The Use of GIS in Determining the Role of Visibility in the Siting of Early Anglo-Norman Stone Castles in Ireland

Lisa Karen McManama-Kearin

BAR British Series 575
2013

Published in 2016 by
BAR Publishing, Oxford

BAR British Series 575

The Use of GIS in Determining the Role of Visibility in the Siting of Early Anglo-Norman Stone Castles in Ireland

ISBN 978 1 4073 1074 9

BAR Publishing is the trading name of British Archaeological Reports (Oxford) Ltd. British Archaeological Reports was first incorporated in 1974 to publish the BAR Series, International and British. In 1992 Hadrian Books Ltd became part of the BAR group. This volume was originally published by Archaeopress in conjunction with British Archaeological Reports (Oxford) Ltd / Hadrian Books Ltd, the Series principal publisher, in 2013. This present volume is published by BAR Publishing, 2016.

Printed in England

BAR
PUBLISHING

BAR titles are available from:

BAR Publishing
122 Banbury Rd, Oxford, OX2 7BP, UK
EMAIL info@barpublishing.com
PHONE +44 (0)1865 310431
FAX +44 (0)1865 316916
www.barpublishing.com

When attempting to make an assessment of what was humanly possible to see....

There is no substitute for the human

ACKNOWLEDGEMENTS

A visual study of this nature could *never* have been undertaken without the excellent resources offered by Ordnance Survey Ireland and Ordnance Survey Northern Ireland. Maps are of particular importance in this research, both modern and historical. Ordnance Survey Ireland's vast corpus of survey material made it possible to bridge the gap of hundreds of years, and to illustrate beautifully the network of roads within the Republic of Ireland. A debt of gratitude is offered to Ordnance Survey Ireland. Many of the images reproduced herein are courtesy of OSI *© Ordnance Survey Ireland/Government of Ireland, Copyright Permit No. MP 0005612*

Ordnance Survey Ireland

National Mapping Agency

Many thanks are extended to the outstanding individuals who helped with this research. Foremost, thanks so much to the two kind gentlemen who have supervised and mentored me, Dr Tom McNeill and Dr Mark Gardiner. Thank you Tom for giving me the 'three-hour version' of the introduction to castles instead of the 'forty-five minute version', and in so-doing changed my life forever. Thank you for the idea of a visibility study based on castles, for your incredible knowledge base and your unflagging enthusiasm for all things 'castle'. Finally, thank you for your continued help and advice, even though you *thought* you had retired. Thanks to you Mark for your wisdom, patience, and your great sense of humour. Thank you for making sure all the 'i's were dotted and the t's crossed', and for facilitating as only a master facilitator can. Thanks so much for sharing your expertise; you are a wonderful tutor who obviously takes delight in the intellectual growth of others. I am grateful that you were willing to be involved in this project from the start, and count myself *incredibly* lucky that you were here to help me finish up.

There are numerous people who have helped to shape this research; grateful thanks are offered to the GIS lab at Queen's, specifically Lorraine Barry, Conor Graham and Jennifer McKinley. Thanks as well to Mike Hernandez from Weber State University, Ogden, Utah, who kindly shared his GIS knowledge. Thanks to Maura Pringle at the School of GAP map library for her help and kindness, to the staff of the Queen's Library Special Collections Department, and to the instructors, administrative and technical staff at the School of Geography, Archaeology and Palaeoecology. Thanks to my peers at Queen's, rock-stars everyone. Thanks to The Queen's University of Belfast which has served me and many thousands of others so well. It was a superior educational experience, I am deeply grateful.

Thanks are given to the government and heritage agencies of Northern Ireland and the Republic of Ireland that allowed access to these wonderful monuments. Thanks especially to Kevin O'Brien at the Office of Public Works in the Republic of Ireland for his time and experience at Trim Castle, Co. Meath. Thanks to all those who have been willing to share their time, knowledge and passion for castles with me, particularly the members of the Castle Studies Group, and all the great individuals who work at the castles themselves.

Thanks to my son Isaiah, who so trustingly put his future into my hands, and *relatively* willingly flew with me to a new country, culture and life. It *could not* have been done without you by my side. To my numerous brilliant children and their spouses, and the growing numbers of *gorgeous* grandchildren, I love you. Thank you all for letting me go for a while so I could achieve this goal. Thanks to my mom and my pop, who have kept the home-fires burning and have supported me with so much love. And finally, thanks to my Thom...my husband, partner, research assistant and academic example. Thanks for encouraging me to follow a dream, even when it meant going 5,000 miles away to get an education; and then supporting me mentally, spiritually and financially for 5 years. I will try to be the best darn retirement-plan you could *ever* have hoped for.

Lastly, thanks to the people of Northern Ireland, and specifically those in the Rosetta and Holywood Road Wards of the Church of Jesus Christ of Latter-Day Saints. I am very grateful for your overwhelming love and support.

Thank you all so much, kare

ABSTRACT

This thesis examines the effectiveness of the use of GIS and GIS viewsheds as tools in the study of medieval castles in Ireland. To date, archaeological usage of GIS viewsheds has centred on prehistoric funerary sites. Little work has been done using GIS in relation to medieval castles, a subject and time-frame which is well documented. To date, no work has tested GIS and viewshed analysis across a wide comparative sample of castles. This study uses GIS to examine the visibility *of* and the views *from* structures about which much is known. A comparable set of twenty sample castles were taken from a particular period in one social/geographical context, the first century of English lordship in Ireland. This has afforded a tougher test of the viewshed than the studies that have already been done; pitting *supposed* purpose and perception against *known* purpose and perception.

Research objectives included exploring the priorities of the first three generations of Anglo-Norman castle builders in Ireland, by determining if there are patterns in site choices. Specifically the project hoped to establish whether visibility may have played a role in the siting of these castles. The concept of 'communities' of viewers was used to separate the different groups of spectators and to humanize the study. Three communities within the castle walls were identified, which would have experienced the projective view (looking outward from the castle): the 'garrison community', the 'public community' and the 'private community'. Viewing positions for these groups were predetermined to be gate-house battlements, great hall and lord's chambers respectively. Two communities were identified outside the castle walls which would have had reflective views (looking toward the castle): the 'stranger community' and the 'local community'. The viewing positions chosen for these groups were the first-sighting upon approach to the castle for strangers, and the parish church or market place for the local community.

The extant castle fabric and the statements the builders made by their choice of building sites were analysed using a combination of GIS, viewshed analysis and an experiential process of personal observation. The results indicate that there is definitely a pattern to site choice. A set of 'rules' or norms have emerged by which a lord or lady might have selected the most suitable position in the landscape for a castle. Generally, the builders of the subject castles placed a high priority on environmental features that could facilitate the successful maintenance of an extensive economic base. These features include access to transportation routes, good land, a fresh and reliable water source, and proximity to a major cross road. While most of these priorities have already been suggested in castle studies, this is the first work that has able to establish an actual pattern involving a large sample population.

The research showed that builders *were* concerned with both projective and reflective visibility, especially within an effective radius of 1km. Prominence, either topographical, architectural or both was also a priority, however, proximity to water and convenience to a cross roads were higher priorities than either visibility or elevation. Ireland provided a landscape uncluttered with feudal manors, a virtual island of 'green-field' sites on which lords could establish their *capita.* Still, slightly over half of the subject castles (11 castles or 55%) appropriated sites or place names of previous (Irish) political, religious or mythological power (*even* when such locations were not necessarily defensively secure.) The study has also shown that the vast majority of subject castles did not meet the two most basic requirements for defensibility: a site that was not overlooked geographically, and an internal/secure water supply. While disregard for these basic tenets of defence has been noted on a site-by-site basis for many years, this is the first research that has been able to supply data in Ireland on a broad scale to confirm it. Essentially, these castles were designed for a formal way of *life,* not for war.

NOTES

Every effort has been made to obtain and provide original sources. However, there were a few instances when information pertinent to this research was cited as a primary source but had been paraphrased beyond the point of recognition. In spite of attempts to track down all original statements, some have proven elusive. When this was the case, the original article was searched for the direct quote. If it was not found, *both* the original source and the citing author are referenced.

To that same end, every effort was made to receive copyright permission for those illustrations that were not the author's own work. In almost every instance, that permission was received; however in a small number of cases, the requests were met with no reply. The origins of those few images are given in-full within the List of Figures, and gratitude is extended to the copyright holders.

LIST OF ABBREVIATIONS

AC	*Annals of Connacht* (1224-1544)
CPI	*Chartae, Priviligia et Immunitates*
COD	*Calendar of Documents relating to Ireland*

TABLE OF CONTENTS

LIST OF FIGURES: Please note that colour figures are available to download from www.barpublishing.com/additional-downloads.html

Part I

Chapter 1 Introduction

Castle studies is a growing field of archaeological research, and the issue of the 'role' of the castle in medieval society has been an intriguing one. The last fifteen years has seen an abandonment of the premise of military purpose as providing the castle's foremost role; a change that has brought an end to the stress that had been placed on the chronology and typology of castle architecture. Instead the move has been to recognize the multiple roles castles played within medieval society. This move has also included an emphasis on the visual impact of castles, and an acknowledgement of the castle as both a central place of political power and as the fundamental hub of estate administration. The result has been that even as the field is growing, it has fragmented. With the loss of a reliance on classification systems, there has been a division into different strands of enquiry. While this can be seen as an advance in one light, after fifteen years the field is still not able to answer the most basic question "what is a castle?" In trying to answer this question numerous aspects have been considered, including the concepts of psychological dominance, social display, economics, domestic comfort, and even recreation or entertainment. One motive for castle siting that has not as yet been researched in depth is that of 'siting for visibility': whether or not builders may have intentionally chosen a site which might have had a psychological impact on the outside observer, been defensively advantageous or even simply visually pleasing for the lord.

Much of the work done in the past decade and a half is a study of castles on an individual basis. Researchers have used something of an anecdotal approach, utilizing examples of castles selected specifically to illustrate whichever aspect the researcher has chosen to highlight. What would be helpful is a systematic line of investigation with a methodology designed to define and compare castles 'across the board', not by simply hand-picking those that fit the model. Several authors (Creighton 2002; Liddiard 2000; 2005) have realized the importance of not isolating the castle from its environment and have used a 'wider lens', taking into consideration the landscape that has surrounded the castle and of which the castle has been an integral part for nearly a millennium. Landscape is a valid piece of any castle puzzle; it is not possible to have a castle without a setting, and every setting (like every castle) has a story to tell. In a landscape context, Ireland (as a theatre for castle studies) provides a unique and valuable challenge. It was the scene of the seizure of large amounts of land in the late twelfth and early thirteenth centuries by Anglo-Norman lords. These were men accustomed to using castles and estates as the building blocks of both their social and their economic structure. But castles and the Anglo-Norman social and economic structures were relatively new concepts for Ireland; it was not (as England was quickly becoming) riddled with parcelled estates and the castles which served as their nodal points. What is seen in Ireland during this period is a one-of-a-kind castle-building event; where lords were largely able to choose their castle sites at will, unconstrained by the density of their peers. This provides an opportunity to examine the statements made by these lords through their choice of building sites.

In the past castle landscapes have been hinted at in the literature by vague comments, such as "the castle is sited to command the pass" (or ford, river, valley, harbour, etc). But castle studies in general have been slow to find effective tools to take an analysis of the landscape further. Liddiard (2008: pers. com) and Creighton (2010) have utilized several cross-discipline methods and have recently experimented with the spatial data management, data visualisation and spatial analysis capabilities (Conolly and Lake 2006: 11) of GIS (Geographic Information Science) as a means of forming a more objective assessment of the landscape surrounding castles.

GIS is a tool that has been experimented with for nearly a decade by archaeologists in the field of prehistory, but in general there has been a distinct lack of interest to embrace this science on the part of archaeologists (Wheatley and Gillings 2000: 2). This may be due to archaeologists as a group being uncomfortable 'embracing' computation (Llobera 2007: 66), but Crescioli et al. have suggested that the diffidence shown by archaeologists towards the use of GIS reflects not only doubt about the value of the technology and concepts in an archaeological context, but perhaps a 'fear' of any innovation in research (2000: 159). This hesitancy to engage the science fully may also have something to do with the fact that, while there are comprehensive texts describing the archaeological applications of GIS available (notably Wheatley and Gillings 2002, Conolly and Lake 2006 and Chapman 2009), at this point, there is not a body of information which clearly points the way for archaeologists who may wish to attempt a visibility study, but lack a background in geographic information science. There are as yet no manuals illustrating the step-by-step process of creating and analysing a viewshed. Any archaeologist interested in exploring the technology must first take a class in geographic information science, or teach themselves the basics of GIS before they can reliably use the various GIS programs. With no 'how-to' guide, the process can seem daunting and is time-consuming.

However, this researcher believes that it might also be the dispassionate objectivity of GIS (its ability to view the landscape virtually, and then analyse it without ever leaving the office), that has been seen as a problem for archaeologists who traditionally prefer a hands-on, 'feet-in-the-mud' approach to an archaeological site. GIS represents a two-dimensional abstraction of a site that at times can be "crude and far removed from reality" (Harris and Lock 1995: 360), and which can lead the observer to think in a static way (Llobera 2003: 39). By removing the human (i.e. the archaeologist) from the scene we are in danger of losing the human meanings of the scene; we can quantify it, reduce it to a plane, de-sanctify it (Tilley 1994: 21), and essentially de-humanize it. GIS on its own creates a picture of and from a castle which would have been completely alien to someone living in the subject time period. The only tool available to the builders of these castles was visual assessment, and as such it is the most important tool that can be used in a visibility study. In any study trying to understand the human experience, retaining that human element is vital. Because of this, Tilley promoted an experiential approach to understanding the use of space, stressing the importance of actually moving within and through the monuments and their landscapes (1994: 73, 81).

This work aims to examine the effectiveness of the use of GIS viewsheds as a tool in the study of castles. This will require the testing of the actual techniques of generating viewsheds in a number of different field situations, and an assessment of their application to a sample of sites. It will also require a consideration of the means by which the tool could contribute to the study of castles. This will be done by taking a sample of castles from a particular period in one social/geographical context, the first century of English lordship in Ireland.

Within a GIS framework were the following aims:

1. Investigate the archaeological application of GIS and viewsheds as wider decision-making tools, not anecdotally through a "high-tech fishing trip" (Lock and Harris 1996: 239), but in a systematic and rigorous manner, across a broad spectrum of physical environments
2. Give an assessment the technical issues related to the use of GIS viewshed analysis on a series of sites requiring comparability, in contrast to single-site studies
3. Explore the use of GIS viewsheds in a valid historical context, using a time period about which much is known and documented, so that we may consider the validity of the conclusions reached by matching it against other sources
4. Assess the results of viewshed analysis when used as a tool to investigate the role of visibility of sites in the landscape of the period

Working within a castle studies framework added the following separate goals:

1. Seek to explore the priorities of the first three generations of Anglo-Norman castle builders in Ireland, through the statements they made by their choice of building sites
2. Determine if visibility may have played a role in the siting of these castles by answering several research-driven questions:
   - Can we detect a prevailing norm, or set of norms, by which a lord fixed on the most suitable position in the landscape for a castle?
   - When faced with the possibility of being able to choose a castle site at-will, unconstrained by the density of peers, did the choice of site address the issue of visibility by:
     a. Taking in a broad spectrum of visibility (projective views large in scope)?
     b. Offering a visually pleasing projective view?
     c. Commanding a militarily advantageous view?
     d. Producing a psychological impact on the outside viewer?
3. While these questions define the general siting aims of the castle, this study aims to refine this further by considering how the views from differing spaces within the castle may have varied according to the specific spectators who were intended to view them
4. Conversely views of the castle from the outside, as seen by residents of communities associated with the castle as settlement, and more distant visitors to the castle, are examined to consider how they may have varied according to the specific spectators that were expected to view them

Twining the two different approaches of GIS viewsheds and phenomenology within the context of castle studies required a final aim:

1. To define a methodology that could combine the two types of analysis to give a better understanding of the historical human communities in question

The study is organised in three parts. Part I comprises an overview of the project aims and method: Chapter 2 gives a brief introduction to GIS terms and principals. It examines the issues relating to GIS usage within archaeology as a whole, and its possible application within castle studies. Next, it explores the concept of 'communities' – as a way of focusing on the viewers within a visibility study. It details the criteria used to isolate the sample castles, and then introduces the sample castle population. In Chapter 3 the boundaries and limitations of both viewshed and phenomenological approaches are considered. Chapter 4 outlines the field research and data-collection processes. It gives a brief introduction to the viewsheds used to illustrate visibility from the prospective of the study 'communities', as well as the format of the site reports.

Part II comprises Chapter 5, and represents the corpus of castle research as individual site reports.

Part III presents the results of the study for discussion, and the conclusions drawn: Chapter 6 highlights the strengths and weaknesses of the GIS and phenomenological approaches. The various issues discovered during field research and lab analysis are examined, and the manners in which the issues were resolved are outlined. Chapter 7 discusses the general siting of the castles, and makes determinations regarding the priorities of the builders through an analysis of the site choices themselves. Chapters 8 and 9 summarize the results of the viewshed and phenomenological analysis in the human context. It looks at the 'communities' of viewers and explores the views upon which each community focused. Chapter 10 gives the conclusions drawn from the research, and discusses the vigour and limitations of combining GIS with phenomenology within the framework of castle studies. Lastly it offers a few suggestions for additional research.

Because this is a visual study, there are many colour illustrations. In an effort to keep printing costs reasonable, all illustrations have been reproduced in a digital format, and are contained on the CD included with this book.

Chapter 2 GIS and Archaeology

2.1 The application of GIS in archaeology

This section will offer a brief synopsis of GIS and the viewshed, and outline some of the issues related to its usage within archaeology in general (see Conolly and Lake 2006 for a comprehensive overview on archaeological applications of GIS). It describes the past anecdotal usage of viewshed technology within archaeology and then relates the technology to castle studies. The concept of the communities related to castles is introduced as a means of focusing the study both visually and phenomenologically. Lastly, the criteria used for the selection of the subject castles are discussed, and the sample population is established.

Gillings and Wheatley have suggested that visibility can be regarded as a key factor in attempting to answer the question as to why a monument is in a particular place, rather than all the other places it might have been located (2001: 2). GIS is a computerized toolset. Conolly and Lake have broken the main tasks of GIS into five groups: data acquisition, spatial data management, database management, data visualisation and spatial analysis (2006: 11). Spatial data (a place, a river, a road, a castle, etc.) and non-spatial data (text and numbers) are captured and stored on both vector and raster data structures. The vector structure is a binary system which utilizes a 0/1 or yes/no system to build lines, points, polygons and arcs. The raster structure is a visual system built using pixels (such as a photograph) which provides a discrete approximation of a continuous field (ibid. 4-5). The resultant stored information can be combined or overlain to create accurate multi-dimensional illustrations of the data. The data is displayed using the Cartesian mapping system (familiar to archaeologists), which makes GIS a flexible and comprehensive analytical tool for visibility studies (Gaffney et al. 1995: 211). Using GIS may help us to “understand the visual impact that a building with some symbolic relevance….has on its surroundings, and to answer questions such as: where can it be seen from? How much of it can be seen at each location?” (Llobera 2003: 31); and we might add, exactly how much of its surroundings were visible from it?

Very simply put, there are two types of spatial data required to use GIS for a visibility study. The first of these are the background information data sets (the digital base map so to speak), the second are the coordinate points of the particular locations of interest. Digital map data for GIS use can be procured in several ways: retrieved from paper maps, aerial photography, and through the use of satellite data or scanned images. This study mostly utilized satellite imagery obtained from both government sources and commercial sources made available through the MIMAS Landmap Project. This is digital mapping data produced by the University of Manchester for the use of higher education in the UK. Occasionally, issues arose which required the use of non-satellite data derived from local frames of reference (paper maps, aerial photography). When needed this information was procured from local government sources (OSNI and OSi). These issues, their solutions and further concerns regarding the digital mapping data utilized are discussed in detail in Chapter 6: section 6.1. Digital maps provide a model of the elevation of the Earth’s surface (Conolly and Lake 2006: 291), and are referred to as digital elevation models (DEMs), or digital terrain models (DTMs). The two terms are basically interchangeable; however a DTM includes information on surface texture as well as elevation. The digital models used herein were DEMs, and were either in a raster format originally, or were transferred to a raster format before use. The DEM portrays land divided into either 25m², or 10m² parcels. This means that all of the ground elevations within those parcels are averaged, and the entire parcel (also known as a pixel) is considered to be all of the same elevation. This accounts for the rather boxy ‘pixelated’ appearance of the viewshed.

The second type of spatial data used with GIS is a coordinate point, for instance, the physical location of the castle on the ground. As with a DEM, these points can be obtained from several sources. Primary sources are those which are hand-taken in the field by the researcher, secondary sources are coordinates taken from paper maps, excavation or site reports, basically any data collected by another researcher. This study intended to use only primary, or personally obtained coordinate data, but in fact, secondary sources of coordinate data were utilized. The issues related to the use of secondary coordinate data and the resolution of those issues are detailed in Chapter 6: section 6.2.

This study utilizes two of the ‘building blocks’ of visibility analysis within GIS: line-of-sight (or LoS) calculation and viewsheds (Gillings and Wheatley 2001: 6). LoS locations are specific points on a DEM that can be connected by means of an uninterrupted straight line (Llobera 2003: 29). Each pixel is a discrete unit that can, or cannot be seen from a central position, and a viewshed is used to calculate which pixels on the DEM can be connected by means of a LoS to the pixel at the viewpoint location. A LoS in GIS cannot be obstructed by artificial or natural encumbrances on the landscape, such as built environment or vegetation. However, even though a location may be connected to an object by a topographically unobstructed LoS, atmospheric conditions may keep that object from being visible from the specified location (Llobera 2003: 29). Effectively the viewshed calculates which pixels or locations are not blocked by natural geographic features (hills etc.) and therefore may be observable from the determined viewpoint. Theoretically, reciprocity (inter-visibility between two points) can be assumed (Wheatley and Gillings 2000: 8). Determining the percentage of pixels visible in relation to the total amount of pixels within a specified radius might be able tell us something about the extent of visibility, both from and of the castle. This process is objective, reliant totally on distinct (and testable) data, and can be relatively easily repeated by others.

Simply put, the viewshed is a visual representation of what can be seen from a specified viewpoint coordinate. Viewsheds are useful to study how visible objects might have been, or from which locations in the landscape an object may have been visible (ESRI 2007: http://resources.arcgis.com/glossary). However, the concept of visibility from and of any point can be confusing (Wheatley and Gillings 2000: 8), and it is helpful to have distinct terms to separate the two views. The term which has been coined within GIS terminology to differentiate the process of looking outward from a specific point (i.e. looking outward from the inside of a castle) is known as the 'projective view'. The process of looking inward toward a specific point (i.e. gazing at a castle from outside it) is known as the 'reflective view' (Loots 1997: 12-16). These terms are used frequently within this study, and interestingly, the simple need to understand projective and reflective positions led almost naturally to a greater reliance on the phenomenological approach.

2.2 The problem of anecdotal study

Even though GIS as an archaeological analytical tool has been available for the past two and a half decades, viewshed analysis in relation to archaeology is still in its infancy. Several studies pairing GIS viewsheds with archaeology have been undertaken over the past few decades (Gaffney et al. 1996:154; Harris and Lock 1995: 349; Kvamme 1995:5; Wheatley 1995:181-182). These have proved beneficial, and have allowed archaeologists to 'cut their teeth' so to speak on GIS technology to determine if it can be useful in the field. But most often archaeologists have simply experimented with GIS as a tool to see what it is capable of, not necessarily with any solid research questions at hand. Such "high-tech fishing trips" (Lock and Harris 1996: 239) have added to the corpus of knowledge of GIS usage within archaeology, but it may not have always enhanced archaeological knowledge as a whole. In this sense it is similar to the fragmented analysis that castle studies has seen; and while anecdotal analysis is usually interesting, it lacks the rigour of a systematic line of investigation.

2.3 Castles and GIS

Most archaeological studies so far that have utilized GIS have focused on predicting where prehistoric sites might be found in the landscape (Gaffney et al. 1996:133); and in assessing the visibility of archaeological sites and their inter-visibility in relationship to one another. With few exceptions (Lowerre 2005; Lilley, Lloyd and Trick 2007), viewshed analysis has centred on prehistoric burial monuments, chosen most likely due to their highly visible site locations. However, Crescioli et al. have suggested that the impossibility of investigating the cultural aspects that were mediated by cultural and social factors surrounding the contemporary prehistoric population represents the greatest problem in the study of prehistoric funerary habits using GIS (2000: 161). In short, the actual events that took place in and around these monuments have been lost to modern knowledge. We simply do not know what happened during a burial ceremony or a solstice celebration, so it is difficult to know exactly what the pertinent questions to ask GIS should be. In contrast, we know fairly well about the typical events and daily life that occurred in and around medieval castles and most of this knowledge is based on historical documentary sources. Castles, which have a known social structure and context also have a firm chronological position; so it is possible to analyse fairly other sites and landscape features (i.e. valuable land or communication and transportation routes) within the environments which surround them. The use of GIS to examine both the visibility of and views from structures about which much is known would afford a tougher test of the science of GIS than the studies that have already been done, pitting supposed purpose and perception against known purpose and perception. It is believed that by applying viewshed analysis to a comparable set of sites within the historic period, we can help take the use of GIS in archaeology a bit further. We can also assess the strengths and weaknesses of GIS as a tool in the study of archaeological sites in their landscape setting.

2.4 The concept of the communities related to castles

The medieval inhabitants of the castles and their hinterlands are the people who will be considered the 'viewers' within this visibility study. (A castle's 'hinterland' is herein considered the rural area surrounding each castle, which includes the area within an 11km radius around the castle.) We might ask the question "who are these individuals?" Within the context of this study the answer includes the indigenous Irish, who up until this point had not seen large stone buildings and from whom the land had been wrung to build the castles. For these people castles were a new type of architecture, strange, perhaps even threatening in their immensity and permanence, at the very least a curiosity. And it also includes the Anglo-Normans; barons, knights, burgesses, transplanted settlers and villagers (each with their own cultural interpretations of the castle), for whom these structures were symbols of authority, wealth and power. The viewers can be separated into two main categories; projective and reflective viewers. The viewers within the castle would have had projective views of the hinterland (viewing the wider environment from within the castle); while the viewers outside the castle would have had reflective views of the castle (viewing the castle in its setting from outside the castle walls.

Several authors have suggested that each castle held within it a collection of assembled communities, each with varying needs and functions (McNeill 1992: 19, 22, 27: Creighton 2002:1). In an attempt to comprehend how the views from a castle may have been experienced by the contemporary medieval human population and as a reminder that these views were in fact a human experience (and therefore often an emotional experience), the idea of assembled communities was expanded to include not only the communities of people living within the castle, but those outside it as well. This concept incorporates both projective and reflective communities. The projective communities were determined to be:

- The 'garrison community'; these are individuals who worked within the castle. They are termed the garrison community, however it could include anyone who had the responsibility of watching for the arrival of visitors to the castle, both friend or foe
- The 'public community'; those who were invited into the castle (Anglo-Norman peers, indigenous Irish guests, elevated community members, burgesses, etc)
- The 'private community'; consisting of the lord, his lady, their family members and their most intimate friends (or councillors)

The reflective communities were determined to be:

- The 'stranger community'; those people travelling through the land which may have approached the castle only occasionally. This loose community may have included any type of traveller including the indigenous Irish (who might be either friendly or antagonistic towards the Anglo-Norman lord living in the castle), an Anglo-Norman peer (a fellow baron, friendly or otherwise), or a client knight of the castle's lord or of a neighbouring lord. These are individuals for whom a reflective view of the castle may have been a rare if not singular experience.
- The 'local community'; those who would have lived daily within sight of the castle, either as part of the village community (or as part of the market-day regulars), and who may never have seen the inside of the castle. As with the stranger community, this group of individuals may have included indigenous Irish (who had been in the vicinity prior to the arrival of the Anglo-Normans, or had moved into the area to take advantage of the settlement), as well as transplanted Anglo-Norman and Welsh peasants brought in by the lord to settle and farm the land, or burgesses and traders persuaded into populating the castle's attendant town. Those from the hinterland who might have come periodically to the castle's settlement are included in the local community because they would have had regular (if not necessarily frequent) reflective views of the castle. For rural castles the local community might only have consisted of those individuals living within the immediate hinterland.

Each of these separate communities would have had diverse world views stemming from their own enculturation, with different expectations of what they were viewing, and for each of these communities the castle would have had different meaning. The castle elements that could serve as representative viewing platforms for the projective communities formed part of the criteria for the subject castle selection and are considered the 'projective prime components'.

General views

We might assume that the higher an element within a castle rose, the more chance it had of not only offering a wide projective view, but also of being seen from a great distance. In every case this assumption proved true; the element that gave the maximum projective view was also the element best or first seen in the reflective view. For this reason the viewing platform offering the maximum potential projective view (gate-house battlements, great tower battlements, or the otherwise highest possible viewing position within the castle) was used to provide a 'general' view. For instance, at Trim Castle the elevation of the battlements of the great tower (24m) afforded the largest projective view possible from the castle. As this tower is also the most easily seen and recognized component of the castle from the hinterland, it was used to provide the general view. The position used for the general view was most often an actual physical location (i.e. battlements of great tower or gate house). For castles which lacked definitive components with elevation, the centre-point of the area bound by the curtain wall was determined, and then the highest probable elevation for that castle (based on the standard component heights used in this study which are discussed below) was applied as the elevation offset. The general view was crucial for establishing all of the other potential views, for once it was established, the views from the separate components were analysed to see what portion of the general view they focused on (projectively); and the degree to which the component was visible from its environment (reflectively).

Prime components and their expected views

For the garrison community the battlements of the gate house were selected as the definitive projective viewing position. Designated observers may have stood at many places within a castle (battlements of the great tower, battlements of a great hall, wall-walks along the curtain, etc). However, the gate interfaces with the outside world, and while it was generally the castle's strongest defensive feature, it can also be considered the most vulnerable area of any castle. It is used here as the representative platform for observation of all approaching traffic but, specifically, for the defensive community or the garrison.

As the great hall would have been the main gathering place within every medieval castle, it was selected as the projective viewing position for the public community within the castle. The private community would have occupied the very inner spaces of the castle, which was often those private areas most deeply embedded into the core of the castle, and hence the most difficult to access: the lord's chamber. As is the case at several of the subject castles, the lord's chamber is often two separate spaces; an outer, more public room or office where the lord's council would have convened (lord's council chamber), and an inner even more private chamber reserved specifically for the great family itself (lord's private chamber).

There are certain things we might expect to see from these three different locations. While it is safe to say that the idea of the castle as being first and foremost a defensive structure has (though lingering) been abandoned, few people would argue the fact that castles were intended to look defensive and, if necessary, to be capable of defence. One of the factors often used to assess defensibility is visibility; in both a general sense and a focused nature. The concept of focused visibility has been behind the recurring theme of observations such as "the castle was built to overlook the ford", which suggests a view targeted on a specific feature in the landscape. The issues of how this 'overlooking' could be turned into control (for instance the constraints of the range of fire on medieval weapons and decreased human visual ability at night) are usually not addressed. This focus can also be seen in some type of managed approach toward the castle itself; the lord would certainly want to know when someone was about to knock on the gate. This management of approach could be for ceremony, or for defence; both would require projective views of anyone approaching. In either case (ceremony/defence) the possible projective views of the observers positioned atop the battlements of the gate house can be revealed in a viewshed. GIS should allow us to assess whether the defensive view did focus on anything in particular and in so doing, introduce a degree of objective assessment into the question about the defensive qualities of a castle.

From the great hall we might also expect to have focused views. In this case however, we would expect the focus to be on a park, or other monument attesting to the lord's seigneurial power, authority, wealth or piety. Such markers might be bridges, religious houses, parish churches or cathedrals, etc. From the lord's chambers we might expect to see recreational views, artistic landscapes, or even views that emphasize economics, such as the town, the market or a harbour. McNeill built a strong case for including the view available from the tops of great towers as a criterion when judging the purpose of any specific castle (2006: 127). His premise was that rooftops could often be accessed exclusively from private chambers and were used as private viewing venues. He even suggests that the towers were planned with that specific activity in mind. While it could not be said that the great towers within the subject castles allowed for exclusive private entry to the rooftops, each of the towers (Trim, Co. Meath; Carrickfergus, Co. Antrim; Dundrum, Co. Down and Nenagh, Co. Tipperary) would have provided an artificial elevation from which to observe the landscape in a way that had not been available to humans in Ireland prior to that time. The panoramic projective views from these heights would have been under the control of the lord, and the opportunity (as well as the ability to extend the opportunity to others) would have been solely within his power.

The locations where a reflective view of the castle could have been seen are termed the 'reflective prime components'. For a traveller or member of the stranger community the reflective prime component would be the spot where the castle was first glimpsed in its setting upon approach. Of course this is an elusive location and could have been numerous places for each castle. To deal with this incredibly open-ended possibility, the likely approach routes to each castle were assessed and then traversed, a process which will be described further below. For the local community there are several points from which we might expect the castle to have been visible within the settlement: market places, some of which were at the very gates of their castles, and the parish church which was often a foundation financed by the lord. These two venues would have drawn both the immediate community living within the urban setting and those living within the hinterland influenced by the castle.

2.5 Sampling strategy: the twenty castles

Castles were erected in Ireland from the time of the arrival of the Anglo-Normans until the 1600s; although the materials used, the shape and the elements that castles contained changed throughout this time. Depending upon the definition used (motte and bailey, earth and timber, masonry, fortified house, tower house, etc), there are thousands of castles in Ireland. Construction of stone castles began c. 1175, the earliest being concentrated in the east, but eventually castles in stone were built all across the island. As the study seeks to understand the priorities of the earliest Anglo-Norman castle builders in Ireland, the sample population had to include only the earliest castles. Unfortunately, for most castles in Ireland, the documentary evidence is fragmentary and gives chronologies which (although helpful), were often less than specific (a recurring theme in the story of Irish castles).

Because of this, much assessment of the construction dates required a 'reading' of the remaining castle fabric and a comparison of the extant architectural elements with similar elements from other Anglo-Norman castles (from Ireland, England, Wales and France), that have already been dated. However, the story of castle building rarely represents a 'one-off' event. Castles are expensive and built to last. They were often built in stages, repaired as needed, refurbished or in some cases almost completely rebuilt to update or modernize them. When trying to read the fabric of a castle, one is confronted with a 'jig-saw' puzzle of intriguing starts and stops, with mixtures of styles and materials that can offer enigmatic and often conflicting clues to determining the dates of the construction episodes. In some of the cases the most likely explanation (when all of the prospects have been weighed) is the best that can be offered (Hodkinson 2003: 32). However, this study uses as many methods as possible to determine the construction date of the subject castles, and these methods are mentioned within the introductory section of each site.

It was necessary to find a corpus of castles which had enough above-ground fabric left with which to build a story, and those chosen needed to be the best examples of such castles. Using this criterion the following twenty castles were chosen:

- Adare, Co. Limerick, constructed c. 1199 by Geoffrey de Marisco

- Athenry, Co. Galway, constructed c. 1240 by Meiler de Bermingham
- Athlone, Co. Westmeath, constructed 1212 by John de Gray / royal
- Ballylahan, Co. Mayo, constructed c. 1240-1250 by Jordan de Exeter
- Carlingford, Co. Louth, constructed c. 1212 by Hugh de Lacy (the son)
- Carrickfergus, Co. Antrim, constructed 1177 by John de Courcy
- Castleroche, Co. Louth, constructed c. 1236 by Rohesia de Verdon
- Clonmacnoise, Co. Offaly, constructed early 1200s, c. 1216 by Ralph Derevaus or Walter Reboth/ royal
- Dunamase, Co. Laois, constructed c. 1181 by unknown builder, but probably Geoffrey de Constentin, Meyler fitz Henry and William Marshall
- Dundrum, Co. Down, constructed before 1203 by John de Courcy
- Ferns, Co. Wexford, constructed c. 1247 by De Valence
- Greencastle, Co. Down, constructed c. 1240 by Hugh de Lacy (the son)
- Kilbolane, Co. Cork, constructed c. 1270 by Maurice de Rocheford (the son)
- Kiltartan, Co. Galway, constructed c. 1264 by Redmond de Burgh (son of Walter)
- Lea, Co. Laois, constructed before 1257 by Maurice Fitz Gerald
- Limerick, Co. Limerick, constructed c. 1210 by unknown royal representative/ royal
- Nenagh, Co. Tipperary, constructed between 1200-1220 by Theobald Walter II (of the Butlers)
- Rinnduin, Co. Roscommon, constructed 1227 by Geoffrey de Marisco / royal
- Swords, Co. Dublin, constructed c. 1200 by John Comyn / Episcopal
- Trim, Co. Meath, constructed 1175-1225 by Hugh de Lacy (the father)

A map illustrating the locations of the subject castles can be seen in Figure 2.5.1; a chart listing the habitation dates and additional information about the prime components can be seen in Figure 2.5.2. While determining the location of most of the prime components within each castle was quite straightforward at many of the sites, it was fairly difficult at a few. When dealing with ruins, gates are generally clear-cut and obvious but it is often difficult to determine exactly where the great hall and the lord's chambers had been located; quite simply it can be very difficult to tell a hall from a chamber. The difficulty in determining the function of a space can be illustrated by the controversy over a room in the lower bailey at Chepstow which has been identified as a hall for guests and their households by Faulkner (1963: 218), as a hall for the constable functioning in the lord's absence by McNeill (1992: 54), and as a kitchen by Knight (1991: 53). Often there are several positions in a castle where either hall or chambers may have been located. In these instances it was a case of narrowing down the possible locations from the options available to find the most likely location for each.

A system of guidelines was needed to help establish the locations of the more elusive components. This was done by listing the activities that would have taken place in the different components, and then by trying to determine if a location would have met the needs for those activities. For instance, the activities within the public space of the great hall would have been different from those in the more private lord's chambers; and with specific functions come specific requirements. Since we should expect that each venue would have been geared to facilitate particular usage, by looking for the elements that would facilitate that usage, we can narrow down the possible locations of the prime components.

The hall was the most public portion of the castle - the assembly point of the great and good of the community as well as honoured guests, courts, councils and clients of the lord. It is important to see this space as primarily an opportunity for display - display of wealth, display of power, and display of hospitality. Since it was the public space and meant to accommodate large numbers of guests, it would need to have been a big room, and would usually have been the largest and grandest room available within the castle walls. It would also have needed to be fairly easy to access by the incoming visitors, and the path to it should not have been hampered by constricted staircases or indirect routes. Access to a kitchen would have been convenient since gatherings within the hall usually included feasting (though cold food was probably a common occurrence). A fireplace would be a significant element for the comfort of the guests and general ambience. This could be located either in the wall or as a central hearth (though having a central hearth could preclude rooms with wooden floors). Latrines were not normally placed next to great halls, but were to be found in more private places. Lastly (and most importantly for a visibility study) would be the windows. We can assume that within the great hall, windows would have been geared to impress those in attendance by framing the wealth and power exemplified by the buildings within the castle ward; or by framing a significant part of the town or hinterland too illustrate the authority and prosperity of the lord.

Using this same approach, we can attempt to determine the location of the more 'private', residential-side of the castle. This would have been the only place in a castle where the lord and his family could actually achieve privacy. Eadie (2009: 132-144) has suggested that there were at least three kinds of privacy required by a medieval lord; privacy for bodily functions, privacy for personal intimacy and privacy for confidential conversations, negotiations or official ceremonies. Because of the need for ceremonial space, the lord's chamber component could be divided into two areas: one or more rooms that are private (to meet the intimate/confidential needs) and a larger, more public locale (a ceremonial venue). The elements we would expect to see in these two components differ slightly, but only in accessibility. Where the need for privacy would be facilitated by restricting access (Dixon, 1996: 53), the need to create a ceremonial scene would require guests to be moved in a dignified, processional manner (ibid. 55), which may or may not be restricted. With this in mind, determining which rooms were not easily accessed could give an indication regarding the location of the residential or intimate/confidential suites of the lord and his family, while chambers that are approached in style might hint at locations with ceremonial functions.

For castles where one of the prime components simply could not be conclusively determined (as with royal castles which may never have had a lord's chamber), an analysis from the component in question was not attempted and that particular component was passed over. In these instances, the examination of the castle states up-front what the missing component is and explains why it is not included. However, the castle is still treated as far as is possible. There were a few castles where more than one component was indeterminate (such as Athlone and Kilbolane). Treatment of these castles was necessarily reduced in scope; a central point was chosen and a viewshed was generated in an effort to tell what the views might have been in a 'best-case scenario'. These castles, though contentious were included in the study in an effort to err on the side of thoroughness, with the hope that an analysis of the extant fabric and its wider setting can still tell part of the castle's own particular story.

A table illustrating each castle's features can be seen in Figure 2.5.3. It should be noted that Athlone and Kilbolane were initially included in the sample because it was thought they might contain a lord's chamber. However, it was later determined that there was no conclusive evidence to support the possible rooms as lord's chambers. While this meant that Athlone and Kilbolane actually contained none of the required prime components, the analysis of these castles offered interesting information pertinent to the research, and they were retained in the study population.

We have now discussed past archaeological applications of GIS. The problems inherent in anecdotal usage of GIS within archaeology have been examined, and the intent to explore the archaeological use of GIS in a valid context (over a large historic sample requiring comparative analysis) has been highlighted. The concept of castle communities and their specific viewing positions have been introduced, and the sample castle population has been established. We will next examine the constraints inherent in the two approaches, and consider how these limitations will be addressed within the study.

Chapter 3 Boundaries of the two approaches

The ability of GIS to strip away aspects of a landscape which might otherwise conceal the presence of a castle greatly enhances the viewing potential offered by viewsheds. This brings up the important point that, without exception the viewsheds within this study represent the maximum visibility possible from and of the study castles. In essence GIS shows what might be visible to the human eye under the best possible conditions. The reverse of this is also true; the viewshed offers an illustration of what the human eye simply would not have been able see, even under ideal conditions. It is reliant on us to realize that during the subject time frame (as now) such conditions may not always have appertained, and in fact, may have been rare. So while a viewshed might suggest that there was inter-visibility between the castle and a point within its landscape, not all human beings would have been able to experience it. This highlights the constraints that form the boundaries around the two different approaches. These, and the concept of the 'maximum view' will be explored next.

3.1 GIS constraints

Past vegetation

In medieval Ireland as now, the presence or absence of vegetation would have had a tremendous impact on visibility, even over short distances. When examining a DEM, and ultimately a viewshed, we are faced with a relatively smooth surface unhampered by trees or other plants and are left wondering if the (often expansive) views suggested by the viewshed could possibly ever have been a reality (Lock and Harris 1996: 222). It has been suggested that at the beginning of the Anglo-Norman presence in Ireland, most of the dense oak forests were gone (although scattered oak remained) and the woodlands were made up of hazel and alder (Pilcher and Hall 2001: 15). Hazel and alder may not be extremely tall trees, but would certainly have been tall enough to curb and restrict views (especially reflective views from a standing position at ground-level); even low trees and bushes could have rendered long-distance viewing difficult.

Chapman and Gearey have suggested that because of the possibly broad impact of vegetation on visibility, this variable should be re-introduced or factored into the viewshed equation (2000:316). This can be done in two ways; the first of these is the use of LiDAR. This is an active sensor mounted on aircraft to measure elevation but which is also capable of detecting and measuring the amount of current vegetation (Conolly and Lake 2006: 305). However, these measurements reflect modern vegetative coverage, which may or may not correspond to past levels. LiDAR data is also quite costly, and this alone placed it out of the reach of this study. Secondly there is the possibility of using environmental data reflecting contemporary medieval vegetation patterns (when available). Lock and Harris have stated that these "missing pieces" in the palaeobotanical jigsaw puzzle are significant and should be taken into account when feasible (1996: 219). Realistically however, Chapman and Gearey had to admit that the "precise spatial and temporal structure of this vegetation cannot be easily established" (2000: 318). Tilley has suggested that even the most refined of techniques (pollen analysis) can give only a coarse assessment of past environments (1994: 73). More pessimistically, Fisher and Farrelly wrote that even armed with palaeobotanical records, it is not truly possible to reconstruct vegetation patterns with enough detail to undertake reliable GIS-based visibility analysis (1997: 587).

Although using palaeoecological studies of the past may allow us to suggest potential vegetative interferences around the study castles, such an undertaking would require specific palynological information for twenty different sites in Ireland from the thirteenth century. Such data is, unfortunately not available. There is not yet a corpus of palaeobotanical data on an island-wide scale for the time frame in question. We should also remember that the camouflaging effects of vegetation are seasonal (with the exception of the evergreen species). Wheatley and Gillings have suggested that even if accurate palaeoecological data were obtained, it might add little to the interpretive method (2000: 6). As the insertion of past vegetation into viewshed analysis is necessarily suppositional (Fisher and Farrelly 1997: 587), it seemed the best course to remove the question of foliage completely. While to some, this may be seen to represent a draw-back, in reality, dismissing the question of vegetation only enhances the maximum potential of visibility.

Distance constraints

One of the most important constraints within a viewshed is the size of the zone of interest, or buffer of land surrounding the component. It is necessary to limit the search field because not placing a distance limit on a viewshed could result in a suggested LoS so distant as to be improbable. Ogburn has suggested that ignoring the limits of visibility can lead to errors in two ways; first an unrestricted analysis could lead to an overestimation of the visibility of an object, implying a larger visual impact than the object merits (a viewshed unlimited by distance could cover tracts of land which would be unrealistic for human perception to handle.) And secondly a highly restricted analysis could result in an underestimation of visibility, implying a smaller visual impact than the object actually may have had (Ogburn 2006: 405). It was necessary to determine the extent of the buffer which reflectively would give the point at which the castle became too distant to appreciate within the landscape and which projectively would suggest the point at which objects seen from the castle were no longer recognizable. There have been several 'visual indexes' developed within the past decade from which to chose, and most have been built upon the work of Tadahiko Higuchi in 1975 (translated into English in 1983).

Higuchi developed his distance index for defining the visual structure of the natural landscape based on the ability of an observer to perceive trees over distances. In Higuchi's long-distance view (approximately 6.6 km and beyond) an observer is able to perceive only major topographical features; in the middle-distance range (between 6.6 km and 360 m) definite tree-tops can be detected; and in the short-range (360 m from the observer), the tree can be recognized as a distinct object, with the observer able to determine separate leaves (Higuchi 1983: 12-14). However, there is a difference between recognizing a tree and a castle in the landscape, and several other methods have been developed specifically for observation of cultural objects in visibility studies (Llobera 2007: 58; Bishop 2002: 718). Most of these methods are nebulous to the point of actually being unhelpful, and an index tailored for the nature of the topography and prolific vegetation in Ireland was needed. Fisher has developed something termed 'fuzzy' viewsheds which suggests that clarity decreases with increased distance (Fisher 1992: 1994; 1996 as referenced in Gillings and Wheatley 2001: 12). While his formula for artificially creating this decrease was not adopted, his two nearest-distance indicators were.

The result includes three distance indicators representing the far, middle and near-distance radii around the prime components. The far-distance index was determined using a conservative interpretation of Brown's 'range of the castle': the area which castle inhabitants could reasonably control (effectively the range that a horse could walk and still be back before nightfall). Brown deemed this to be between 8-10 miles or roughly 12-16km (2004: 148). Although a travelling distance of 10-15 miles or 16-24km for a horse is actually closer to reality (http://www.wwwestra.com/horses/history_travel.htm), Brown's 'range' is more realistic in regards to the limitations of human acuity. This was reduced slightly to 11km, and can be seen in Figure 3.1.1. The middle-distance view from 4-1km, (Figure 3.1.2), and the immediate or near-distance view of 1km surrounding the prime component, which Fisher refers to as a 'zone of perfect clarity' (Gillings and Wheatley 2001: 12) (Figure 3.1.3) were both suggested by Fisher. Almost all viewsheds produced within this study are illustrated using this tailored visual index of 11km, 4km and 1km radii. It should be noted that although the views from the differing radii of each component are run independently of each other, they represent the same elevations and fields of vision (discussed below). This allows the visible pixel percentages within the various distances to be measured and compared with each other and against the components of other castles in easily comparable increments.

Elevation

The DEM provides the X, Y and Z coordinates of the topography surrounding each castle (Z representing the elevation of the ground at each site). In order to determine what can be viewed from a certain position in the DEM, the elevation of that specific position must be determined (i.e. the height of the tower/gate/hall from which you wish to visualize a view). This is entered into the data-set as the 'elevation offset' (an elevation above ground-level). As many of the prime components are damaged and no longer stand to their original height, conventional elevations were used at all castles (except where extant fabric provided the actual elevation of a component). The following heights were assumed for specific floors: 1.7m for ground-floor (which is a compromise between the 1.6m and 1.75m suggested by Gillings and Wheatley as the optimum height of an adult (2001: 13); 6m for a first-floor, 9m for second-floor, 12m for third-floor etc. Hence, the first-floor of a great hall (the majority of medieval halls were at first-floor) was given an elevation offset of 6m, and a fourth-floor great tower was offset at 15m. The battlements of great halls, smaller towers and gate houses were given a standard elevation of 10m above ground-level, unless an extant gate tower was obviously higher than 10m.

Azimuth

For projective views from a component, it is important to know what contemporary medieval anthropogenic structures may have hindered a view; i.e. a gate tower blocking the view from the lord's chamber, or a great hall which might have had windows along only one wall. This numerical 'view' is entered as an 'azimuth'. An illustration of how the addition of azimuth data can shape the viewshed can be seen in Figure 3.1.3; in this viewshed the windows of the great hall were exterior to the castle compound only to the south-west, and the resultant projective view reflects this curtailment. Initially, an attempt was made to enter the exact azimuth from every external opening from a prime component, while subtracting from the azimuth the obvious contemporary obstructions. However, it eventually became obvious that applying this level of accuracy resulted in a spurious precision; requiring suppositions to be made about how far into a window embrasure an observer might have been standing, how close to the window, etc. It was determined by the project supervisory panel that this was an unproductive avenue and that only basic directions should be utilized, i.e. if there were external openings to the south-west, an azimuth taking in the entire south-west should be used (minus any contemporary obstructions). The resulting viewsheds are therefore, once again an illustration of the maximum amount of visibility from a location, and are not meant to provide an exact representation of any one person's projective view. The azimuth utilized for the general view mentioned above (which was taken from the highest possible element of the castle) was a full 360° to imply that from those positions topography alone would have been the major visual obstruction.

Visual artifacts

Most of the satellite imagery available today is from declassified government intelligence sources (Connolly and Lake 2006: 68, 72). Most of it is of a very good quality. However, it is worth noting that erratic features (unofficially known as artifacts) (http://resources.arcgis.com/content/search-result?searchKeyWord=artifact&clearCacheTime=1279122123&searchTaxonomy=&searchProduct=&searchRC=All&searchCollectionType=0) do occasionally crop-up, especially (though not exclusively) where large bodies of water may interfere with satellite coverage. These are not natural features and can usually be recognized as such. Figure 3.1.1 shows a good example of linear artifacts to the west, east and south of Trim Castle; these particular artifacts are the edges of the satellite photographs or tiles. (When obvious artifacts are seen in the DEM for a castle within this study, they are noted in the text portion of the analysis.) The final constraint within GIS is the fact that the interpretation of the viewshed analysis has a human dynamic. While the methods used are objective, the analysis made from a DEM and an ensuing viewshed is open to individual interpretation. This creates an odd conundrum: an objective analysis with an element of subjectivity to it. While that human element cannot and should not be removed, it can be acknowledged. Obviously the viewshed cannot offer the actual experience of the past; though certainly there is the possibility that these views could have been experienced. The opportunity to explore the projective and reflective views of the subject castles which the medieval inhabitants of Ireland would have found familiar allows us to be one step closer to them phenomenologically. But while GIS can remove the obstructions and camouflaging abilities of vegetation and man-made structures, humans in the real world must constantly contend with optical interference.

3.2 Human visibility constraints

Constraints on visual and recognition acuity

It has been demonstrated that if not bound by topographical constraints or curtailed by arbitrary boundaries, GIS could extend the possible LoS from a castle to improbable distances. In contrast, the ability of the human eye is quite restricted, and there are many factors which further diminish our visual ability. This finite human characteristic constitutes the major drawback within a visibility study and brings to light the consideration that visible is not necessarily the same as noticeable. While a castle in the distance may not be obscured by topography, vegetation or built environment, it still might not be distinguishable. Both the structures and the environments surrounding them have a large part to play in whether or not a castle can be seen within its setting. Simply because you can see an object far-off in the distance does not guarantee that you can recognize what it is you are seeing (Gillings and Wheatley 2001: 12). Sight is the sense that provides most humans with the majority of their information about the world around them and, the importance of vision to human perception is apparent in the size of the visual cortex of the brain (Llobera 2007: 52) (Figure 3.2.1). Ogburn has suggested that visual acuity has the most direct influence on how far away an object may be seen (2006: 406). Recognition acuity (the ability to recognize and identify an object) is the most widely known measure of visual acuity, and the most relevant in identifying a distant object within a landscape (Ogburn 2006: 406). Fraser has suggested that having some measure of prior knowledge as to the existence and position of a target point is very helpful in determining what that object is once it is sighted (Fraser cited in Gillings and Wheatley 2001: 12). The phenomenological portion of this study has suggested that if one expects (or hopes) to see a particular object in the distance; one is more likely to recognize that object when it is first spotted. But, as Ogburn (2006: 405) has asked, how visible is visible? It might be supposed that castles, which can be relatively substantial anthropogenic features would have been relatively easy to recognize within the landscape, but that was certainly not always the case.

Natural and environmental constraints

One dynamic that that can limit visibility is seasonal atmospheric conditions (Gillings and Wheatley 2001: 12). The presence of precipitation, haze, fog or pollution can increase the optical density of the atmosphere, decreasing visual acuity and recognition acuity in turn. This is why views appear so much clearer in dry climates, deserts or elevated areas, and why landscapes in Australia or the western deserts of North America seem to have such high levels of clarity compared to parts of the eastern seaboard of North America and Europe (Bishop 2002: 713, 717), which without doubt includes Ireland. This said, even though atmospheric conditions can reduce visibility, some objects can be recognized over great distances, even under less than perfect conditions; Ogburn cited the Golden Gate Bridge in San Francisco, California which can be seen, even in overcast or hazy conditions from as far away as 20km (2006: 410-411). This suggests that at times, the nature or colour of the object itself significantly affects its own visibility.

Cultural objects in the past may have had slight contrast with their surroundings, due in part to the use of locally available materials. In order for these objects to be highly visible, or even to stand out at all, special measures may have been needed. It may have been for this reason that William the Conqueror imported Caen stone from Northern France to dress the White Tower (Tower of London). Salzman (1952: 157) has suggested that the use of a lime-wash or other surface treatments were applied with “a lavish hand” throughout the medieval period; Corfe Castle (Dorset) was white-washed both externally and internally in 1243, as was Guildford Castle (Surrey) in 1255. And it would appear that the Caen stone alone did not achieve the full desired effect at the Tower of London, for the tower was apparently “resplendent” with white-wash (ibid). The White Tower, which is depicted in a book of poems by Charles, Duke d'Orleans (1391-1465), can be seen in Figure 3.2.2. This illustration implies that the tower was called the White Tower because it was indeed white. We have ample reason to suppose that castles were often treated externally in some fashion (Ogburn 2006: 407), and can point to numerous examples of castles portrayed in medieval art which suggest this, such as the examples within the Duc de Barri's illuminated 15th century Book of Hours shown in Figures 3.2.3 - 3.2.8. This treatment is still being used on castles in Scotland and in Europe as shown in Figure 3.2.9, a modern photograph of the great hall at Stirling Castle. During the subject time period in Ireland, which had a shortage of stone (freestone) which cuts well (hence the use of rubble walls instead of ashlar stone work on many castles in Ireland), we might expect such rendering, and it is certainly found on later tower houses.

Along this line, contemporary depictions of late medieval castles frequently portray flags and pennants flying over gate house battlements and/or cap-houses of the great towers. Although the images shown above (Figures 3.3.3 – 3.3.8 above) have been reduced greatly in size making it difficult to see, flags and pennants are present on many of these castle illustrations. The use of such devices could have aided in visual recognition in three ways: the use of colour would help a traveller to distinguish the solid (often dark) masonry of the castle from a vegetative monochromatic theme within the surrounding environment; the movement of a fabric flag in the wind atop an otherwise stationary structure could call the eye; and the elevation advantage of the increased height of the pennant's pole over the top of the castle could increase the chance of a reflective view by someone pressing for a first-sighting of the castle. The first impressions of Sir Gawain regarding Bercilak's castle found in the medieval poem Sir Gawain and the Green Knight, suggest that not only were the turrets gleaming white with painted pinnacles, they had intricately carved ornamental tops (Anderson 2005: 200).

Similar to the reduced visibility caused by atmospheric conditions is that caused by a competing background environment. For instance, although a dark building may be a potent visual marker perched atop a hillside and silhouetted against the skyline, when it is located on the side of a hill nestled within a backdrop of verdant foliage it may not stand out well during certain seasons. Likewise a castle in a valley may have to compete visually with its environment (vegetation or man-made structures), and without distinguishing aspects such as flags, banners, pennants, or as Fraser suggested, a beacon (as quoted by Wheatley and Gillings 2000: 6) to set it apart from its surroundings, it can easily blend-in or seem to melt visually entirely into its environment.

We have seen that there are inherent boundaries placed upon the research by the very nature of the approaches being tested. Most of these are of a technical nature and are relatively straightforward. Some, as regards human visual ability under diverse conditions, can be anticipated, but cannot be resolved. Instead, the constraints must simply be acknowledged and dealt with when, and if they occur. Next, the process of data collection, collation and analysis is summarized, and the site reports and viewsheds will be introduced.

Chapter 4 Data used

4.1 Data collection and recording

Assessment of castle fabric and personal observation of the projective view

Field research took place over two summers, (2007 and 2008). Subject castles were visited, each site visit lasting between two to four days. A determination of the prime components within and outside each castle was done by a careful examination of the castle site compared with published castle plans. Geographical coordinates of these prime components, as well as around the extent of the curtain walls were taken using a hand-held GPS unit (Garmin eTrex ® H GPS Personal Navigator >10m (33 ft) RMS: accurate to within 3-10m on the X and Y axis). After establishing the locations of the prime components, the openings of extant windows and exterior doors within each were assessed; specifically the directions the openings faced and the azimuth of the projective views. Measurements of all these openings were taken using a standard metric tape-measure and an extendable 5m measuring rod (often only approximate measurements were obtainable due to safety concerns), and the elevations of the openings were determined. Then any extant special features (mouldings, sills, draw-bar holes etc) were noted. Next, photographs were taken of the prime components, and when possible interior and exterior views were taken of each window and door opening. When safety permitted, the projective views from each component were assessed and photographed to verify if the visual expectations mentioned above were met. These photographs were later compared against the viewsheds from those components to see if the viewsheds were in fact realistic.

Assessment of the wider environment and documentary cartography

When the site examination of the castle fabric was complete and projective views had been assessed, the wider environment of the castle was investigated. This was done by noting the presence or absence of topographical features as well as the proximity to a fresh water supply, transportation routes and villages or towns. Circumstantial evidence, such as street and townland names as well as landscape evidence of seigneurial markers such as parks, fishponds, warrens, dovecotes, bridges, mills, religious establishments and planned settlements (Liddiard 2000: 2, 60-62, 67, 78, 84, 97-98) were noted if found. Documentary evidence of these lordly aspects was subsequently searched for in literature and then included in the data. The possibility of a managed or structured approach to the castle described by Liddiard (2005: 10, 142) was also assessed.

Subsequently, the extended reflective views were pursued. To test the views which might have been experienced by the stranger community, the possible likely approaches to each castle were determined by a comparison of the joint Ordnance Survey Northern Ireland (OSNI) and Ordnance Survey Ireland (OSi) Complete Road Atlas of Ireland (1998) with the Taylor and Skinner 1778 Maps of the Roads of Ireland itinerary of Ireland. Those routes deemed similar enough to have offered a contemporary sighting of the castle were travelled, starting generally at an 11km distance from the castle and proceeding towards the castle (this was not a hard-and-fast rule, especially when obvious topographical features would have obscured the castle for much of that distance). When the castle came into view the first time, the position coordinates were recorded using the GPS hand-held unit, and the castle was photographed from this location. This reflective view is known herein as the 'first-sighting', and is seen in the site reports on the viewsheds generated from the gate house battlements.

It must be stressed at this point that reconstructing the transportation routes in and around early medieval castles throughout all of Ireland required a heavy use of later medieval cartographic evidence. The range of early maps of Ireland are limited and "of variable quality" (Reeves-Smyth 1983: 119). The documentary cartography used almost exclusively in this study is the all-island itinerary published by Taylor and Skinner in 1778. While this is substantially later than the subject time frame, these are the only older comprehensive maps covering the entire island. These strip road maps included details of settlements, the transportation routes that connected them and often the names of the landowners in between the settlements (Reeves-Smyth 1983: 126). Before settling on the Maps of the Roads of Ireland as a likely standard for determining medieval transportation routes, a comparison was made between these 1778 maps and several of the oldest maps of Carrickfergus. As a royal castle, Carrickfergus was given almost preferential treatment by early cartographers, prompting the creation of several of the earliest maps either suggesting or actually showing roads in Ireland. These early maps (which are not shown in this study) include 'Kragfargus Towne', c.1560 (British Lib., Cotton MS Augustus I ii 42); Carrickfergus, c.1596 (Public Record Office, London, MPF 98 (ex SP 64/1/31); and Carrickfergus, 1685 by Thomas Phillips (National Library of Ireland, MS 3137 (42). The roads suggested in these early maps supported the routes depicted by Taylor and Skinner in use around Carrickfergus 200 years later. Interestingly, these same roads are still in use (with slight alterations) at Carrickfergus, and personal observation suggested that most of the roads depicted by Taylor and Skinner in 1778 are reflected in the modern roads in use in Ireland today.

Finally, the reflective views of the local community were determined by locating the seigneurial aspects of the market place and the parish church (or cathedral or religious house founded by the lord). In some cases it was difficult to identify the exact location of these components. Often the original market area for an urban castle has been fossilized in the standing village or town, but this is not always the case. Many towns have simply engulfed the market place, and at castles (which may once been urban but are now quite rural) the discovery of the locations of the market awaits excavation. Determining the position of the parish church was a little easier; when it was not possible to locate the original seigneurial urban parish church established as part of the 'castle package', attention was given to the town's present-day Church of Ireland parish church. The Church of Ireland often simply appropriated the already established medieval places of worship within communities; many of which have been in continuous ecclesiastical use from very early in the Christian period right up to the present day (Roulston 2004: 1, 157; Hamlin 2008: 187). Once located, the reflective views from these prime components were photographed (when possible) and GPS coordinates were taken.

4.2 Data analysis

After the field work had been completed the coordinates of the prime components taken on-site were transferred from the hand-held GPS unit to the computer using a transfer software program provided by DNRGarmin (Department of Natural Resources, State of Minnesota), version: ArcView Extension 5.4.0. (http://www.dnr.state.mn.us/mis/gis/tools/arcview/extensions/DNRGarmin/DNRGarmin.html). The data was placed in an Excel spreadsheet format that could be used with the GIS program. The program used at the Queen's University of Belfast is the ESRI (Environmental Systems Research Institute) ArcGIS version 9.3; the international-standard in GIS software. The field coordinates were then added to a DEM within the ESRI program.

As the ability to visualize the terrain around a monument is one of the most beneficial aspects of the GIS DEM, a topographical map of each castle was created and is included as the initial illustration in the castle's illustration section. This allows the wider environment surrounding the castle to be understood graphically. An example can be seen in Figure 4.1.1. Likewise, determining the location of a monument on a cultural map can fix it spatially in our minds, while binding it to the land gives it a 'place'. This allows us to 'see' it mentally within its wider environment and to visualize it in relation to the features (both topographical and man-made) around it. For ease of this mental positioning of a castle, and because it is a simple process in GIS, a current digital road map was overlain onto the topographic map of each castle, placing the castle within the context of its current urban or agrarian setting. A 2km radius around each castle is used so that the roads approaching the castle can clearly be seen. This is shown in Figure 4.1.2. The maps are from two different sources: OSNI 1:50,000 raster maps, information made freely available under the Northern Ireland Mapping Agreement (NIMA), and OSi 1:50,000 raster maps (which had to be purchased). The sample viewsheds in Figures 3.1.1 – 3.1.3 illustrate the differing styles of digital road maps provided by the two different sources. Figures 3.1.1 and 3.1.3 used OSi digital road maps, and it is obvious that they offer a good amount of detail in road names and route numbers, place names, monuments and even topographical changes. Figure 3.1.2 is an example of OSNI road data, which although distinct and easy to see, offers no extraneous information and is somewhat less helpful.

These modern road maps turned out to be a useful tool in several other ways; by overlaying the existing road systems with the topography we are able to see how the modern roads come in and out of LoS with a castle. This helps illustrate how the castle's topographic setting could have interfered with both the projective and reflective views. Also a comparison of the recent maps with the older maps makes it possible to see the degree of change these transport routes have undergone since the publishing of the Taylor and Skinner itinerary. It was possible to determine that the modern approaches were similar enough to those depicted by Taylor and Skinner in 1778 to offer basic comparable reflective views of the castle.

Finally, constraints were set, (distance of views, elevation offsets and azimuths) and a general viewshed for each castle and its prime components were generated. Examples of viewsheds have already been introduced in Figures 3.1.1 – 3.1.3. These offer samples of the three different DEM sources as well as the two different pixel sizes used in the study. Figure 3.1.1 is an example of the Landmap DEM (accurate to 25m), Figure 3.1.2 was created using the OSNI DEM (accurate to 10m), and Figure 3.1.3 utilized the OSi DEM (also accurate to 10m). A table illustrating the various tools and programs utilized in this study can be seen in Figure 4.2.1.

Brief tour of the viewshed

In the sample viewsheds above, the areas which are visible are depicted in black, and those areas which cannot be seen are not obscured in anyway (coloured). Although portraying the visible area in black and fully illustrating the portions of land that are not visible may seem counter-intuitive, this reverse portrayal of visibility allowed the DEM to illustrate best the topographic reasons why areas are not visible. In many cases the percentage of land that could not be seen was much larger than the percentage visible and to do it the other way around (showing only what is visible) would have been less informative. Because theoretically inter-visibility between two points can be assumed (Wheatley and Gillings 2000: 8), the same viewsheds are used to portray both the projective and the reflective views. Consequently areas in black could have been visible *from* the component, and/or could have had a view *of* the component. Areas illustrated in colour are places that would not have been visible from the component, nor would they have had views of the component.

The site report

A site report for each castle establishes the known construction history and the subsequent owners of the castle throughout the subject time frame. It then looks at the building sequence of the castle, pin-pointing the order of construction and locations of the prime components (gate house or gate tower, great hall and the lord's chamber/s). Proximity to natural amenities, such as sources of fresh water and fords, as well as to anthropogenic features such as transportation routes, towns and seigneurial features are assessed. It also looks at the possible significance of each castle site to the indigenous Irish prior to the arrival of the Anglo-Normans in an effort to determine if the sites had previously been associated with authority (power-bases either religiously or royally within the pre-Anglo-Norman Irish population). The model then establishes what the battlements or the openings (windows and doors), of these components might have looked out upon. Finally the projective and reflective views are analysed and compared to views seen at other Anglo-Norman castles within the study. The following illustrates the format of the site report:

- **Introduction:** a review of the documentary sources of the castle and an introduction of the history behind the castle's construction.
- **Location of the principal components:** a discussion of where the projective viewing positions were located in relation to the castle in general.
- **General view:** a discussion of the maximum possible views (both projective and reflective) that could be obtained for each castle.
- **Projective views:** an analysis of each of the prime components within the castle, specifically the directions that the windows and doors of the components faced; with an analysis of the projective viewsheds from these places. This section ends with a comparison of these projective views with each other and with the prime components of other Anglo-Norman castles.
- **Reflective views:** an analysis of the reflective viewsheds from the hinterland as a whole, but specifically from the prime components outside of the castle: the various roads approaching the castle, the parish church and/or market place (if these locations are known). This section ends with a comparison of these reflective views with the locations of reflective views from other Anglo-Norman castles.

Having now described the process by which the data was collected, viewsheds created, and the model which was used to analyse the results, we will move on to the compilation of site reports and comparative analysis of the sample castle population.

Part II

Chapter 5 Corpus of castle research

5.1 Adare Castle, Co. Limerick

Note: Because of the large amount of stone remaining at the castle site, the GPS satellite signals 'bounced' and became 'scrambled'. This simply means that when they were projected onto the DEM, the locations were incorrect, and did not reflect the true location of the prime components on the X and Y axis (this issue is discussed in detail in Chapter 6: section 6.2). In place of the faulty primary coordinate data, OSi orthophotographic coordinates were used for all primary components: outer gate, inner gate, early and later hall and parish church. These were found online at: http://ims0.osiemaps.ie/website/publicviewer/main.aspx#V1,546949,646800,7.

Introduction

Adare Castle, Co. Limerick was built on low ground against the northern bank of the River Maigue approximately 18km south-west of the city of Limerick (Figure 5.1.1), at a position along the Maigue that was both fordable (the name Ath-dara means 'ford of the oaks') *and* navigable (Dunraven 1865: 160). It is believed to have been built upon a substantial Irish ringfort (Dunraven 1865: 105; Adams 1904: 3; Leask 1977: 34; Dunne 2007: 167-170). This belief was challenged by Rynne in 1961 (193-202), and later by Sweetman in 1980 (1-6), who suggested the foundations of the castle were entirely Anglo-Norman. Neither documentary nor archaeological evidence for this 'ringfort versus ringwork' debate is conclusive, but regardless of its origins, the location of Adare Castle was chosen by Geoffrey De Marisco, who had been granted the land by 1199 (McNeill 1997: 38, 50).

De Marisco began his stone castle at Adare soon after he received the land (McNeill 1997: 38), but possibly as early as 1236 the castle and land had come into the possession of Maurice Fitz Gerald, the elder (Dunne 2007: 155). It became the possession of Maurice Fitz Gerald the son at the time of his marriage to Agnes de Valence in 1266, and was improved by the Fitz Geralds in the second half of the 13th century. The castle and lands were held by the Kildare portion of the Fitz Gerald family for the next 300 years (Dunne 2007: 155-6). Much of this castle has been repaired, first by the Dunravens in the 19th century and more recently by the Office of Public Works. The effect of this modern repair work made it difficult for this researcher to determine some of the modern rebuilds from the 19th century work. Fortunately a Master's thesis done by McCleary (1997: 77-127) of Queen's University Belfast includes a skilful photo documentary of the castle before the recent OPW's repair work was conducted. His efforts are gratefully acknowledged.

Today Adare Castle sits on the outskirts of what is termed 'Ireland's prettiest village'. However the position of the current town is relatively recent (19th century). Thomas states that the town had already been moved from its original location sometime before 1656 (1992: 2-3) and suggests that the most likely spot of the original town is the north bank of the River Maigue, surrounding the castle on three sides and encompassing the parish church (Figure 5.1.2). Although there was a tradition of a town gate being located approximately 18m from the castle's north gate (Dunraven 1865: 127), again the archaeological and documentary evidence for this specific location is inconclusive. That there *was* an early town associated with Adare Castle is incontestable, but the exact position is unproven.

Seigneurial aspects of Adare Castle are suggested by both excavation and documentary sources and includes the following: a mention by (The Countess) Dunraven of an early mill at Adare (1865: 128); the original castle plan of Adare drawn by J.R. Jobbins in 1865 labels the stream north of the castle as the 'Old Mill Stream'. This same plan also illustrates a linear channel extending from the Maigue northwards and Dunne suggests that had this channel connected with the stream, it would have extended the moat and would not only have "greatly enhanced" the protection of the castle, but it could also have helped to power a mill (2007: 159). This idea is supported by the fact that in the 19th century the probable foundation of the mill and a millstone were uncovered to the west of the castle, between the moat and the semi-circular tower in the north-west curtain (ibid).

Documents cited by Dunraven suggest that a hospital was established either before or during the subject time frame and its presence on or around land granted to De Marisco would have added to the prestige of the manor. At some point in the study time frame (and perhaps even before-hand), the manorial parish church of St. Nicholas of Myra was built just north of the castle (Killanin and Duignan 1967: 54), however the earliest documentary reference to the parish church at Adare is from taxation records dated between 1291 and 1302 (Dunraven 1865: 94). Although eventually Adare included three priories (Trinitarian, Augustinian and Franciscan), only the Trinitarian Abbey was founded during the study time frame. It was built during the 1230's across the river approximately 700m south-west of the castle. Early documentary sources do not mention a deerpark, but there are two later references to a deerpark covering hundreds of acres in sources mentioned by Dunraven; one dating from 1767 and one from 1776-78 (1865: 2-3).

Because the location of the early village is unknown, we cannot tell how the road might have passed through it or what seigneurial markers may have been displayed along its route before the travellers arrived at the castle's outer gate; giving us no *current* evidence of a managed landscape. Certainly however, the bridge itself would have been a potent marker. Figure 5.1.3 is a reproduction of the Ordnance Survey Ireland digital raster map which gives an illustration of the network of modern-day roads surrounding the castle.

Location and discussion of the primary components
The primary components of gates, great hall and lord's chambers are to be found in four principal features at Adare Castle; the outer gate, the inner gate, and the two great halls (an early one, and a later one built 50 to 75 years after the first). The outer gate allows entry to the castle's outer ward, and the inner gate guards a moated inner ward in which a tower is located. The two halls are positioned along the south curtain wall, the oldest at the west end, and the newer hall to the east (see Figure 5.1.4). The location of the lord's chamber was difficult, for although a good case can be made for the location of the lord's chambers during the later part of the study time frame, the exact position in the early stages of the castle is unclear. It was suggested by Adams in 1904 (2-3) and has long been assumed that the central building in the inner ward was a 13th century keep (owing no doubt to the fact that the ruins now are of a four storey tower). It is now agreed however that this building only had a ground-floor and a first-floor (reaching to attic level) (Sweetman 1991: 67; McNeill 1997: 41; McCleary 1997: 46) until the 15th century when another two stories raised it to tower height. Subsequent writers have perpetuated the idea of this building as a keep or great tower even though this is not technically true for the study time frame (Leask 1977: 34-35; Sweetman 1980: 1; Sweetman 1991: 67, 74; Dunne 2007: 157). Though this structure may well have taken on the role of keep or great tower *eventually*, its use as lord's chambers during our time period cannot be substantiated and it would be inappropriate to consider it as such here. Unfortunately this leaves us without a location for the lord's chambers for the early phase of Adare.

Nevertheless, after the construction of the later great hall at Adare, the early hall may have filled the role of lord's chambers (McNeill 2009: pers.com.). Indeed, all that seems to have been missing from the early hall to turn it into a comfortable set of apartments was a garderobe and fireplace. McCleary suggests a fireplace may have been present in the now somewhat restored west end (1997: 33), and of course it is possible the space had a central hearth, but the truth is we may never know. What we do know is that at some point in the late 13th century (around the same time the later hall was constructed), an elaborate latrine tower was built attached to the south-west corner of the early great hall, extending out over the Maigue. Because of this significant addition to an already fine venue, it seems more than likely that the early hall did in fact become the lord's chambers and it was examined in both the context of great hall and lord's chambers.

The later great hall and the latrine tower provided something of a challenge to this examination of Adare Castle. Because there are no clear clues by which to date these structures, they have been given a rather unspecific construction date-range of sometime within the 13th century (Adams 1904: 2; Leask 1977: 35), or even of the *late* 13th century (Sweetman 1991: 67; Dunne 2007: 161). If they were built towards the tag-end of the 13th century, they *may* not fall within the time perimeters of this study, but as there is no way of conclusively establishing their construction dates, for the sake of thoroughness they have been included here.

The first component built at Adare Castle is thought to be the early great hall. It has been dated by its fine two-light windows (which are comparable to a window of this date extant in the Cistercian monastery of Monasternenagh (Leask 1977: 35; McNeill 1997: 38), to the years immediately before or after 1200. A sequence of construction has been detailed by McNeill in the castle plan seen in Figure 5.1.4, which suggests approximate dating from the first construction to the stabilization efforts conducted by Lord Dunraven in the 19th century.

General views
As illustrated by the topographic map, Adare Castle sits on very low-lying land. What perhaps is not clear from this map is that this position, on rich ground immediately adjacent to the River Maigue is somewhat bowl-like; with the area north of the castle rising slightly, and then dipping again within 500m. The effect of this position on the projective views from Adare can be seen in the general viewshed created from the position of the later lord's chambers in Figures 5.1.5-7. It was generated with an elevation offset of 10m (approximating the wall-walk level of the latrine tower), and represents the most extensive of all the projective views seen from Adare Castle during the study time frame. For this reason it is included as the general viewshed.

If we use the analogy of a clock face, the farthest reaching views from the lord's chambers are to the south-west between 6:00 and 7:00. Towards this direction, there is a fairly wide area visible approximately 9 to 11km away. From 4 and 9km from the castle, there are small patches of visibility between 4:00 and 5:00, and between 8:00 and 10:00. Within a 4km radius, the densest visibility is to the south-west between 6:00 and 7:00. The best over-all views are seen in the south half of the first kilometre, but there are no uninterrupted or far-reaching projective views from Adare Castle. It must be noted however that as the latrine tower was not built until the late 13th century, the views seen in Figures 5.1.5-7 were only available at the *end* of the study time period.

Projective views: viewshed and personal observation
Gates
The current castle at Adare has four gates; the outer gate, the inner gate, and the north and east gates. The later north and east gates are little more than basic openings through the curtain wall fortified only by wooden doors; similar to the original gate at Carrickfergus Castle. There was also a water-gate, which was a simple door-way leading off the Maigue and opening directly into the later great hall. Because the north and east gates are chronologically later than our study time frame, and the water-gate was a simple postern door geared to accommodate river, and or service traffic, only the main outer gate and the inner gate were considered here. The viewsheds from both these gates are sparse and fall mostly within the first kilometre surrounding each gate.

Outer gate

The outer gate was a simple round opening, which faced west and opened into the outer ward. McNeill suggests it was built after the early great hall, but before the inner gate; and only later strengthened with a tower and portcullis (1997: 38). This is evident in the clear straight joint between the tower and the early great hall. The viewshed for the outer gate gives a startling example of the limitations of Adare's defensive projective views in general. This viewshed was generated using a 10m elevation offset to approximate the height of the outer gate tower, and the obstructing effects of the great hall, the inner gate and the central building were factored into the calculation. The result can be seen in Figure 5.1.8 which shows how very little would have been visible from the top of the outer gate tower. Indeed, *all* of the land visible from this gate falls within the immediate 1km radius around the castle, so an 11km and 4km radius view are not shown.

The Maigue would have brought river traffic from the Shannon to the north, and the road from Limerick would have conveyed overland travellers from the largest town on the west coast of Ireland (and its royal castle). These approaches would have been closely observed by look-outs stationed atop the gate tower. It is noteworthy that the river does not come into view from the outer gate until it is less than 200m away, and the road only crosses the line-of-sight from the gate tower at one point; approximately 250m north of the castle, then it drops from view again and reappears only 100m or so from the outer gate. Essentially, traffic (both friend and foe), coming from the north or the west would have been unobservable until it was at the front door. The projective views of traffic approaching from the south are no better; this road does not come into the sight from the outer gate at any point until it is within 150m of the castle.

What *would* have been visible from this gate is the original fording point of the river, which is located about 75m closer to the castle than is the current stone bridge, and would have been well within bowshot. This perhaps gives us a better indication of where the main transportation route might have ran during the subject time frame. The likelihood is that the road actually passed within 10 - 40m of the outer gate before crossing the river at the ford. This would have allowed approaching traffic to be visible within about 133m of the outer gate, but this is only slightly better than the modern road. This is perhaps a good place for a reminder that the viewshed represents only topographical obstructions to vision, and does not take into account vegetation or built environment. If the area surrounding the castle had not been kept clear of vegetation and encroaching buildings, the projected view from the outer gate would have been even more limited. These approximate views (taken from the wall-walk just north of the gate) can be seen in Figures 5.1.9 and 5.1.10, and a photograph of the gate from the exterior is seen in Figure 5.1.11. The fording point on the Maigue is visible in the lower left portion of Figure 5.1.9, suggesting that from the outer gate, the ford would have been completely observable. Personal observation from this point supports the viewshed well, and highlighted the problem that vegetation causes in projective views.

Inner gate

The inner gate house and ward were built sometime within the first 75 years of the castle. It is slightly more technologically advanced than the outer gate, having a drawbridge which protected its outside, opposing loops on its east and west side walls to provide flanking fire onto the base of the inner curtain walls, and a stout wooden door on the inside (Dunne 2007: 161). The upper parts of this structure were stabilized by Dunraven in the 19th century, but are likely to be an accurate representation of the original. The two main gates are separated by approximately 27m and are at roughly the same elevation. The viewshed from this gate can be seen in Figures 5.1.12-14 and was generated using a 10m elevation offset, and by taking into account the obstructions caused by the central building and the great hall. The later great hall was not factored into the calculation because for most of the study time frame it may not have been in existence. However, including the later hall in the equation did not alter the available projective view much, as very little is visible even *without* that as an obstruction.

The projective view of approaching traffic is also limited from the top of this gate. To the north, the Limerick Road is not visible until approximately 100m from the castle, and about that same distance can be observed along the road to the south. It was not possible to ascertain whether the fording point would have been visible, but the location and height of the outer gate and curtain wall may well have made a view of the ford impossible. Unfortunately access was not available to the top of this gate, and no photographs from this vantage point are available. Figure 5.1.15, which was taken within the inner ward, looking out through the inner gate illustrates that this view would have been almost entirely of the interior of the castle. Figure 5.1.16 is a photograph of the exterior of the inner gate.

Great halls

Early great hall

As mentioned there are now two buildings identified as great halls at Adare. The earlier was a first-floor hall, at the angle of the west and south curtain walls. This is a rectangular building measuring 17 by 9.4m internally (Dunne 2007: 159), its long axis running east to west. The main entrance is in the north wall, directly south of the outer gate. Although many fine features have been lost to time and much repair work has been conducted, several windows are original and provided both dating evidence and the pattern for reconstruction. The viewshed in Figures 5.1.17-19 illustrates the projective views possible from this hall. It was generated using an elevation offset of 6m to approximate first-floor level. The visual obstructions caused by the outer gate, inner gate and the central building in the inner ward have been factored into the calculation by adjusting the azimuth offsets. What is slightly surprising was the *lack* of projective views, either immediate or far-reaching from this first-floor hall.

The rising and falling ground around the castle curtails visibility to what is basically the local neighbourhood. Using the analogy of a clock face, the farthest reaching views from the great hall windows are to the south-west between 6:00 and 7:00. The 11km viewshed revealed that there are some far-reaching views from this direction (visible approximately 9 to 10km from the castle). But between 4 and 9km, there was virtually nothing visible in any direction. Within the 2 to 4km radius the slice from 6:00 to 7:00 is again visible in splotches, but little else. The best over-all views then are to the south-west, within the first 4km from the castle. But there are no uninterrupted projective views from the early great hall at Adare Castle.

The most accessible and easily viewed projective views from the first-floor hall would have been of the river flowing directly along the south curtain wall, the ford across the Maigue and the south bank of the river opposite the castle. The narrow west window would have been able to view the ford, a very small portion of the Limerick Road as it directly passed the front of the castle, and the approach to the main gate. To the north and east, the projective views would have been almost entirely of the interior of the castle (although before the later hall was built there may have been some views of the immediate countryside to the east). It is suggested by the viewshed that the possible views from the early great hall were not extensive and would not have been either especially impressive or informative.

The hall had a total of six windows, five of which have twin-lights; two in the north wall east of the entrance (the west window in this wall is a modern replacement, based on the east window), one centred in the east wall and two more in the south wall. These two south facing windows pierce the south curtain wall at first-floor level, as does the narrow single-light window which was inserted in the north-west angle of the west wall in the late medieval period (McCleary 1997: 33). Both windows on the north wall currently have window seats; however, a photograph (not shown here) taken by McCleary (ibid. 109) seemed to suggest that the window to the west may not have had one before the recent renovations. The entrance and north-facing windows can be seen in Figure 5.1.20, and the window in the east wall can be seen in Figure 5.1.21. The windows on the south wall (Figures 5.1.22 and 5.1.23) also have window seats, and though they pierce the south curtain wall, only the west window has a draw bar hole. The single-light window in the west wall is shown in Figure 5.1.24, and though it also pierces the west curtain wall at first-floor level, it does not have a draw bar hole. Still extant in this hall are parts of the exterior mouldings on the windows of the north wall, the fine red sandstone casing with roll moulded sill on the east window, and the continuous roll mouldings across the top of the interior embrasure on the west window in the south wall. Unfortunately the windows were inaccessible and photographs of the views from them are not available.

The later great hall

The second great hall at Adare (like the earlier hall) is built against the south curtain wall. It is located just 15m to the east of the first hall, and is a rectangular building, the long axis is east to west, and the ends are gabled (only the west end is extant). It is larger than the early hall, measuring 22.7 by 11.2m (Dunne 2007: 161), and incorporated twin service rooms at its eastern end. Unlike the earlier building it was a ground-floor hall and the gables are windowless. The Dunraven plan shows that this later hall had a porch and two aisles (1865: 321). It was certainly a grand space, with a wide entrance, four stone pillars, a water-gate doorway, and three large pointed two-light windows. Sweetman suggests there were also a pair of two-light windows in the north wall which has unfortunately been reduced to foundation level (1991: 67).

The viewshed from the windows of this structure can be seen in Figure 5.1.25-27. It was generated using an elevation offset of 1.7m (approximating eye-level on a ground-floor), with the curtain wall and the central building of the inner ward taken into consideration as visual obstacles. We might expect that the projective views (at least to the north and south) from the two halls would have been similar, perhaps differing only slightly due to the later hall's position being approximately 15m farther to the east. The fact that it did not have the height advantage of a first-floor elevation might suggest that the views would be even more limited, and indeed the ground-floor location did rob this hall of any views to the north beyond the curtain wall. But what was discovered is that even with the seeming elevation disadvantage of a ground-floor location, the projective views to the south from this hall are considerably *more* extensive than from the earlier hall. This may be helped by the fact that the ground on which the later hall was built is roughly 2m higher than the early hall (25m versus 23m above sea level respectively). But the shift to the east is apparently enough to decrease the limiting effects of the lip of the depression on the projective views to the south-west. While views do extend to the 11km radius around the castle, the best views are within 4km, to the south and west.

The large main entrance, which can be seen in Figure 5.1.28, is on the north wall. The east end of the hall has a standard service-end plan (Figure 5.1.29) with two side doors opening onto service rooms, and a centre door which opens onto a passage. The passage (which is much restored) is approximately 8m in length. It runs between the service rooms to an outside entrance and Sweetman suggests this may have lead to a kitchen at the east end of the hall (1991: 67). (This centre door was interesting in that it has opposing draw-bar holes, supposedly for security, which are located in the passage, on the *outside* of the door.)

Three of the five openings in the south wall (which faces the Maigue) have been altered by restoration and repair work. This wall now contains two doors to the east end, seen in Figures 5.1.30 and 5.1.31, and three window embrasures seen in Figures 5.1.32-34. We can be fairly certain that the west-most door which is now partially blocked (Figure 5.1.31) was the water-gate. The east and centre embrasures seem to be original and contain two-light windows with pointed arches; however ogee-headed lights were inserted into the west window at some time in the 15th century (Dunne 2007: 161). All three windows have window seats and draw-bar holes.

A photograph of the view to the north from the later hall can be seen in Figure 5.1.35, it is clear that the curtain wall cuts out any distant view in this direction. A panoramic view of the River Maigue and the opposite river bank (taken from the door/window at the east end of the south wall), can seen in Figures 5.1.36 and 5.1.37. From these photographs we can see that though the exterior projective views to the south are sylvan and aesthetic, they look out at ground-level and directly into vegetation which completely obscures the far-reaching view suggested by the viewshed. This serves to underscore the value of GIS in this context. Although the hall would have had views of the ford across the Maigue (demonstrating the power of the lord to control traffic at this point in the river), whether or not the views of the other bank displayed lordly wealth would have depended upon what had been located directly across the river. Fortunately, although we have no hints at what that might have been, we now know from the viewshed that the scene may have been somewhat informative about the hinterland.

Lord's chambers

The location of the lord's chambers during the first stages of the castle is unknown, but as mentioned above the early great hall makes a fairly convincing and accommodating lord's chamber after the addition of the latrine tower over the river. The view from the first-floor of the early great hall has already been discussed, and projective views to the south-west would only have decreased with the construction of the latrine tower. However, the latrine tower has an added storey which increased the available viewshed from this portion of the lord's chamber considerably. Although it is doubtful that the lord and his family would have gone to the garderobe on the first-floor of this tower for the express purpose of taking the view, the latrine chute does not appear to extend to the second-floor and it is not impossible that this storey was used as a salon. McCleary suggests that an interesting offset in the width of the extant south and west walls at first-floor level may indicate that a battlement-walk ran along the roof-line of the tower (1997: 33-35). A photo of this structure, taken from outside the later great hall can be seen in Figure 5.1.38.

The windows in the south and west walls of the tower are extant. Those on the first-floor (garderobe level) are narrow single-lights and appear to be original. The windows on the second-floor have been altered with the insertion of ogee-heads (similar to those inserted in the west two-light window in the later hall), but both of these extant second-floor windows have original window seats. These windows can be seen in Figures 5.1.39 and 5.1.40. If we take into consideration the added elevation of the second-floor of the tower, the window seats in the embrasures at that level and the wall-walk above it, it seems quite conceivable that the extra storey and wall-walk above (possibly as an observation platform) were intended to provide additional projective views from the lord's chambers. With this in mind a viewshed was generated from the position of the wall-walk using an elevation offset of 10m. The obstruction of the outer gate and the central building in the inner ward were taken into account. The result, which was introduced in Figures 5.1.5-7 was a surprise. The projective viewshed from this location not only exceeded all other views from the castle, it actually *doubles* the view possible from the later great hall. The viewshed shows that a larger portion of the area outside of the 4km radius is visible from this location; an entire slice between 9:00 and 10:00 on the clock-face has been revealed. Although it is still patchy, this projective view is certainly more extensive than from any other prime component at Adare Castle.

Comparison of these projective views with each other and the views from other castles

The overall projective views that could be had from Adare Castle during its first half century were limited to the immediate localized area. The major factor in this was the choice made to build on low, undulating land along the banks of the Maigue, but this is compounded by the rich land and the amount of vegetation that richness affords. It is not that there were no higher elevations upon which to build in the area, because it is obvious from the topographical map in Figure 5.1.1 that higher positions were available quite nearby. Other than the obvious advantage of observing the ford on the River Maigue, there seems little to recommend this subtle site. It has been suggested that Adare was built on a ringfort, appropriating not only a prepared site, but the prestige and authority already inherent in that site. Unfortunately current knowledge, the lack of documentary sources and the available archaeology simply does not provide enough information to determine if this is true. We may never know. But for *some* reason, De Marisco picked this particular piece of ground upon which to build his castle, and the Fitz Gerald family settled in after him, content to remain in an inconspicuous location.

Other castles which have drastically reduced projected views are Swords, Co. Dublin and Lea Castle, Co. Laois. These castles were also built next to rivers in low-lying areas. Probably the best comparison would be Swords, for (like Adare), Swords had higher ground in the vicinity which might have been chosen for the castle site, but was not. As a result extensive views were not found at either site, both castles are able to observe little more than the land directly adjacent to their curtain walls (which, at both castles were curtailed by vegetation, and in the case of Swords by built environment). Unfortunately the reasons for the choices of sites in either case are unknown.

This can be contrasted with Carrickfergus, another low elevation castle, built next to water (along the Belfast Lough). Like Adare, Carrickfergus was built in the bottom of a bowl-like depression. However, at Carrickfergus the bowl extended out in almost every direction for many kilometres, the ground sloping gently up and away from the castle. This had the effect of providing the castle with a three-sided 'amphitheatre' type of projective views. At Adare the bowl rises around the castle only in the immediate area, and then falls back again, the 'rim' of the bowl effectively cutting off extensive views.

While exterior projective views from the great halls may not have been expansive, they could have been bucolic and serene. It seems significant however; that the *interior* windows of the great halls were able to offer views of the inner gate and its drawbridge. A curious aspect of the construction of the early great hall is why they did not initially go just a bit higher with the hall. We have seen how the addition of just one more storey (onto the latrine tower) basically increased the projective view from this same location nearly four times. Although the first great hall at Carrickfergus was at or very near sea-level, the accommodation went 'up'; the lord supplied *himself* with far-reaching views even if he did not supply his wider circle of *guests* with those views from the hall. This seems to have been attempted at Adare during the late 13th century re-fit of the great hall into the lord's chambers; perhaps the construction of this extension to the lord's chambers was intended to correct the problem of poor projective views.

So we can ask, what were the priorities of De Marisco and the Fitz Geralds as they built the castle at Adare? Both viewsheds and personal observation have shown that it cannot have been visibility. An assessment of the construction sequence of the extant castle fabric gives some hint. First there was a great, unprotected hall. Then the west and north-west curtain wall was built (and presumably some sort of wall completed the circuit to the east). At this time, and ever after the early great hall formed part of the curtain, its two large windows piercing it (and therefore weakening its defensive capability), at first-floor level on the south side. We can only assume that defence must not have been a large concern. Next in the sequence is the central building in the inner ward, followed by the latrine extension to the early great hall (which then likely became the lord's chambers), and the construction of an even larger ground-floor great hall. It is interesting that the Fitz Geralds chose to build a ground-floor hall rather than a first-floor hall, even though a first-floor location would have greatly increased the projective view. However, by the later 13th century, ground-floor halls were popular (as seen by the construction of the later great hall at Trim). Finally was the construction of the wall around the north-east, east and south of the outer ward. The sequence and type of construction at Adare suggests that throughout the study time frame the priority was on making the castle a commodious and fashionable place. These priorities certainly seem to have held greater importance than either visibility from the great hall or defensibility of the castle as a whole.

Reflective views: viewsheds and personal observation
Approaches

The Taylor and Skinner map of the roads approaching Adare (shown in Figure 5.1.41 confirms that in 1778 the town of Adare had one main road travelling through it, allowing approach from two directions; the north and the south. First-sightings were assessed from the north using the Limerick Road (the N20) coming from Limerick town, and from the south using the Rathkeale Road (the N21). We should remember that the River Maigue was navigable to this point from the mouth of the Shannon and can certainly be considered as an approach, but unfortunately assessment of siting from the river was not possible and is not included here. However, it is interesting to note that none of the viewsheds generated for Adare suggest good reflective views of the castle from downstream (the direction that would have been most navigable), although the later lord's chambers and later great hall can be seen from approximately 500m upstream of the castle. The problem with obtaining a reflective view travelling along the approach from the north is mainly topographical (the castle is visible a bare 195m away). Although the viewshed in Figure 5.1.27 suggests Adare could have been seen from the road 100m before the spot it was actually visible, because of vegetation and a stone wall, it was *not* visible. But even without those obstructions, since the road curves around the north-west side of the castle, this would still have placed the first-sighting only about 195m away as the 'crow flies'. The photo of the first-sighting from this direction can be seen in Figure 5.1.42 and was taken at the modern public Adare Golf Course.

Approaching from the south is only slightly better. The castle was visible from 200m away (at a point just as the road begins to cross the stone bridge). The photo in Figure 5.1.43 illustrates roughly what a first-sighting from this direction would be like. Unfortunately the road at the bridge was too heavily congested and dangerous to stop and take a photo at the exact spot of first-sighting, so a position on the south bank of the Maigue, just west of the bridge (which can be seen in the photograph) was used. Figure 5.1.26 suggests that the castle should be visible from as far away as 2km at several points along the road coming from the south, and possibly in the past this may have been true. Today however, this road travels through the centre of the town of Adare, where built environment and vegetation completely obscures the castle until the traveller reaches the bridge. A summary table of the approaches can be seen in Figure 5.1.44. Personal observations of the approaches to Adare Castle have suggested that the viewshed as portrayed to the north is accurate. To the south however, current built environment and vegetation mean that the viewshed is hopelessly optimistic. The castle is by no means dominant in its surroundings.

Community

Today the outer gate at Adare Castle appears as little more than a sturdy doorway into the castle ward. We cannot know how much of a presence it had in the daily lives of the people of the town without knowing where the town sat in relation to the castle. And, as mentioned above, the location of the medieval town and its market place is uncertain. We can however attempt to envisage it in an urban context, as part of a town, or on the periphery of a town. And, in so doing try to determine if it may have been seen as overtly intimidating, over-bearing or psychologically daunting by those who saw it daily. The fact is that this outer gate does not come across as imposing or threatening in any way. It makes no convincing statement of overt power, or even of reliable defend-ability. It is so benign as to appear residential, even inviting. Adare Castle has all the trappings of a defensive castle: a moat, crenulations, inner and outer wards with gates. But for some reason, perhaps because it is built on such low ground (and the gates and walls are not tall by any means), it has more an air of romance than of intimidation about it.

What we do have at Adare is the location of the parish church, still standing only 120m from the great tower. So, although we are unable to assess reflective views of the castle from the site of the market, we can do so from the church. The castle as seen from this vantage point is illustrated in Figure 5.1.45. As with the reflective view from the outer gate, from this vantage, the castle is picturesque, but not inspiring or intimidating. Currently the most impressive portion of the castle from this reflective point is the central building, which is of much later date than the subject time frame. For contemporaries, the roof of this building would have barely been evident above the curtain walls. The visible castle for an individual standing at the parish church between 1200 and 1275 would have consisted of some type of curtain wall and the roofs of the great halls. Had there been buildings (even of modest height), between the church and the castle, the castle would have been obscured. Today, even with the tower rising above the curtain, the sense of the castle is benign, companionable, more like the residence of a wealthy neighbour than of a lord; it can only have appeared more unpretentious in the 13th century.

Comparison of these reflective views with each other and the views from other castles

The reflective views of Adare Castle suggest that the castle simply did not stand out in its environment in any way during most of the time frame this study is examining. The outer gate captured only 1.31% of the reflective view of its closest hinterland, the inner gate was slightly more visible, but could be seen by only 2.13% of its hinterland. The early great hall could be seen by 1.39%. Clearly the castle was not originally sited to be seen. It became more visible towards the end of the study period with the construction of the later great hall and the latrine tower. The later hall, built on a slight rise within the outer ward could be seen by 2.78 % of the surrounding area, and the latrine tower extending over the Maigue captured the most reflective views at 5.5 %; almost doubling the castle's visibility within the hinterland.

Adare was built on an obviously low site, with multiple points within a 200m radius that were of a higher elevation. In an effort to determine how the visibility might have been affected by utilizing a higher position, a viewshed was generated from the location of the highest of these points. The elevation offset used was 10m, with an azimuth of 360°. The resulting percentages of pixels visible from this alternate site were surprising: 9% in the 11km radius, 18.6% in the 4km radius and 40.35% in the 1km radius, compared to 5.5%, 10.34% and 28% respectively from the later lord's chambers. This is an increase in visibility of 61% in the far-distance, 56% in the middle-distance and 70% within the near-distance. Clearly visibility was not a major consideration of siting at this castle.

Actually the inner gate makes a greater statement of power and defence than the outer gate. This gate would only have been seen by those invited into the castle or by those who had gained entrance by force. And perhaps those were the very people the inner gate was intended to impress...individuals able to size up the defensive capabilities; either casually as guests or by default of aggression. The military strength of Adare was in the inner gate and curtain wall, and it was almost wholly passively defensive in nature.

The castle that most closely compares with the lack of command seen at Adare is Swords. Neither one of these castles stood out in their environment, and almost had to be 'stumbled upon' to be noticed. Like Adare, the ecclesiastical castle at Swords has an almost friendly 'good neighbour' aspect to it with the chapel windows piercing the curtain wall at ground-level directly east of the gate itself. But Swords still makes a *pretext* of control with its gate tower. Adare's gate tower blends into the curtain wall so effectively as to seem to be an after-thought, tucked behind the entrance so as to not be aggressive in any way. With the exception of being within arrow's reach of the main road and the ford across the Maigue, Adare could not physically control any aspect of its environment.

By ruling out defensibility, intimidation and physical control as the main concerns at Adare, we are left with the possible priorities of social display, economics (basic administration of the manor) and commodious habitation. The site chosen and the type of castle built suggest that the builders were relatively disinterested in making an *overt* social display. For instance, on this low-lying site a great tower similar to that built by De Lacy at Trim or by De Courcy at Carrickfergus would have been more appropriate if reflective views had been important. That such a tower was not built might imply that this was not a very high priority. Still, there is unmistakable social display at Adare in an inward way; both great halls and the inner gate were fine, fashionable and aesthetic structures, but an observer needed to be within the curtain walls to really appreciate that fact. So although ostentation does not seem to have been a driving priority, it was obviously important. What we see at Adare was a successful manorial administrative centre which was continuously inhabited up until the time it was rendered uninhabitable in 1657. The durability of this as both economic hub and habitation speaks to its position along a major transportation route and the importance placed upon the accommodation within the walls. It was simply a good place to raise a family. The priorities suggested by the reflective views support those suggested by the projective view; commodious habitation, economic base and social dominance.

5.2 Athenry Castle, Co. Galway

Note: OSi orthophotographic coordinates were used for the following primary components: great tower, east curtain building, market place and parish church. These coordinates were found online at: http://ims0.osiemaps.ie/website/publicviewer/main.aspx#V1,550392,728001,8.

Introduction

Athenry Castle, Co. Galway is situated on east Connaught lowlands, at a natural fording point beside the small River Clareen (which becomes the Clarinbridge downstream (Givens 2008: 21). Although the area where the castle and town are located is relatively flat, the land directly to the west, north and east rises considerably within 2km of the castle. The castle itself is sited on a small mound which the OPW suggests was once swampy and entirely surrounded by the Clareen (OPW nd: 1). It lies 29km east of Galway, 55km south-west of Athlone and 200 miles west of Dublin. The topography of the 11km radius surrounding the castle can be seen in Figure 5.2.1. It is not known if the site had been previously used by the Irish, either secularly or religiously. However as it lies along both the most westerly north-south route on the island and an east-west esker, it may have been a regional crossroads (Givens 2008: 21). The OSi map for the area immediately surrounding Athenry can be seen in Figure 5.2.2 (OSi map).

The land was granted by Richard de Burgh to Peter de Bermingham in 1238 as a reward for his participation in the successful invasion of Connaught in 1235. Peter's son Meiler de Bermingham appears to have been the first lord of Athenry, and consolidated his lordship with the erection of the castle (Orpen 1920: 211-213). Leask suggests that construction was *c.* 1250 (1936: 156; 1977: 36), but Sweetman stylistically dates the plunging arrow loops of the great tower in the early 13th century (1991: 70). It was most likely begun immediately after the acquisition of the land by Bermingham (around 1238) and was established to such a point by 1241 that De Bermingham felt confident in pledging his resources to found a Dominican priory just south of the castle (MacAlister 1913: 197; Rynne 2000: http://www.athenryheritagecentre.com/history_castle.htm).

The OPW suggests that the castle may have originally been surrounded by an earthwork and timber palisade (OPW nd: 3). Initial construction provided only a first-floor great hall, but the great tower was probably built to its full height by the late 1240's. Athenry is an urban castle, set against the outside north-east angle of the walled town of Athenry (see Figure 5.2.3). The town walls were constructed in the final decades of the 13th century (Givens 2008: 27) and may be considered a symbol of seigneurial influence. Other seigneurial aspects at Athenry include St Mary's Collegiate Church founded in 1240 (ibid. 20), the Dominican priory founded in 1241 (which may have been visible from the windows of the hall/chamber on the south-east curtain wall), and mills along the Clareen to the south. There is no documentary evidence for a deerpark, but the amount of land included within the town walls suggests that the area was either prosperous, or Bermingham was an optimistically ambitious lord (ibid. 11). The presence of a hospital is hinted at by the extant south town gate, 'Spittal Gate', though the hospital and the road it sat on are both gone. As this southern approach passed the hospital, it might be circumstantial evidence for a managed landscape by a manipulated approach.

Location and discussion of the primary components

Because the location of the original gate is unknown, only the primary components of great hall and lord's chambers are extant at Athenry. These are seen in two principal features: the great tower and a separate large rectangular building set against the south-east curtain wall, which are illustrated in the castle plan seen in Figure 5.2.4 (castle plan by McNeill). The location of the current gate into the castle was thought to have been the position of the original gate (OPW nd: 4; Rynne: http://www.athenryheritagecentre.com/history_castle.htm), and Leask suggests that it was still standing at the end of the 18th century (1936: 156). However excavations in 1989 and 1990 failed to find any trace of a gate (McNeill 1997: 131-133; Papazian: http://www.excavations.ie/Pages/Details.php?Year=&County=Galway&id=3343). As there was not an exact geographical spot from which to generate a gate viewshed, it was determined that the gate component would not be treated in this study.

The first structure built in stone at Athenry Castle was a squat building of two storeys which soon rose to become the great tower. This contains the original first-floor great hall, but it is an unusual great hall in that it contains a garderobe (set diagonally from the main entrance). It is possible that the great tower was intended to eventually become the Lord's Chambers, and that the inclusion of a garderobe was in preparation for that usage. If that is the case the tower may have held the great hall only briefly because at sometime around 1250 Piers de Bermingham (son and heir of Meiler) built what has been referred to as both chambers (McNeill 1997: 134) and banqueting hall (Givens 2008: 24) directly against the south-east curtain wall. This building has three fine (though small) pointed arch windows with relieving segmental arches that pierce the curtain wall at what is essentially the first-floor level on the exterior of the castle. The windows (all of which have seats), and the ground- floor location (which was a popular trend in great halls during the latter half of the 13th century) suggests that it certainly could have been a great hall similar to those added at Adare and Trim in the late 13th century. However, the use of this structure is in doubt, and because of the possibility that both buildings could have contained either hall *or* lord's chambers at various times, this study examined each structure within both contexts.

General views

The 11km viewshed (Figure 5.2.5) was generated using an elevation offset of 12m to approximate the elevation of the battlements atop the great tower. In it we see that the most extensive far-reaching views are to the south-west, south and south-east (although there is a smattering of visibility to the far north). There are no far-reaching views to the west or east. This is caused by the fact that hills which surround the castle to the west, north and east rise rapidly to the west and east, but more gradually to the north. This rising ground effectively reins in the projective views to the immediate hinterland, while the area to the south slopes downward, spreading out before the castle. The slope increases the amount of visibility to that direction. Closer to the castle the views are more densely packed. The 4km radius of this same viewshed (Figure 5.2.6) shows that areas up to 3km away can be seen (though somewhat patchy) to the north-east between 12:00 and 3:00 on the clock face; and between 4:00 and 6:00 within the 2km ring. But we see that the most interesting and complete projective views are within the 1km radius (Figure 5.2.7) which has a line-of-sight with over half of this area.

Projective views: viewsheds and personal observation
Gate

As mentioned, the projective views from the gate at Athenry was not addressed in this study, as the location of the gate has not been determined. However, defensive projective views *can* be determined from the viewshed seen above; indeed no better views could have been seen from the castle than from the top of the tower. In Figure 5.2.7 it is seen that each of the approach roads to the castle are covered by the viewshed to some degree. Traffic approaching from Tuam in the north-west (modern road R347) could have been viewed from the top of the tower up to 900m away. In Figure 5.2.6 we see that traffic from the north-east (modern road L3107) could have been seen (although sporadically) nearly 2km away. Traffic coming from the east along the old Loughrea Road (road number unknown) could have been seen as far out as 1km, although there are long stretches where that road is not visible from the castle. Approaching from the south (along the modern R347) visibility extends to at least 200m, and then patchily from that point to the 1km radius. Defensively Athenry was in a good location to be able to spot approaching traffic (either friend or foe) once it was within 1-2km.

Unfortunately the top of the tower was not accessible for personal observation. However the third- floor of the tower was accessible, and it was possible to take photographs from the north-east and south-west facing windows. Although they are not as far-reaching as photos taken from the battlements would have been, they do give a good indication of the possible projective views from those points. These photos can be seen in Figures 5.2.8-10, and would seem to confirm the views suggested by the viewshed.

Great hall and lord's chamber

As mentioned, the great tower *and* the building along the south-east curtain wall were examined in the context of both great hall and lord's chambers; we will address the great tower location first. They will be referred to as the first-floor great hall/lord's chamber within the great tower, and as the possible later great hall/lord's chamber respectively.

Great tower's first-floor great hall/lord's chamber

The great tower sits within the north half of the ward. It is rectangular in shape, roughly 16m by 10m externally, with the long axis running north-east to south-west. The main entrance faces south-east and was probably reached by way of a timber stairway. There are six windows on the first-floor level, one on each of the four main walls, and two narrow lights in the garderobe (which was not treated in this study). All of the main windows have window seats and three of the four have draw-bar holes (the south-west window being the exception). Photographs showing the main entrance and the projective view from that point can be seen in Figures 5.2.11 and 5.2.12 respectively. Each of the main windows and most of their available views can be seen in Figures 5.2.13- 21 (note: the photo of the projective view to the west-south-west from the south-west window did not turn out, however the parish church *is* visible from this window; also the photograph of the view from the south-east window did not turn out, but the projective view from it would have been similar to the view seen from the main entrance in Figure 5.2.12).

Several examples of elaborate decoration are still extant in this building and include a pointed moulded arch atop the main entrance that is supported by columns with delicate floral capitals (Leask 1977: 36-39). This door also has an interesting stone slab projecting over the exterior, apparently a remnant of a ruined fore-building. Two of the windows (south-west and south-east) have columns with carved capitals (each one distinctive) and moulded bases. Three of the windows have graceful trefoil-pointed windows (north-east, south-east and south-west).

The projected views from a hall or lord's chambers in this location could have been both interesting and informative about the hinterland, and similar (though more limited) than those available from the battlements of the tower. Views in all directions would have been hampered to some degree by the curtain wall. Unfortunately we do not know to what height the curtain walls rose or how much they may have obscured visibility from the first-floor of the great tower, so it should be remembered that the projective views seen in the viewshed are the *best possible* views that would have been available. The farthest views should have been to the south-west, which slopes away from the castle. The viewshed in Figure 5.2.22 generated from the great tower using an elevation offset of 6m shows that this is true. However, close-up versions of this same viewshed using a 4km and a 1km radius (Figures 5.2.23 and 5.2.24 respectively illustrates that within the *immediate* hinterland visibility is best to the south-east; this is owing to the rise of the hills in that direction. Although vegetation makes it difficult to be certain, the photographs taken from the first-floor level of the great tower suggests that the viewshed seems accurate.

The possible later ground-floor great hall/lord's chamber

The second possible location of the great hall or lord's chambers was within the rectangular building in the south half of the ward. This ground-level building is represented by three foundation walls and the south-east curtain which formed its fourth wall. It spans the length of the south-east curtain, stretching between the two circular towers and measures roughly 24 by 10m externally, with the long axis running from the north-east to the south-west. The building may have risen above ground-floor level, but nothing remains today beyond the height of the curtain wall. The location of the main entrance is not readily apparent, but there are three extant windows piercing the curtain. The north-east and central windows are the largest, nearly twice the width of the south-east window. These three windows have pointed arches with segmental arched embrasures and window seats. They overlook the River Clareen, framing a sylvan landscape of meadow and trees. But the view is neither vast nor sweeping; it is in fact quite limited by the vegetation which grows almost to the castle walls. These windows and their respective views can be seen in Figures 5.2.25-31.

Interestingly the viewshed from this location (seen in Figure 5.2.32) suggests that even without the vegetation as obstruction, the projective view from these windows would be limited. This viewshed was generated using an elevation offset of 1.7m. It is remarkably thin on projective views; a patch of visibility is seen within the 2 to 3km radius between 5:00 and 6:00 on the clock face, and between the 10 to 11km radiuses the upward slope of a hill can be seen towards that same direction. But it is in the close-ups of this viewshed seen in Figures 5.2.33 and 5.2.34 that the surprising *lack* of projective views is best illustrated. These poor projective views stem from three factors, the first being that the windows are along only one wall (hence only one direction can be viewed). Originally of course there may have been windows along the north-west wall facing into the ward, but realistically at ground-floor level such windows could have only offered internal views of the castle. Secondly, the low elevation of the room obviously hampers the amount that can be seen. And finally the general topography beyond the castle wall rises and then falls, effectively cutting off any projective view until an even higher hill rises about 10km away. The view seen today is attractive, but it is not informative or even all that interesting. If a deerpark had been located beyond the small river directly south-east of the castle, the view from this building may have been both sylvan *and* impressive, but as mentioned this research found no hint of a deerpark in either documentation or place names. The viewshed and the town plan (Figure 5.2.3) suggest that the Dominican priory might have *initially* been visible from these windows (before the erection of the town walls). But unfortunately the current vegetation made a comparison of the viewshed with personal observation of the priory impossible.

Comparison of these projective views with each other and the views from other castles

Although the far-reaching projective views from the battlements atop the great tower were only fair (4.95% of hinterland was visible); views within the castle's immediate area were much better. In the 4km radius around Athenry 12.25% of the hinterland was visible, and within the1km radius around the castle more than half the available area was visible (51.71 %). The reasons for the excellent close-range views are the topography around the castle and the height of the battlements. The castle is on a slight rise in an otherwise low lying area. Here the surrounding land slopes gently upward to the west, north and east, and slopes gently downward to the south. The effect is basically that of a three-sided bowl whose inner sides are visible up to the point of the rim of the bowl. Other castles that sit within bowl-like depressions are Athlone, Adare and Carrickfergus. Of the three, Athlone is most similar to Athenry because it shows a similar trend (low far-reaching views but substantial closer views). At Athlone the 11km viewshed suggested that while only 6.27% of the extended hinterland was visible, within the 4km radius 20.40% of the surrounding area could be seen. This percentage nearly doubles within a 1km radius (where 39.29% is visible). Both Athenry and Athlone were urban castles in contested areas, located on raised ground next to rivers (though the Shannon is certainly much larger than the Clareen), and both were well able to keep their towns under surveillance. Adare and Carrickfergus were also urban castles sited next to water (Adare on the River Maigue and Carrickfergus on the Belfast Lough). However, neither of these castles sat on raised ground. Like Athenry, Carrickfergus had a great tower that elevated its occupants high above the surrounding area which meant that it allowed for extensive projective views (interestingly however, its great tower was not poised to survey the town but focused on the harbour, the lough and the open sea instead). (Adare is perhaps the least good comparison, for it had a much less extensive projective view, was built low without a great tower, and the location of the original town is in question.)

The projective views from the two possible locations of great hall and/or lord's chambers are interesting to contrast. Neither position gave far-reaching views; the first-floor location allowing views of only 2.34% and the ground-floor location a mere 0.27%. However of the two, the first-floor location definitely offered more extensive *immediate* projective views; these windows could see 27.22% of the area within the 1km radius compared to the ground-floor location's view of 0.63%. The view from the first-floor included the parish church to the south-west, but interestingly the viewshed suggests that the church would not have been visible.

The builders of Athenry Castle seem to have been cautiously optimistic of their acceptance by the Irish and the possibilities for success in their new lordship, and almost torn between meeting the needs of comfort, display and defence. This is shown by a number of aspects. Athenry was in a turbulent area in a marcher region, but it was not settled by large amounts of transplanted Anglo-Normans or Continental immigrants. Instead it attracted mostly an Irish population which became quite active in the development of their town (Killanin and Duignan 1967: 75-76; OPW nd: 1; Givens 2008: 24). Givens suggests the De Berminghams might have expected some hostility from the immediate Irish neighbours, and offers this as an explanation for the castle being separated from the town (as well as the hinterland) by both wall *and* moat (ibid. 23).

It is difficult to accept that De Bermingham felt greatly threatened by the Irish living within his town; he seems to have been genuinely dedicated to the success of his barony and his immediate neighbours. He may have been more cautious when it came to the Irish *outside* the town's walls however. Indeed it is quite possible that the De Berminghams felt it necessary to defend themselves and their townsfolk from a hostile *external* Irish population because towards the end of the study time-frame they threw up a wall and moat around the perimeter of the town that was so large (28.5 hectares) that it encompassed twice the area of the town of medieval Galway, and was similar in scope to the Hightown circuit of Kilkenny (Givens 2008: 26-27). There was plenty of room within the town's walls to protect livestock and inhabitants from casual raiding and represents a major commitment on De Bermingham's part to his Irish townsfolk. Realistically however, such a perimeter (2,000 m) was very long and would have been nearly indefensible.

Having said this, it is interesting that at about the same time (or slightly before) the construction of the town defences, Piers de Bermingham followed the fashion of the day by piercing the curtain wall to create the windows of the large ground-floor rectangular building (whether hall or lord's chambers) similar to those built during that time at both Trim and Adare (see Figure 5.2.31 above). Like Trim, these windows were effectively at first-floor level on the exterior of the castle (at Adare the windows were still ground-floor on the exterior of the castle). Admittedly the windows were protected by a moat, and not extravagantly large (as are the hall windows at Trim or Adare), so perhaps their modest size represents a compromise between defence and display. Still these windows were certainly wide enough (even the smallest of the three) to allow a grown man access to the castle. It seems unlikely that anyone who felt truly threatened in their environment would intentionally breech their own defences in such a blatant manner simply to be stylish. So, perhaps the risk was not as grave or as continual as has been thought in the past.

Other than the fact that the parish church was visible from the first-floor room but is not within line-of-sight from that location in the viewshed, personal observation did support the GIS. In general, personal observation of the projective views combined with the viewshed does suggest that visibility had a role to play in the siting of Athenry Castle. Priorities of the builder appear to have been carefully balanced at Athenry Castle between commodious habitation site, administrative centre for the barony and defence.

Reflective views: viewshed and personal observation

Reflective views of Athenry Castle have been established by the viewsheds in Figures 5.2.5-7, where we have seen that topography allowed a good possibility for travellers in the near hinterland (within the 4km radius) to see the castle; and for the immediate community of Athenry itself (within the 1km radius) an excellent possibility of viewing the castle. Barring obstructions from built environment and vegetation, approximately half of the town (51.71%) would have had reflective views of the castle. However the same topographical restriction that bore on the projective views influenced the reflective views (namely the shape of the bowl-like depression in which the castle sat). Anyone living or approaching the castle from beyond the rim of the basin (farther than approximately 1 or 2km to the west, north or east) would not have had many opportunities to see the castle, while those living or approaching from the south-east, south or south-west would have had a better chance to catch a reflective view.

Approaches

The Taylor and Skinner 1778 map shown in Figure 5.2.35 shows four main approaches to Athenry. These earlier roads have solidified into modern roads, with the addition of one other main approach from due south. The viewshed seen in Figure 5.2.6 and 5.2.7 suggest that almost all the routes into Athenry should allow for reflective views from about the 1km radius, and some roads travel in and out of the line-of-sight as far out as 2km (Caheroyn Road and the Loughrea Road). Without man-made structures and vegetation to contend with, the castle is well placed visually and could have figured strongly within its landscape. However the fact is that current far-reaching reflective views are simply not possible due to built environment and vegetation, no matter *how* conducive the topography is; and even the more immediate views are curtailed by these same obstructions. Unfortunately all photographs of the first-sightings were inadvertently unavailable due to issues with digital equipment.

Perhaps the worst approach for reflective views of the castle is from what presumably was the most important road: the Galway Road (modern 348 north-east bound). This route enters the town from the south-west, and a traveller on this road cannot actually view the castle until it approaches the turn onto Ct. Lane (modern R347) on which the castle is located, only approximately 120m away. The built environment of the town is the main cause of this poor view. When travelling north-west along the Loughrea Road (the modern R348) the viewshed in Figure 5.2.6 suggests that at least sporadic views should be available from almost 2km away. However, topography, vegetation and built environment compete successfully for visibility until the modern traveller turns onto Ct. Lane less than 200m away. This becomes the default approach from north-bound R347 as well, the most modern addition to Athenry's transportation arteries. Traffic on this route is diverted through the one-way roads in the centre of town and must come to the castle in a circuitous manner. The viewshed suggested this approach might have been able to see the castle several places within the 2 to 1km radius, the reality is that the castle is not visible until the viewer is on Ct. Lane, again less than 200m away. The problem here is built environment.

The Tuam Road (R347) travelling south-east is slightly better. The viewshed shows that this road does not enter the line-of-sight to the castle until well within the 1km radius, but built environment hides any view of the castle until it is approximately 350m away. The Caheroyn Road (L3107) travelling south-west is within occasional line-of-sight to the castle from at least 2km away, but built environment and vegetation only allow views of the castle from 500m out. The best reflective view of Athenry Castle was from the country road directly east of town (old Loughrea Rd). Coming west this road is within sporadic line-of-sight at about the 1km radius mark, but first-sighting of the castle was from 580km; again, built environment and vegetation restrict visibility. Currently Athenry Castle is neither dominant nor intimidating in its landscape, nor is it friendly or benign as are the castles at Swords and Adare. The curtain and great tower simply appear to huddle cautiously at the periphery of the town. A table listing the initial sighting points on the approaches to Athenry Castle can be seen in Figure 5.2.36.

Community

The medieval town of Athenry grew up directly south-west of the castle, separated by the curtain wall and a moat. The original parish churchyard now has a Church of Ireland building, and the market square (or triangle in this case) was only a few dozen metres to the south-west of the parish church. The viewsheds generated from the locations of the parish church and market place (Figures 5.2.37 and 5.2.38 respectively), which were created using elevation offsets of 1.7m suggest that the ground upon which the great tower at Athenry sat was not visible from these vantage points. This is an important place to note that though often the *ground*-level of a building might not be visible from the ground-level of another location, the *top* of an elevated object such as the great tower of Athenry (which rises 12m or more) might easily have been visible. This was most likely the situation at Athenry, and is illustrated for the parish church (although in reverse) by Figure 5.2.39 (photo of church from third-floor of tower) which shows that St. Mary's church was certainly visible from the elevation of the third-floor of the great tower. It was unfortunately not possible to prove that the great tower could be seen from the market place.

Such a view of the great tower from the church or the market, if not hampered by built environment or vegetation might have seemed rather formidable, though perhaps not threatening. There is little about Athenry that comes across as threatening. It would certainly have been the tallest man-made structure the townsfolk were accustomed to seeing, perhaps even the tallest building they would ever see, but it does not give off the squatting aggressive stance that Athlone's tower does. Athenry generates more of a 'hunker down', watch-and-wait feeling than a sense of imminent attack. In fact, it may have even seemed protective to those in the town who might have looked towards this defensive tower and its inhabitants (the same family who had built their town wall after all) as shields between themselves and those from outside the town walls.

Comparison of these reflective views with each other and the views from other castles

It is impossible to know to what degree the main gate of Athenry Castle imposed itself into the lives of the community of Athenry and the surrounding hinterland, or even to know how much of a presence it had in their daily lives. The fact that it has disappeared so completely that not even excavation has been able to find it suggests that the gate may not have been substantial (McNeill 1997: 134), though Rynne suggests otherwise (Rynne 2000: http://www.athenryheritagecentre.com/history_castle.htm).

Currently the castle cannot be seen particularly well within its surroundings due to competing built environment and vegetation. It is very nearly swallowed up by the modern town. Due in part to the lack of a gate tower (or any gate at all) the castle does not intimidate or dominate. The great tower itself is not exceptionally tall (as compared to Trim or Carrickfergus), and though it does seem to govern (or at least seem to *try* to govern) its surroundings, it does not intimidate or dominate. Perhaps the best comparison to make with Athenry might be Swords, whose (extant) gate was not particularly menacing. Swords Castle comes across very convincingly as first and foremost an administrative centre, and secondly as a benign or even neighbourly entity in the community. Though Athenry seems neither benign nor neighbourly, it is a very credible administrative centre.

The best comparison of reflective views would be with Athlone Castle, which sits in a similar location and like Athenry commands the attention of its immediate surroundings quite effectively. But, interestingly Athlone does not do quite as well as Athenry (when total pixels visible from the hinterland are compared). Athenry is actually very well sited for reflective views within its immediate community. The physical choice of location for the castle was brilliant, and (unless it was simply a fortuitous accident) suggests that De Bermingham was not a fool and that he intended to be successful in his barony.

Using personal observation the reflective view of Athenry seems moderately defensive, modestly pretentious, but very liveable. It feels like the home of a determined, but cautious administrator who took no great pains to either impress or intimidate his neighbours. Givens has suggested that Meiler de Bermingham and his son Piers were either unduly optimistic, or were locked in some type of seigneurial competition with their overlord De Burgh (Givens 2008 26). Of the two options, the reflective view would support unduly optimistic. The prudence displayed in the modest sizes of the windows in both the great tower and the curtain wall suggests that De Bermingham, although *influenced* by fashion, was not a slave to it. In general both the reflective viewshed and personal observation tend to agree with the priorities suggested by the projective views; that Athenry was a carefully liveable balance of administrative centre and defence.

5.3 *Athlone Castle, Co. Westmeath*

Note: OSi orthophotographic coordinates were used for the great tower and parish church. These coordinates were found online at: http://ims0.osiemaps.ie/website/publicviewer/main.aspx#V1,603947,741399,7.

Introduction

Athlone Castle, Co. Westmeath was built along the gently sloping west bank of the River Shannon, in a river valley extending from the southern tip of Lough Ree. The area is quite low-lying, surrounded on three sides by higher ground. Because the castle rises quite high (6-7 m) above the surrounding land, it has been suggested that it was built upon an already existent motte, and it is possible that the builders incorporated an esker knoll into the motte, scarping it to shape (Orpen 1907: 264-265; Leask 1977: 42). The geography of the site is illustrated by the topographical map seen in Figure 5.3.1.

The spot was previously owned by the Priory of St. Peter, which was compensated for the land by the crown (Orpen 1907: 263). The original motte is believed to have been built by Geoffrey de Costentin sometime before 1210. But the land was appropriated from Costentin by King John around 1210 for the site of a royal castle; and John de Gray, bishop of Norwich (who was the current justiciar of Ireland), was entrusted with building a royal stone castle (Orpen 1907: 261-265). The original stone tower collapsed in 1211, killing at least nine Anglo-Norman men. Undaunted, De Gray immediately rebuilt, although he apparently revetted the motte with masonry to prevent another such collapse (ibid).

The town of Athlone is separated into two halves (east and west) by the River Shannon. It grew quickly and by 1251 plans to enclose the town had commenced. Orpen suggests the castle was built on the Connaught (west) side of the river to protect the bridge from attack from the west and to 'overawe' the Irish on the northern frontier of Connaught (Orpen 1907: 261). It was built in several stages and has undergone much change since 1210, the most extensive alterations being the enormous reconstruction efforts of the 1790's (Kerrigan 1995: 227-232). In fact, the only aspects of the castle left dating from the subject time period are the base of the polygonal (10 sided) great tower, two ¾ round towers built along the frontage of the Shannon and the substantial wall between those towers (Leask 1977: 42; McNeill 1997: 129; Sweetman 1999: 86). The medieval town plan can be seen in Figure 5.3.2 (town plan by Givens).

The extant seigneurial aspects of the royal castle at Athlone consists of a graveyard from the original Franciscan Abbey sited to the north-east of the castle across the Shannon, founded in 1241; and portions of the wall around the medieval town. A street in the west half of the town; Deer Park Road, which runs east to west approximately 500m south of the castle, is taken here as circumstantial evidence that a park once existed. There are references to a mill at either end of the original bridge, a pre-Norman Cluniac priory founded in 1150, a kiln and fulling mill, and Orpen mentions that the town itself had a meadow-pool and fisheries. By 1293 the selling of eels at Athlone had become fairly lucrative (Orpen 1907: 271). A look at the Athlone medieval town plan by Givens suggests that there was probably a managed approach, at least from the Dublin direction, which appears to pass St Mary's Church, going through the market place to the medieval bridge, and directly to the castle on the west side of the Shannon. The OSi map of the modern roads surrounding Athlone can be seen in Figure 5.3.3.

Location and discussion of the primary components

The Athlone Castle plan can be seen in Figure 5.3.4. Today the castle seems fairly well intact, but the truth is that very little of what is extant is medieval. We can only say with any degree of certainty that the base of the now relatively short great tower in the centre of the castle, and the two angle towers facing the Shannon are pertinent to this study. It is impossible to know where the original main gate stood, or where the hall and the lord's chambers may have been located. The castle has been altered to the point that there is not one certain primary component left for us to examine. Although we might be safe in assuming that the lord's chambers would have been within the great tower (which was probably the first element of the castle built in stone), and thereby try to gain some kind of a story from that component, to do so would be supposition (there is also the possibility that as a royal castle, Athlone may never have had a lord's chamber component). In contrast, determining the primary components *outside* the castle gates (the approaches and the location of the market and parish church) was relatively straightforward.

General views

It is possible to gain a fairly clear understanding of the *general* projective views from the castle as a whole through the viewsheds generated from the position of the castle's great tower (the tallest original point left from the subject time frame) seen in Figures 5.3.5–7, these are the 11, 4 and 1km viewsheds respectively. This view was created using an elevation offset of 9m (approximating the tower's second-floor level), with an azimuth of 360°. Of the 608,222 pixels visible in the 11km radius, 38,154 pixels could be seen from this vantage point (6.27% of the hinterland was visible to some degree from the great tower). Figure 5.3.6 illustrates that though the views were not focused in any specific direction, a great deal of the immediate 4km radius was visible, (including many places along the approaching roadways), and the 1km viewshed in Figure 5.3.7 confirms that the majority of the town of Athlone could be seen from the great tower. Although the River Shannon (at least to the north) does not fall generally within the line-of-sight, certainly the fording point, the bridge and much of the 1km south of the castle were visible. The castle was clearly in a very good position to observe its near hinterland, and would have had excellent all-round views of the immediate locale.

Projective views: viewshed and personal observation
Because the locations of the internal prime components are unknown, and in an effort to be as accurate as possible in this examination (and to not fabricate a story just for the sake of *having* a story), determining the projective views from the prime components within Athlone Castle were not attempted. With no projective views to discuss, a comparison of the projective views is not feasible and is not attempted here.

Reflective views: viewshed and personal observation
Using the same illustrations mentioned above (Figures 5.3.5-7) we can gain an understanding of how much of the hinterland could see the castle. We can see that the majority of the 4km radius around the castle had some measure of reflective views. Past the 4km ring, the castle was visible sporadically to the south and south-east, and only intermittently to the north, east and west. At 3: 00 and between 5:00 and 6:00 on a clock face there are some views from distances as far away as the 11km radius, but in general, far-reaching views *of* the castle are patchy and not widespread. It was possible, as mentioned above, to establish three external prime components at Athlone. These include the approaches, the market and the parish church within the immediate community.

Approaches
The Taylor and Skinner map seen in Figure 5.3. 8 shows that in the 1770's the castle had three land approaches (from the north, east and west), and could be approached from both the north and the south on the River Shannon. Athlone Castle sits relatively elevated in an otherwise low-lying area, and the viewshed suggests that this allowed irregular pockets where reflective views were possible. Note that two places in the 4km viewshed (Figure 5.3.6), seem to have more of a concentrated view of the castle than anywhere else; towards 2:00 (along the Ballymahon Road) and towards 8:00 (along the N6). Unfortunately the castle was not visible from either of these spots on these approaches; vegetation and built environment have effectively obstructed views and in this case the viewshed does not reflect reality in the near distance. Coming a little closer to the castle we see that the situation does not improve with proximity.

Today the old roads illustrated in the Taylor and Skinner map are fairly well represented by modern thoroughfares. From the east, several roads converge to enter the east half of the city (the Ballymahon, the Dublin and Mullingar Roads). The resultant single street leads west to the bridge over the Shannon and it is only at that point (approximately 250m away) that the castle is actually visible coming from this direction (see Figure 5.3. 9). The close-up viewsheds in Figures 5.3.6 and 5.3.7 suggest that if there was no visual interference the castle would be visible along this approach as far out as 500m. This does not reflect reality however, as the reflective views coming from the north, east and south are clogged with built environment which completely blocks sight of the castle until it is directly across the river.

The same pattern occurs on the west approach, though the resultant view is slightly better. The Galway, Tuam and Roscommon Roads converge into R362 before entering the city limits. This road approaches from the west and then bends sharply to the south-east, where it parallels the Shannon and becomes Grace Road which runs directly towards the castle. Coming from this direction, it is not until the bend at the Shannon that the castle comes into view, approximately 500m away (see Figure 5.3.10). Figure 5.3.7 illustrates how this road is only intermittently in the reflective line-of-sight of the castle. Indeed the point of first-sighting is actually the *first* point at which the castle can be seen to any extent along this approach road. In this case, the viewshed quite closely reflects reality. The interference in this case is topographical, as well as built environment. The castle, though almost malignant in its imposition on the immediate surroundings, is not highly visible from any great distance. This is owing to topography in the first instance, but where topography *does* allow reflective views, vegetation and built environment contrive to block them. A summary table of the approaches can be seen in Figure 5.3.11.

Community
The town of Athlone is now split into two halves by the Shannon, and is connected by a bridge. Initially only the east side of the town may have been walled (Thomas 1992: 16), and as the original market place and parish church (St. Mary's) were located just within the walls of the Dublin Gate in the east town, the first community seems to have grown-up on the Meath (east) side of the river. At some point in history the Cluniac priory on the west side of the river (in the south-west corner of the town walls) also served as a parish church and a market was established south of the castle at the west end of the bridge (ibid). However, development of the town on the west side of the river cannot be reliably established as part of the study time frame, so only the east town was considered as the castle's community for the purposes of this study. According to Figure 5.3.7, the parish church and market would have had good reflective views of the castle. A viewshed generated with an elevation offset of 1.7m from the site of the parish church illustrates that the castle does indeed fall within line-of-sight from this vantage point (Figure 5.3.12). If nothing had competed with that line-of-sight, the castle would have been a potent statement of royal authority and power from these prime components.

The castle today has an overwhelmingly martial aura, and commands the environment surrounding it. The dark stone and the height of the walls with the huge angle towers rising up from the bank of the river are impressive and seemingly unassailable; the refurbished great tower perched atop the massive motte still expresses rule. There is no doubt that the presence of the king of England was felt in the town of Athlone. Today however, this overwhelming sight is hidden by the infrastructure of the town. Personal observation determined that built environment makes a current view of the castle from the location of the market place or the parish church impossible. There is simply too much built environment between the castle and these prime components to allow for any reflective views.

Comparison of these reflective views with each other and the views from other castles

The gates of the castle have been altered to the point that we cannot be at all certain where they were originally placed. The current entrance is to the north, up a switch-back ramp overlooked by gun-loops and angle towers. It seems likely that the original gate would have had a strong martial theme and would have reflected the almost belligerently aggressive nature of the entire structure. But it is impossible to know which side of the castle the main gate might have been on, or how that gate appeared to the community living in Athlone, or to the people from the hinterland. Unfortunately this makes it impossible to compare with the gates of other subject castles. However, even though we do not have an extant gate to give us a visual 'feeling' of the entrance, we *can* at least take those medieval portions which we know fit our time frame into consideration while trying to determine how the castle was perceived by the outside community. This can probably be summed up in four words; 'threatening, and no nonsense'. Today the traveller comes upon the castle as a surprise when rounding a corner, or at the east base of the bridge. It is impressive, undeniably military in character and dominates the immediate surroundings. In the past (and particularly within the study time frame) Athlone Castle would have been the largest structure in sight and probably appeared overpowering and unassailable to the contemporary Irish on either side of the Shannon. It certainly would have been a strong reminder to the citizens of Athlone of exactly who was in charge of the town, if not the region. In this Athlone can be compared to Trim, another urban castle. But, unlike Trim, Athlone was not the personal home of a baron, and unfortunately the amount of change that has been imposed upon Athlone makes it impossible to know if the castle exhibited any softening domestic aspects.

The viewshed implies that the castle was at least dominant in its immediate settings (the 500m radius), and (barring man-made or vegetative obstructions) could have been visible from most places within the town on both sides of the river. The viewshed can also help us determine how Athlone Castle may have been perceived by contemporaries. The extent to which the past environment blocked reflective views within the town is unknown, but the site tends to suggest that buildings, houses and trees of even modest height could effectively have closed views of the castle. Personal observation backs this up, and has shown that the castle does not currently figure largely in its wider environment.

This may partly be explained by the fact that since Athlone was built using dark grey granite stone, viewers of the castle would have had the 'Carrickfergus phenomenon' to contend with; the dark, low lying castle basically blends into its environment, the dark green vegetation of the surrounding landscape only serving to camouflage it even further. The dark tower simply does not show up well, as is illustrated in Figure 5.3.10. Because the castle was within bowshot of the ford across the Shannon, it was perfectly placed to *physically* control it, as well as to control the medieval gated stone bridge built at the castle's south-east angle. It was able to guard the city on the eastern river bank, and to allow Anglo-Norman soldiers to venture at will into the Connaught territory (Orpen 1907: 259). All of this seems to suggest that Athlone Castle was strategically sited first and foremost for defence and aggression. It is not thought that the current stone castle was built directly upon the site of the earlier wooden castle built by Turlough O'Conor in 1129 (Leask 1977: 6), but it was certainly built *at* Athlone which had been a regional seat of authority and the key to the communication between Connaught and Meath (Orpen 1907: 258). This was an important crossroads, and a place of power. Establishing a royal castle here figuratively placed the English crown in the geographical heart of the island; this was not only politic in dealing with the Irish, but it sent a powerful message of control to every Anglo-Norman lord staking a claim in Ireland. The reflective views of Athlone Castle, as seen in the viewsheds and by personal observation support the suggestion that aggressive military defence, control and psychological dominance were the priorities of the builder of Athlone Castle.

5.4 Ballylahan Castle, Co. Mayo

Introduction

Ballylahan Castle, Co. Mayo is situated only 57m from the Cloonlee River, and 340m due south of a ford on the River Moy (a major route-way in north Connacht) (Malcolm 2007: 198). It was approximately 12km north-west of Athlone, 73km north of Galway, 65km south-west of Sligo, and 225 north-west of Dublin. This is an area of rolling countryside where elevations fluctuate between 0 and 373m above sea-level. The topographical map for this castle can be seen in Figure 5.4.1.

It is not known if the site had been in use prior to the building of the castle. The land, which became the barony of Gallen was first granted to Richard de Burgh, then granted to Hugh de Lacy (the son), and finally granted to either Jordan de Exeter I or his son, Jordan II (Knox 1908: 307). Dates for this castle are somewhat nebulous, but Orpen gives Jordan de Exeter II credit for being the first de Exeter to occupy the land which became known as the manor of Athlethan (the broad ford), between 1240 and 1250 (1920: 197). It is not known if the initial castle was built over a period of time or all in one effort, but we can tell by the use of hand-gun loops in the north-west curtain wall that the castle was utilized for several centuries. McNeill suggests that the nearly circular curtain walls seen today (measuring approximately 40m across), probably follow the lines of the original castle enclosure (1997: 136-7).

Currently the castle is rural, laying 2km to the north-east of Strade. Evidence of seigneurial aspects is both documentary and circumstantial. A reference in the *Calendar of Documents relating to Ireland* grants a weekly market and a yearly five day fair to the "town" of Adleen (CDI 1252-1284: 250). Jordan de Exeter established the monastery of Athlethan (now known as Strade Abbey) *c.* 1252 (Knox 1908: 95). Apparently it was built for the Franciscan order, but by 1253 De Exeter had "turned out the Franciscans" and installed the Dominicans' (Knox 1908: 307). Orpen suggests that the incorporated town (which was the site of the fair) and the location of the abbey were not far apart (1920: 198). There is also a reference in the *Calendar of Documents relating to Ireland* to De Exeter being granted the right to have free warren in all his demesne lands in Ireland (CDI 1252-1284: 250). Today the only detectable seigneurial aspect of the castle *is* the abbey. The fact that the road leading from Westport and Castlebar passed through Strade and therefore passed the De Exeter monastery and market place is considered here as circumstantial evidence of a managed approach. The OSi map showing the immediate 2km radius surrounding the castle can be seen in Figure 5.4.2.

Location and discussion of the primary components

What we see at Ballylahan Castle today amounts to the remnants of a twin-towered gate house and portions of the curtain wall. The gate house opened to the south-east, facing the road which came from Castlerea (ultimately Athlone), toward a hill which rises approximately 18m above the level of the castle. There is a moderately sized fireplace extant in the north curtain and there is evidence of stone buildings mixed with the tumbled curtain walls beneath the grass in the inner ward. Without more fabric, identification of the position of the great hall or lord's chambers must only be speculative, and analysis of these components was not attempted in this exercise. There does not appear to be any documented plans of the entire castle at this time, but a plan of the gate house can be seen in Figure 5.4.3.

General views

The viewsheds for Ballylahan Castle are seen in Figures 5.4.4–6. These were created from the position of the gate house, using an elevation offset of 10m to simulate battlement level, with an azimuth of 0° - 360°. There is a smattering of visibility between 12:00 and 2:00 stretching from the castle to the 11km radius (Figure 5.4.4. There are also far-reaching views towards 5:30 and 9:00 on the clock face. The best over-all views are illustrated by Figure 5.4.5, and are to the north between 11:00 and 2:00 on the clock face within the 4km radius, and to the south (6:00 on the clock face) within the 1km radius. There is no long-range uninterrupted visibility from this castle, the maximum uninterrupted distance being only 300m to the north-east. This is best illustrated in Figure 5.4.6 which clearly shows how the gentle hills surrounding the castle to the east, south and west could severely limit both projective and reflective views. An interesting aspect of the views at Ballylahan is that the town of Strade where De Exeter built his seigneurial abbey can neither see, nor be seen from the castle.

Projective views: viewsheds and personal observation

Gate

The extant stone gate had two D-shaped towers, with a narrow entrance passage between them. The inner walls of the towers at ground-level were quite thin, and only the north-east tower had arrow loops at ground-level. The entire structure measured roughly 15m by 12m; the longer axis running south-west to north-east. The nineteenth century sketch by an unknown artist shown in Figure 5.4.7 (courtesy of Duchas) shows the towers rising above first-floor height (Sweetman 1999: 111), but we cannot be certain how many floors the gate house originally had. Currently only the front portion of the north tower, and a small part of the south-east wall of the south tower remain. The extant fabric of the gate house can be seen in Figures 5.4.8 and 5.4.9.

The viewsheds introduced above suggest that an individual attempting to monitor approaching traffic from this gate house would have been at a great disadvantage. There were at least three approaches to the castle: from the south-east, the south-west and the north. These are represented by the modern N58 and R321. Figure 5.4.5 illustrates these approaches and suggests that along the northern stretch of N58 there is a portion of the road as far away as 1.4km that could be seen from the castle. Though the projective view to this direction was patchy as can be seen in Figure 5.4.6, the ford across the River Moy was visible. To the south-east and south-west visibility is not as good. Only a very short portion of R321 can be seen (approximately 250m), and while the southern segment of N58 is better (with projective views to approximately 500m), it is still not a substantial distance. This is especially interesting in that this is the direction of the monastery and the town of Athlethan (Strade). The problems with projective views are mainly that of topography. Although the castle sits in a sylvan location with a river, pasture and green hills spreading out to every side, the hills effectively block almost all far-reaching views, and whatever projective views there were, had to compete with vegetation. The top of the gate house was inaccessible and personal observation from this position was not possible, however the views from ground-level seen in Figures 5.4.10-16 would tend to support the projective views suggested by the viewsheds.

Great hall and lord's chambers

As mentioned above, the locations of the great hall and lord's chambers are unclear and these components were not treated in this exercise.

Comparison of these projective views with each other and the views from other castles

It is difficult to tell the whole story of the projective views from Ballylahan as we have only one principal castle component available. But although we do not have exact locations or heights of the windows in the great hall or lord's chambers, we can still make some comment about the quality (degree and scope) of views the missing components *might* have had. A clue has been left to us by the nineteenth century sketch seen in Figure 5.4.7 which suggests that until that period Ballylahan's curtain walls stood to at least 9m, (3m higher than a typical contemporary first-floor level). Unless the hall or lord's chambers were in a tower rising to second-floor level they would not have had projective views over the curtain walls. This seems unlikely, judging from the lack of tumbled fabric within the castle ward, unless of course the tower had been of timber. If the tower did have three storeys, the second-floor (at approximately 9m) would have had no greater elevation advantage than that used for the viewsheds from the gate house (10m).

Had De Exeter been greatly concerned about defending his castle, he might have placed it in a more defensible position. Though it was built on a rise of land, elevated above the most immediate ground, the site is over-looked by several hills; to the south-east, the south, the south-west, west and north-west, all within 200m (and several of them *much* closer). The site choice he made does not seem to reflect any anxiety over attack. The percentages of far-reaching projective views available from the gate house battlements were very low (only 2% of the hinterland). Views closer to the castle were slightly better, in the 4km radius around Ballylahan 6.5 % of the hinterland was visible, and within the1km radius around the castle 23% was visible. Clearly the near-distance views were a higher priority; but, that said, it is not unreasonable to say that the overall projective views from the gate house at Ballylahan are rather weak. Indeed Ballylahan Castle has one of the worst projective views in this study.

This weakness has everything to do with the type of topography De Exeter chose to build in. But although the constant rise and fall of land almost guarantees patchy views, if the amount and extent of the projective views had been an issue, the castle site could have been chosen better. For instance, if the builder had chosen to build on the hill less than 200m to the north-west; the projective views would have more than doubled in each of the views. This alternate site would have allowed an almost uninterrupted view of the River Moy stretching 1.5km (the chosen site offered a *very* patchy view of approximately 1.3km of the river). But interestingly, the site that was chosen was half the distance of the alternate site to the main north-south transportation route.

The castle which offers the most similar projective views from the gate house is Clonmacnoise in Co. Offaly. It was built on low lying ground next to the *Esker Riada*, the ancient east-west 'highway' across Ireland. Sandwiched between the River Shannon and the esker, it had a very truncated projective view (4.6% at 11km, 12.47% at 4km, and 35.32% at 1km). Although Clonmacnoise did not have an extensive far-reaching view, the view it did have was of the esker, and as such, of approaching traffic. At Ballylahan, the views of approaching traffic from the south-east and south-west were quite limited (approximately 200m and 500m respectively). But the view to the north (though interrupted), is still fairly good at 1.4km. It was interesting to discover that even if De Exeter had built upon the hill to the south-east, his views of the approaches would not have been significantly better. To the north there would have been fewer interruptions in visibility, but the projective distance would have been the same at 1.4km. To the south-east and south-west, while gaining some distance, it would not have been a significant amount (350 and 700 total metres visible respectively). So, although building on the higher location might have offered 360° panoramic views within at least a 3km radius, observation of the approaches would not have been appreciably better. If the ford across the Moy was of interest (even to monitor if not control) building upon the hill would have placed the castle 20m nearer, and would have increased the quality of the views considerably. This suggests that while observing approaching traffic was of some interest, he was willing to trade the greater projective views the alternate site would have offered for the sake of convenience. Building where he did (on the gentle rise next to the Clooneen River), allowed him to monitor the approaches and the ford across the Moy to some extent, be above flood level and still take advantage of the Clooneen as a water source; a very practical choice. Generally, this seems to suggest that visibility probably did not play a role in the choice of siting. The location for the site was most likely determined simply the fact that it was a pragmatic habitation site; the priorities appear to have been economics and comfortable habitation.

Reflective views: viewshed and personal observation

When analysing Figures 5.4.4-6 from a reflective viewpoint, we are led to the conclusion that this castle, which could see but little of its hinterland, could likewise not be well seen from it. Field observation determined that unless an individual knew exactly what they were looking for, the castle is very difficult to see. As mentioned, the topography (rising and falling hills) provided the major obstacle. This is a rural area offering very little built environment to compete with the reflective views, but there is no doubt that vegetation played a large part in obscuring the castle.

Approaches

The 1778 Taylor and Skinner map can be seen in Figure 5.4.18. This map is awkward in a few respects; there are roads illustrated which today are no longer through-ways, and the castle is depicted in the wrong position (the approximate correct location of the castle has been added to the map in the form of a star). But this map is valuable as it provides the main routes from the north, south-east and south-west and shows that they are almost identical to those in use today.

The 1km radius viewshed seen in Figure 5.4.6 suggests that traffic approaching north bound along modern N58 (from the direction of Strade and the Abbey) could have seen the castle when they were 500m away from it. This was not the experience of this researcher; casual vegetation effectively hid the castle until it was only 135m away. A photo of the first-sighting from this direction can be seen in Figure 5.4.19. Coming south bound on this same road (N58) the castle was visible from much farther away, but only when the researcher was aware of what to look for. At the location of the ford across the Moy the castle comes into view (350m away). Although the viewshed suggests it might have been seen much earlier (from as far away as 1.4 km), vegetation obscured reflective views until the point where the road meets the river. This sighting is illustrated in Figure 5.4.20.

The approach which allowed the furthest reflective view was that from the R321 north bound. On this road, which weaves through the hills to the south-east, the castle was sighted from 360m away, just as the road comes around the base of the hill that the castle faces. In this instance, the viewshed was not accurate, for it suggests the castle would not be visible until the traveller is within 200m. Indeed, in general the viewsheds did not give accurate indication of the possible first-sightings of the castle; in two cases the cause may have simply been competition by vegetation, but in the final instance the representation of the line-of-sight itself was inaccurate, well beyond the supposed 25m accuracy limit. Unfortunately the narrowness of the road, combined with its winding and busy nature meant that obtaining a photograph from this reflective viewpoint was too dangerous to attempt (even early in the morning), and there is not a photograph to illustrate this sighting. A summary table of the approaches can be seen in Figure 5.4.21.

Community

As mentioned above, there was indeed a medieval town, but we are safe to say that there could have been no reflective views of the castle from anywhere within the town. As far as the inhabitants of such a town would have been concerned, the castle would not have been a daily sight. Indeed, according to the viewshed (Figure 5.4.4) most of the residents of the south-western quadrant of the castle's hinterland would not have had a reflective view of the castle. What the view was like for those who *were* able to see Ballylahan Castle is impossible to know; we cannot (from the ruins available for us to inspect today) determine if the castle had been aggressive or benign in nature. What can be said is that at present, Ballylahan Castle is almost a non-entity within its environment.

Comparison of these reflective views with each other and the views from other castles

In general this castle cannot be seen well in its present surroundings; it does not stand out, dominate or intimidate. 'Fleshed-out' with walls and gate towers to their full height and buildings within the enclosure it may have been reasonably impressive, especially if the castle had maintained a *cordon sanitaire.*

The reflective views of the castle, as suggested by the viewshed and noted by personal observation certainly seem to support the idea that visibility did not play a role in the choice of site for Ballylahan Castle. What we see here is a practical, pragmatic habitation site, close to a steady supply of water, with the added benefit of having a reasonable ability to monitor approaching traffic. The priorities of the builder suggested by the reflective views support those suggested by the projective views; economics (management and administration of the manor with immediate access to the main transportation route), and a habitation site made commodious by the proximity of the Clooneen river.

5.5 *Carlingford Castle, Co. Louth*

Note: Due to an anomaly in the digital elevation model data for this area (the DEM data freely available at Queen's GIS lab which was used to generate the 25m accuracy viewsheds for all of the other subject castles), contour data for the Republic of Ireland portion of the study had to be purchased. This took the form of OSi 10m resolution contour data in vector format. This data was combined with OSNI 10m resolution data available for Northern Ireland under the Queen's University license, and a digital elevation model in raster format was generated. This is the industry standard method of interpolating data, and thanks are gratefully given to Conor Graham, GIS Officer at the School of GAP at Queen's who helped in this process. As a result, the GIS visuals for Carlingford are slightly different than other illustrations within this study. Namely, most of the area in the Irish Sea along the lower half of the illustrations at the 11km radius level appears to be 'missing'. To put it simply, the portion of the radius shown in white is all ocean, and not relevant to the study.

Introduction

Carlingford Castle, Co. Louth was built on a neck of greywacke sandstone (elevation approximately 15m above sea-level) which juts into the south-west end of Carlingford Lough. It is approximately 90km from Dublin, sitting roughly 27km south (downstream on the Newry River) from Newry, and 27km north-east of Dundalk. Carlingford is located on the Cooley Peninsula, nestled in the south-west edge of the Mourne Mountains. The surrounding land is quite rugged to the west, north and east with peaks rising to between 300 to 580m above sea-level. To the south and south-east the land is low and flat. The topographic map is illustrated in Figure 5.5.1. The town of Carlingford sits just above sea-level along the coast, south and south-west of the castle. During the study time period, the eastern coastline (aided by a ferry between Carlingford and Greencastle on the south-eastern shore of the lough) provided a safe route between Ulster and Dublin.

The lough is somewhat protected from the Irish Sea, and the harbour at Carlingford was at one point deep enough to provide a haven for ocean going vessels. The name Carlingford suggests the lough had been in use by Vikings at one time, and although tradition holds that a great naval battle took place there (Givens 2008: 43; Killanin and Duignan 1967: 137), no archaeological evidence for habitation (either by the native Irish or the Vikings) has been found at Carlingford.

John de Courcy granted revenues arising from the ferry crossing the lough to the Abbot of Downpatrick Priory *c.* 1184 (Dugdale 1846: 1124; Leask 1941: 3). The area was granted by Prince John to Bertram de Verdon *c.* 1189-91 (Givens 2008: 43; Gosling 1992: 1), and there were already burgages in Carlingford by this time. De Verdon died in 1192, and the manor and part of the Cooley Peninsula were granted to Hugh de Lacy (the younger) upon his marriage to De Verdon's daughter Lescelina (Gosling 1992: 3). By 1210 De Lacy had become embroiled in an upheaval between the king and William de Braose and was forced to flee. He may have put the castle to the torch before leaving because the Irish Pipe Roll for 1211-1212 shows King John who seized it in July 1210, ordering repairs to be made (Davies and Quinn 1941: 64-67). By 1212 a castle existed in Carlingford which is probably, at least in part the present one.

The castle was administered by the crown and passed through the hands of several royally appointed constables over the next decade and a half, but was eventually restored to De Lacy in 1227 (Adams 1904: 247; Givens 2008: 43). De Lacy granted the castle and town of Carlingford to David, Baron of Naas upon his marriage to Matilda (De Lacy's daughter) in 1229 (Gosling 1992: 5; Givens 2008: 47), and De Lacy heirs held Carlingford until *c.* 1315.

Carlingford is an urban castle, sitting on the far north-eastern edge of its medieval town. The town plan is illustrated in Figure 5.5.2. Seigneurial aspects of Carlingford Castle include the town, the market and a mill (Davies and Quinn 1941: 64-65) which was apparently the only mill in the area (Gosling 1992: 22), the harbour, a ferry across the lough between Greencastle and Carlingford, and the parish church (Church of the Holy Trinity) located to the south end of the town. Though Carlingford was probably founded to be the *caput* of the younger Hugh de Lacy's lordship, it soon became a castle which was never the principal residence of its lord (like Dunamase in Co. Laois and Dundrum in Co. Down). While it is possible that a park may have been included in the layout of the *caput* from the beginning, this researcher found no early reference to one in the area; however there is mention of a park "without the town walls" *c.* 1609, owned by Bagenal who had been granted the manor and castle at Carlingford in 1559 (Adams 1904: 247-8). The evidence of a managed approach is visible even today as the road from the south (from Dublin, Drogheda and Dundalk) passes the parish church, goes through the market place before flanking the harbour and its ferry, eventually passing directly in front of the castle's main gates. Coming from the north the castle is the first (and potent) indication of an Anglo-Norman settlement. An OSi map of the area immediately surrounding the castle can be seen in Figure 5.5.3.

The castle as seen today was built over several periods, with a large portion of the original castle being extant (the west side). The original fortress consisted of a many-sided curtain wall surrounding an almost oval courtyard. Presumably all of the original buildings within the ward were made of timber, and may have been quite elaborate (McNeill 1997: 43; Sweetman 1999: 71-2). The castle plan can be seen in Figure 5.5.4.

Location and discussion of the primary components

The primary castle components are to be found in only two principal features at Carlingford; the gate house and the great hall. The gate house, which was almost certainly the first component to have been built, is located along the western curtain wall and faces west. The north half of the gate house still stands, however the south half was destroyed during the construction of the rail lines in the 19th century.

McNeill suggests that the first (timber) great hall was most likely located within the original castle ward, just to the south of the gate house, along the curtain wall (2009: pers. com.). His belief is based on the fact that this segment of wall appears to have been the only straight section of the early oval curtain. Medieval convention required a rectangular hall in order to provide upper and lower ends, and as no *non*-rectangular medieval halls have yet been identified in Ireland it would be unusual, though not impossible, to find one at Carlingford. Unfortunately with the exception of the western curtain wall and its three arrow loops, no trace of this building is extant today. Several authors have suggested that the added eastern section of the castle represents a later hall constructed *c.* 1261-2 (Gosling 1992: 41; Leask 1941: 3; Leask 1977: 61 & 63; Sweetman 1999: 72), with further refinements being made to its south-east end as an accommodation complex in the 15th and 16th centuries. McNeill however suggests that the monies spent in 1261-2 on Carlingford would have been nowhere near enough to construct the massive rectangular structure on the east side of the present castle, and is doubtful that this portion of the castle was constructed as early as the end of the 13th century (McNeill 1997: 43, 193). While the original hall may have been located within the east portion of the early castle, the later addition masks any trace of such. Excavation is currently underway at Carlingford which may help settle some confusion, however with this controversy in mind only the possible original great hall suggested by McNeill along the south-west curtain was considered for this study.

Unfortunately the later additions to the east side of the castle also make it impossible to tease out the location of the lord's chambers. These were either destroyed, or having been built in timber have left no archaeological footprint. Since we cannot pinpoint even a reasonable possible location of the initial lord's chambers with anything close to certainty, this component was not addressed in this exercise.

General views

The 11km radius viewshed shown in Figure 5.5.5 was generated using an elevation offset of 10m, with a full 360° view. The centre point is the gate house on the west side of the curtain wall. It should be noted that the viewshed is plagued by an artificial limit along the south-east, south and south-west, which demarcates the extent of land. However, even with this difficulty it is clear from the viewshed that the castle could have had a vast projective view. With the exception of the mouth of the Newry River it takes in uninterrupted views of the lough, the entire east coast of the lough, and projects south-east right out into the open water of the Irish Sea.

The most far-reaching views are to the north-north-west, the east and to the south-east, where in each case there are views extending to the 11km radius (although not shown to the south-east). The best over-all views are of the lough itself and the area within the 1km radius. Not a great deal of the Cooley Peninsula can be seen from the castle for two reasons, the shoreline rises and falls in elevation and then cuts back to the west just north of the castle, restricting the projective views to the north-west (see Figure 5.5.6); and the rise of Carlingford Mountain and Slieve Foye effectively cuts off views beyond 1km to the west (though a bit further can be glimpsed to the south-west). However the full shoreline down to Greenore Point is visible, and, perhaps most interesting of all, almost the *entire* medieval town could have been seen from this vantage point (see Figure 5.5.7).

Projective views: viewshed and personal observation
Gate

Today the gate at Carlingford Castle is greatly diminished (see Figure 5.5.8). It faced west and would have consisted of either a single tower or a true double towered gate house (McNeill 1997: 41; Sweetman 1999: 35). If a single tower, dimensions would have been approximately 10m by 14m; twin towers would have been approximately 10m by 5m each. Figure 5.5.6 illustrates the extent of the visibility of the approaching traffic from the gate house. Although traffic coming from the west would have been mainly local inhabitants returning from the common farming and pasture lands on the side of Carlingford Mountain, it is interesting to see that they could have been observed from the castle a full half a kilometre before they reached the town limits. Traffic coming from the north, parallel to the Newry River (modern-day R173) could first be seen from the gate house when they were 1.3km from the castle. Travellers coming overland from Dundalk (across the peninsula, modern-day R173) may have been visible from the castle as far away as 2km. The best viewed land-based approach is to the south-west along the coastal road to the tip of Greenore Point (modern R176). From this direction individuals were in the line-of-sight for a full 5km. However, the best over-all projective view was of water-based traffic approaching via the lough (with the exception of the north-west portion), which could have been completely visible to the extent of human vision.

The main problems with projective views from the gate house would have been the topography along the north shore line and in the centre of town (which is evident by the two blind spots towards 7:00 on the clock face in Figure 5.5.7. Vegetation could have played a role by cloaking approaching travellers, and built environment (any structure rising above ground-floor) would have interfered with the view into the centre of town. However the unlimited view of the lough would have been hampered only by inclement weather. Unfortunately, access was not available to the top of the gate house. Still as seen in the photos in Figures 5.5.9-14, the projective views are extensive. Although these photos were taken at ground-level and lack the elevation that the gate house would have offered, it is clear that they do back up the extensive views suggested by the viewshed.

Great hall

As mentioned above, for this study the location of the original first-floor hall is assumed to be immediately inside and to the south of the south gate tower. This rectangular structure appears to have been approximately 14m by 8m in size (external dimensions), with the long axis running north-west to south-east. The only extant wall of this structure is the south-west curtain which is pierced by three first-floor arrow loops. The curtain wall and arrow loops can be seen in Figure 5.5.15. As there is no trace of the timber portion of the hall, we have no external doors to examine, but the main entry would most likely have been in the eastern wall at first-floor level. Any windows within the timber portion of the building would have been internal to the castle ward and would not have offered external views. The existing south-west facing arrow loops were narrow and would have offered a fairly restricted viewing range. Access to the interior of the castle was not allowed, so it is not known if the arrow loop embrasures had window seats.

The 11km viewshed from the great hall can be seen in Figure 5.5.16. It was generated using an elevation offset of 6 m. A visual range between 170° and 255° on the compass was factored into the equation, and though this is a rather optimistic portrayal of the viewing range (a 'best case scenario'), it would not have been impossible. With three openings (though each was limited in width) the overall scope of the projective view could have been fairly extensive, covering most of the 1km radius south-west of the castle. Within the extent of this view would be portions of the market place, most of the medieval town and a large portion of the common agricultural area along the side of the mountain. With this in mind we can be certain that the scene from the great hall was impressive (though perhaps not sylvan), and definitely interesting, in that it would have offered a wealth of information about the town and the immediate hinterland. However the projective view from the hall would not have been very far-reaching, extending approximately 2km in distance. Interestingly, what could *not* be visible from the hall according to the viewshed was the parish church.

Problems with the projective views illustrated in the viewshed involve patches of invisibility within the otherwise nearly blanket-like coverage of the town. These can best be seen in Figure 5.5.17 and Figure 5.5.18. Figure 5.5.17 shows a strip of land, widening towards the south-east that should not be visible just beyond the 1km ring. Figure 5.5.18 gives greater detail of the two spots of land almost dead-centre in the 1km viewing radius (7:00 on the clock face). These patches are related to the topography of the peninsula which dips and rises between the coast and the mountain.

As access to the arrow loops was not possible, photos were taken directly below them and are seen in Figures 5.5.19 and 5.5.20. These give a fairly good indication of the possible projective view from this hall. It is obvious that not only might the hall's view have been restricted, it could have easily been obscured by vegetation. However, in general the photos do back up the viewshed from this location.

Lord's chamber

As mentioned, due to the uncertainty of the location of the chambers during the study time frame, this component was not treated in this exercise.

Comparison of these projective views with each other and the views from other castles

We can certainly see that the siting of Carlingford Castle was advantageous as far as projective views were related. But it was interesting to note that the location of the parish church offered excellent projective views as well; a viewshed generated from the church site (which will be introduced later) showed that an observer from that point was able to see 64.67% of the immediate hinterland. Built on a rise on the south end of the medieval town, it would have also made a good site for a castle. Perhaps the parish church had already been erected when De Lacy laid out his plans for Carlingford Castle. Though moving the church would not have given Hugh de Lacy (the father) a moment's hesitation, as he was beheaded attempting to build the castle at Durrow on church property, it is possible that his son was more circumspect about appropriating religious land. In any case, the spot chosen for the castle turned out to be even better than the church site for projective views. The gate house at Carlingford was able to observe 27% of the extended area within the 11km radius. In the 4km radius the projected view was 52.9%, and within the 1km ring around the castle, slightly over 90% of the surrounding area was visible. Actually, owing to the choice of site the projective views from Carlingford's gate house are among the most extensive of any castle in this study.

The view available from the great hall was comparatively quite low, capturing less than 1% (0.62%) of the long-range projective view possible in the 11km radius, placing it on a par with Limerick (0.24%), the later great hall at Athenry (0.27%), and Trim (0.5%). However, when the percentage of the visible immediate hinterland (1km) is considered, Carlingford does better at 22.30%, placing it close to the later great hall at Adare (19%), the early hall in the great tower at Athenry (27.22%), and the early hall at Dunamase (29.89%).

In many ways the site where Carlingford was erected is much like the site at Carrickfergus Castle, Co. Antrim. Interestingly, although Carrickfergus was built by John de Courcy, it was later owned by Hugh de Lacy the son. Both fortresses were built on small spits of stone projecting into an open lough, and nestled in a bowl-like manner by the surrounding hills. At both sites the shape and size of the sites determined to a large extent the shape and size of the castles. Both castles began relatively small, basically oval in shape, and were extended over time. As to the quantity and quality of projective views offered by each castle, their settings adjacent to kilometres of water open their respective projective visibility impressively; and both castles seem to focus the projective view towards the water.

However, we can find contrasting points between the two castles. De Courcy's Carrickfergus with its great tower, small gate and large windows piercing the curtain wall at first-floor level seems to place higher priority on personal comfort, and social and economical factors than De Lacy's Carlingford. Carlingford did not have a great tower similar to that at Carrickfergus, and the early buildings within the enclosure seem to have been made of timber not stone. Carlingford's massive gate house paid more attention to security, as did the mural towers along the curtain. All the openings in these towers and the curtain wall were narrow to the point of being arrow loops instead of true windows. From extant physical features alone, security appears to have been a higher priority initially at Carlingford Castle than it was for Carrickfergus. However, if we compare the reality of the political situation and the siting of the castle to the projective focus of the castle, we see another picture, one *less* concerned with defence. The main threat to the castle would have come from the Irish; from the direction of the north, and north-west. The Anglo-Normans basically controlled the sea, and there could have been little threat from the open water. The projective views from Carlingford were *not* toward the direction of greatest threat. This might suggest that the perceived Irish threat-level was actually quite low, or that defence was not as high a priority as the castle style suggests.

In general, the extensive projective views suggested by the viewsheds and supported by personal observation, definitely seem to suggest that visibility played an important role in the siting of the castle. Finally, one interesting discovery was that the market place was visible from the hall while the church was not; however the castle was clearly visible from the church but not easily seen from the market.

Reflective views: viewshed and personal observation

As seen in the viewsheds generated from the gate house (Figures 5.5.5-7), Carlingford Castle was capable of commanding a relatively expansive reflective view. It does in fact capture one of the highest percentages of any of the reflective views from a gate house/tower in this study. The problems affecting reflective visibility would have been related to the 'usual suspects': topography, vegetation, built environment and as mentioned above, weather conditions.

Approaches

The Taylor and Skinner map in Figure 5.5.21 shows three arteries entering the town which are fairly well represented by the modern roads previously mentioned in the gate section of this chapter. Two other approaches are important at Carlingford; a local road to the west of the town which once lead to the common agricultural land on the slope of Carlingford Mountain, and to the east, the lough itself. The small road would have been little more than a track-way during the study period, but it did have its own gate into the city. Because this path would have regularly been traversed by members of the immediate community who would have routinely glimpsed the castle from it, it is important to include when trying to discern the role of the castle's visibility in the community. River Street as it is now called, runs south-west directly off the market place, parallel to the river (little more than a stream actually) which once supplied much of the village water supply. The road begins a steep climb almost immediately after leaving the market, heading to the sloping pastures west of the town. According to the viewshed (Figure 5.5.7) from almost any point on the side of Carlingford Mountain within the 1km radius of the gate house, the castle should be visible. Personal observation showed that it was clearly visible from River Street at 620m distance from the castle (see Figure 5.5.22). However it was interesting to discover that as an observer neared the town (within .5km) a ridge creates some problems with reflective views of the castle. These problems are not portrayed on the viewshed.

Coming from the north along R173 the castle was visible at almost the exact distance suggested by the viewshed in Figure 5.5.23, approximately 1.3km away. Until this point topography hides the castle, and after this point although the viewshed suggests the line-of-sight between the road and the castle remains more or less constant, vegetation makes the castle impossible to see at times.

R176, the coastal road from Greenore appears on the viewshed to have continuous reflective views of the castle from Greenore Point right into Carlingford town. The reality is that built environment and vegetation hid the castle until approximately 1.7km away (see Figure 5.5.24). Surprisingly the most far-reaching reflective view on an approach road was found coming across the gently undulating land at the south-eastern edge of Slieve Barnavave; the Dundalk Road (R173). Again, almost at the very coordinate that the viewshed suggests a reflective view of the castle could be possible (at 1.8km away) it is (see Figure 5.5.25). As the road approaches the town from that point, the reflective view of the castle is infrequent, hampered mainly by vegetation and built environment, although at one stretch a ridge also obscures visibility of the castle. This blind spot is portrayed in the viewshed.

The final approach to Carlingford Castle would have been by water, either by ferry from Greencastle or from other vessels plying the lough. Unfortunately an approach via water was not attempted, although personal observation showed that the castle can be seen from several spots along the A2 roadway on the eastern shoreline.

In general (with the exception of the blind spot along River Street which is not shown in the viewshed), personal observation supported the GIS at Carlingford. Personal observation of the reflective views available along the approaches to Carlingford Castle suggests that visually the castle was and is an undeniably strong element in its environment. It is obviously powerful, though not militarily aggressive in nature. From every direction when a reflective view was possible the castle portrayed a no-nonsense, 'hunkering' strength. In its wider landscape Carlingford is dominant but not necessarily intimidating. A summary table of the approaches can be seen in Figure 5.5.26.

Community

The medieval town of Carlingford developed directly adjacent the castle along the south shoreline of the lough. Givens suggests that it was well developed, with the streets, harbour and marketplace all well laid-out (2008: 47). The functioning ferry terminal was a major cog in the Anglo-Norman transportation net-work, and the Carlingford market would have been the chief market for the surrounding region. The waters of the lough proved commercially profitable for the local inhabitants as well as fishermen from England. Gosling suggests that by 1250, Carlingford ranked among the top ports of Ulster along with Dundalk, Strangford, Carrickfergus and Coleraine (1992: 7).

The market place has been fossilized in the current street pattern approximately 300m to the south of the castle, almost dead centre in the medieval town (see Figure 5.5.2 above). It is rectilinear in shape, and opens onto the town's main roads. The present street level of the market square is higher than it would have been medievally, having been raised in the 19th century when the river which runs down from Carlingford Mountain was piped (Gosling1992: 30). A viewshed generated from the *centre* of the market place (not shown) suggested that from 1.7m above the current street level the castle is not visible. The viewshed seen in Figure 5.5.27 however, also generated at a 1.7m elevation but from the *top* of the market place (at the head of the street leading to the castle) shows that the castle is *just barely* visible. This viewshed was used because it was created from the same coordinate as the photograph in Figure 5.5.28 (market photo), and shows exactly how tenuous even the best view of the castle from the market place would have been due to topography. The photograph looking towards the castle (which can just be seen above the roof-top of the last house) illustrates how modern built environment completely obscures the castle from the vantage point of the market. But remember that this is a fossilized medieval street, and the forerunners to these same houses could have blocked the medieval view just as effectively as their modern counterparts still do. It is quite obvious that the castle did not loom large over the market place. Indeed, the gates of Carlingford Castle faced west, towards Slieve Foye and Carlingford Mountain, and would not have been squarely visible from the market or the church. The castle gate would have been 'side-on' to the town. So, though the walls of the castle *may* have been visible from the market place, the gate itself could not have been easily viewed, and so may not have seemed imposing or threatening.

The much renovated parish church (the Church of the Holy Trinity) sits another 100m to the south-east of the market and acts as the visitor's centre of the Carlingford Lough Heritage Trust today. Built on an area of rising ground (approximately 15m above sea-level), it sits at about the same elevation as the castle itself. The viewshed seen in Figure 5.5.29 suggests that though there is a low lying area sandwiched between the castle and the parish church, the castle was clearly visible from the church. This is supported by personal observation as shown in the photograph in Figure 5.5.30 (castle from church). In this photo we can clearly see the low area between the two structures, as well as determine that even with built environment and vegetation the two are inter-visible (something which is not always the case). As seen in this photo, the castle stands out clearly against the skyline, and although it does not seem threatening (at least today), it certainly imposes itself on its environment.

The last photo of Carlingford Castle, seen in Figure 5.5.31 was taken from a position that during the study time frame would have been part of the harbour (the land has been claimed from the lough and now extends the town eastward approximately 100m). This photo was added simply because a view of the castle from this location would have been a common sight for vessels nestled in the medieval harbour of Carlingford. Although nearly obscured by the B & B to the left in the middle-ground of the photo, silhouetted against the skyline as it is, the castle (even in ruins) is potent and impressive.

Comparison of these reflective views with each other and the views from other castles

Because Carlingford was built on a promontory running out into the lough, many of the land-based reflective views silhouette the castle against water. Carlingford Castle is built of greywacke; a dark sandstone, which logic might suggest would blend into the backdrop of the water. That does not seem to be the case. Unlike Carrickfergus, which is also built in a dark stone, and which fairly disappears against the water of the Belfast Lough when viewed from land, Carlingford not only 'holds its own' visually, it actually seems to stand out. (The nature of the loughs themselves may have a part to play in the visibility of the castles. Belfast Lough which extends north-eastward appears to be a slate-grey at almost any time of day. Carlingford Lough extends south-eastward and generally has the appearance of being blue, or even turquoise.) On the other hand, when the dark stone of Carlingford is viewed in competition with the surrounding landscape against hills, houses and trees for example, it tends to lose the battle for visibility and simply blends in with its surroundings. Embedded in its environment it is much more difficult to see. Basically, Carlingford and Carrickfergus are both similar and contrasting in their reflective views.

It is also possible to compare the gate at Carlingford to two other De Lacy gates; one built by Hugh (the son) and the other by his brother Walter. The gate built by Walter de Lacy is seen at Trim Castle; built prior to 1224 (O'Brien, 2008: pers. com.) the West Gate was the public entrance to the *caput* of the liberty of Meath. It was perfectly placed to observe the *entire* medieval town, located only 100m from the market place; it was very much a presence in the daily life of the community. The largest and final gate constructed at Carrickfergus was probably built by Hugh de Lacy (the son) after 1226. This was a massive gate house with a portcullis and two completely round towers which jutted outward into the junction of the port and the market place. This De Lacy gate positioned itself firmly (almost overwhelmingly) within the nexus of the town. Of the three, Carlingford (although not benign) seems by far the least daunting. These three gates trace a pattern of gate building by the De Lacy brothers that seems to go from being respectably strong at Carlingford, to outwardly intimidating at Trim, and then on to extraordinarily imposing and overtly threatening at Carrickfergus. This may reflect an increasing harshness in the political climate, or simply the growing egos of the De Lacys.

Carlingford Castle can be compared to Dunamase whose many gates were on the far side of the rock, and away from the town, and so were not visible from the town. Placed this way, the gates simply could not impose visually on the daily lives of the town's community. However, anyone coming to the Cooley Peninsula or to the Carlingford market (specifically from the north which was territory not yet held by Anglo-Normans) would be forced to pass the castle and its gate directly. To these travellers or traders, the strength of the castle would be constantly re-enforced as they approached. Carlingford Castle may well have been sited specifically for reflective views; and this supports the idea that the role of visibility was important in the siting of the castle.

In general, the 'feel' of the castle from almost every reflective position is one of strength, but also of mystery. Carlingford is very like Dunamase in that sense; it comes across as unquestionably tough, commanding, and obviously in control of the region. It *owns* its landscape and psychological dominance must have been a priority. Its position overlooking the harbour and the ferry suggest that like Carrickfergus, at Carlingford overseeing the administration of the economics of the manor were very important. But the main priority of functional defensibility as security suggested by the projective views is certainly supported in the reflective views.

5.6 Carrickfergus Castle, Co. Antrim

Note: OSi orthophotographic coordinates were used for all prime components, and were found online at: http://www.nimap.net/Mapviewer/Default.aspx.

Introduction

Carrickfergus Castle was built on a rocky promontory of dolerite which projects into the Belfast Lough from its western shore. The castle sits just above the high tide mark, along an edge of relatively flat shoreline, on a site surrounded on three sides by low rising hills. Consequently, Carrickfergus sits in a 'bowl-like' setting. The topographical map can be seen in Figure 5.6.1. On the west of this small peninsula, the bay is deep enough to enable relatively large water craft to tie up, providing some cover from the north-westerly wind (McNeill 1981: 2). This safe haven would have been an attractive feature to trading ships. The sturdy foundation of the peninsula allowed the castle to be built in masonry, and a natural spring rising from the rock guaranteed a constant supply of fresh water (Robinson, 1986: 1). It is believed that the natural spring was considered a holy well and that the promontory may have held religious significance to the Irish before the arrival of the Anglo-Normans. Carrickfergus, having been a royal castle for much of its history, has more documentation than most other castles in Ireland. It vies with Trim for the distinction of being the first Anglo-Norman castle built in stone on the island. The construction of Carrickfergus included at least ten building periods (McNeill 1981: 41-51); but the first three periods were the main building phases of the castle in stone, and coincide with the time frame of this study.

In period I (1177-1195), John de Courcy erected the great tower, a great hall, a small gate and curtain wall (which now forms the inner ward) on the southern half of the peninsula. In period II (1216-1224), under the direction of King John of England, the curtain wall was extended to the east encompassing the rocky ledge that had been the previous entrance to the castle. Two towers were placed along the eastern wall, the castle entrance was moved to the north, and a proper gate house was installed. During period III (1226-1264), Hugh de Lacy (the son) doubled the size of the castle, taking advantage of the rest of the rocky headland by extending the curtain wall to the north, encompassing the entire peninsula. He added a large gate house with portcullis and two completely round towers facing north-west (McNeill, 1981: 44). There is also the possibility that he might have erected a substantial great hall, located along the east curtain wall in the outer ward during this time period (ibid. 45).

Carrickfergus was an urban castle; the nexus of the town during the middle-ages was directly north of the castle where port, market and main approaches all met in front of the massive gate of the castle. A plan of the medieval town can be seen in Figure 5.6.2. At Carrickfergus we find seigneurial markers of lordship related to urban life (Liddiard 2000: 17-18, 57-62) such as the associated settlement of Carrickfergus town itself (whose ancient layout has been reasonably preserved within the modern town), a prosperous port and several religious establishments endowed by the earliest earls. Although the *Plan of Carrickfergus* (British Magazine, May 1760) which is not included in this study clearly shows a 'Lord's' garden, and the James O'Kane map of 1821 shows a deerpark at Carrickfergus, these seem to have been post-medieval additions, associated with the great Jacobean house, Joymount, built by Sir Arthur Chichester, and within the modern environment there is not conclusive evidence of either an extensive garden or a deerpark having been initially associated with the castle. Available documentary evidence in the form of maps (which are not included here) consists of: 'Kragfargus Towne', *c.*1560 (British Lib., Cotton MS Augustus I ii 42; Carrickfergus, *c.*1596 (Public Record Office, London, MPF 98 (ex SP 64/1/31); and Carrickfergus, 1685 by Thomas Phillips (National Library of Ireland, MS 3137 (42). These maps all suggest a managed landscape in the form of structured approaches from the west and north which led travellers directly past the religious houses, into the market place, where the main gate of the castle was clearly visible. The OSNI map illustrating the castle in its modern setting can be seen in Figure 5.6.3.

Location and discussion of the primary components

At Carrickfergus the prime components can be found in six features; three gates (one from each building period), an early great hall, and the great tower which held both the lord's council and his private chambers. The castle plan can be seen in Figure 5.6.4. The original entrance was a simple gateway opening to the east along the edge of the promontory (the castle at this time was compact and quite constricted, the great tower taking up almost a quarter of the available space (McNeill 1997: 24). King John's gate (the second or intermediate gate) faced north. While the construction of an outer (now middle) curtain to the north and east greatly increased the defensibility of the castle, it did little to increase the space available within the castle itself. Unfortunately the gate to this middle curtain is not extant, though the outline of the curtain wall remains. The later and final gate, in the twin-towered outer gate house, also faces north, and is still standing today.

The castle's first great hall was not located in the tower, but was in the inner ward, north of the original entrance and against the east curtain wall. This location is marked by fine windows which are still visible in the original curtain wall. After 1226, when the castle was extended northward to take in the whole of the peninsula, the outer ward along the east curtain may have been the site of a later first-floor great hall (McNeill 1981: 45). However, because the actual location of this hall is uncertain, it was not treated within this study. The second-floor of the great tower at Carrickfergus Castle was a fine space, with a fireplace and single latrine, but relatively poorly lit in comparison with the floor above. Although it is touted as a banqueting hall in the modern interpretation of the castle, it should be considered part of the lord's chamber, and was examined here as a council chamber. The third-floor of the tower afforded the best accommodations, and was almost certainly where De Courcy, King John (during a brief stay in 1210) and De Lacy would have had their private chambers.

General views

The general viewshed for Carrickfergus Castle was created using the location of the battlements of the great tower with an elevation offset of 24m, and an azimuth of 360°. It can be seen in Figures 5.6.5-7. The percentages of visibility in the three views are: 17.15% in the 11km radius, 62% in the 4km radius and 80.88 in the 1km radius. The topographic map seen in Figure 5.6.1 shows several slight linear edges of the DEM tiles at about 10:00 to 1:00 on the clock-face and from 11:30 on the clock-face to approximately the centre point. These are very minor digital anomalies, and caused only minor artifacts in the resultant viewsheds (most clearly seen in Figure 5.6.6).

The 11km view in Figure 5.6.5 suggests that the most far reaching views from the great tower would have been those to the east, the south and the south-west. The views directly west are hampered by the rise in elevation of the hills to the west, as are the views northward. The viewshed illustrates the way in which the hills to the north and west 'reign-in' visibility to these directions within the 4 – 5km radii. Much of the visible area around Carrickfergus is water. In the DEM used here, the water in the bay has been given a *nodata* value (it simply does not exist as far as the DEM is concerned). It seems obvious when looking at the viewshed in Figure 5.6.5, that if land all along the shoreline of the lough has inter-visibility with the castle, then the water linking castle and shoreline should *also* have inter-visibility. This is indeed the case, even though this visibility is not represented by the viewshed. Likewise, to the north-east the open lough stretches all the way to the Irish Sea, and visibility between 1:30 and 4:00 on the clock-face may have stretched much farther than even the 11km radius; to the limits of human visual acuity. Photographs of the projective views from the battlements of the great tower can be seen in Figures 5.6.8 -11.

Projective views: viewshed and personal observation

Gates

Early gate

The earliest gate at Carrickfergus was a very simple opening in the curtain wall. Although this gate is no longer functional, the curtain wall is extant and the position of the gate along the east edge of the peninsula is clearly discernible. The viewshed from this location can be seen in Figures 5.6.12–14, and was generated using an elevation offset of 10m with an azimuth of 350°–320° to simulate the obstruction in the projective view caused by the great tower. The percentages of visibility from this site were 14.25% within the 11km radius, 44.3% in the 4km radius and 55.1% within the 1km radius. The maximum area that the initial Carrickfergus castle was built to dominate and exploit was located within a triangle encompassing the English towns of Larne to the north, Antrim to the west and Carrickfergus and the Belfast Lough to the east. The key transportation and communication routes in the earldom roughly formed the triangle seen in Figure 5.6.15 (Carrick transportation triangle). As two of these routes traced the path of least resistance (Larne to Antrim and Antrim to Carrickfergus), is not surprising that those routes carry the modern transportation infrastructure in the area; between Antrim and Larne, the broad rich lowland of the Six Mile Water Valley would have acted as corridor, as it still does today in the form of the A57/M2. The road between Larne and Carrickfergus came almost directly south from Larne and today is a third class roadway (undifferentiated on OSNI road maps), still maintained and regularly used. Between Carrickfergus and Antrim town, the early road followed the Three Mile Water River through the Carnmoney Pass; today the modern regional roads of B90 and B59 still follow this route. Figure 5.6.13 suggests that the road from Larne might have been visible for approximately 1.5km from the early gate. Likewise, the road from Antrim follows a natural depression along the foot of the escarpment which is also best seen in Figure 5.6.13 and comes into line-of-sight approximately 1.75km from the early gate. The projective view to the east (across the lough towards open sea) has the best visibility from the early gate, stretching quite possibly to the 11km radius or beyond (Figure 5.6.12). The top of the early gate house was not accessible and no photographs are included from this position, but photos of the gate can be seen in Figures 5.6.16 and 5.6.17.

Intermediate gate

This gate no longer exists and its location was approximated from the foundations of its curtain wall which have been exposed by excavation. The viewshed from this position can be seen in Figures 5.6.18–20 and was created using an elevation offset of 10m, with an azimuth of 230° – 180° to simulate the obstruction in the projective views caused by the great tower. The percentages of visibility from this position are 13.5% within the 11km radius, 55.8% in the 4km radius and 67.3% within the 1km radius. Figure 5.6.19 suggests that the road from Larne might have been visible from the intermediate gate throughout nearly the entire 4km radius. However, the road from Antrim (also best seen in Figure 5.6.19) has only a very patchy line-of-sight with the intermediate gate until it reaches 1.75km from the castle. As with the early gate; the projective view to the east (across the lough) has the best visibility, stretching to the 11km radius (Figure 5.6.18). As this gate is no longer extant, no photographs of the projective views from this location were possible.

Later gate

The viewshed from the main gate house at Carrickfergus Castle is illustrated in Figures 5.6.21–23 and was created using an elevation offset of 15m, with an azimuth of 180° – 160° to simulate the obstruction in the projective view caused by the great tower. The percentages of visibility from this location are 14.54% in the 11km radius, 58% within the 4km radius and 74% in the 1km radius. Figure 5.6.22 suggests that the road from Larne could have been monitored from the later gate throughout the full 4km radius; but the Antrim Road is still only observable for

1.75km from the later gate. The projective view to the east again stretches across the lough to the 11km radius (Figure 5.6.21). Unfortunately, access to the top of the gate house was not available, but photographs of this gate can be seen in Figures 5.6.24–26. The main feature of the gate house is the way it dominates the market and town.

Great hall

The great hall at Carrickfergus Castle was located against the east curtain wall of the early castle. It was a low first-floor hall, rectangular in shape, with the long axis of the building running roughly north to south. There were two large exterior windows to the east, both of which are extant. These two stone windows seem to have been fine openings; the exteriors have chamfered sills, jambs and heads. The head of the north window is round while the south window is bluntly pointed. This blunt point is wrought in sandstone while the rest of this window's dressings are in Cultra stone. The rear arches are round-headed and quite large. The north window still has a portion of a window seat (on its north side), and originally probably had seats on both sides of the window (McNeill 1981: 20). The viewshed from the location of the great hall can be found in Figures 5.6.27–29 and was generated using an elevation offset of 4m (a low first-floor level), with an azimuth of 18° – 180° to represent the fact that the hall's only exterior projective views were to the east. The percentages of the three views are 5.1% within the 11km radius, 11.4% in the 4km radius, and 3.33% within the 1km radius.

We might expect the view from this hall to reflect the power, authority and even the piety of John de Courcy who was very generous to the church. But because the great hall had little elevation, what would have been seen at Carrickfergus (at least until the second period of construction when the rest of the headland east of the original entrance was enclosed by the middle curtain) would have been the coastline to the north-east, and the wide, far-reaching vistas of the lough. Photographs of these windows and the extant east wall of the great hall can be seen in Figures 5.6.30-35.

Lord's chambers

Lord's council chamber

This chamber is located at second-floor level within the great tower, which is basically square in shape, as seen in Figures 5.6.39–42. Although it was divided by a solid masonry wall in the 16th century, it was built as a single space, possibly divided by a timber arcade along the line of the later wall. The viewshed from the position of the lord's council chamber was created using an elevation offset of 12m and an azimuth of 20° - 290° to simulate the fact that windows are found only in the east and south walls of the chamber. These viewsheds can be seen in Figures 5.6.36-38. Percentages of visibility from this location were 13.3% within the 11km radius, 25.26% in the 4km radius and 31.19% in the 1km radius.

A fireplace is set more-or-less in the centre of the west wall, and a door giving access to a latrine is in the south wall. This chamber originally had five window openings, two on the east wall and three on the south wall. The north window in the east wall is a round-headed single opening, with chamfered stone jambs and a sill of Cultra stone. The embrasure splays to a round-headed rear arch also of Cultra stone. The south window on this wall is set in a large rectangular embrasure with a round-headed rear arch. This is a two-light window, with a later pillar of red sandstone between the two lights. A single block of Cultra stone heads both windows (McNeill 1981: 25), and there is a drawbar hole in the north splay and socket in the south splay to close a single shutter. The three main windows on the south wall are displaced slightly to the east, possibly due to the position of a mural stair from the first-floor to a double latrine (ibid. 26). Two of these are round-headed windows with segmental rear arches and between them is a narrow loop-sized window with an external round-headed opening of Cultra stone which splayed to a round-headed rear arch (although this is now partially blocked by the later insertion of the cross-wall) (ibid. 26). At the western end of this wall is a small window placed high (to avoid the latrine stairs) with a segmental rear arch; this window has not been treated photographically in this study. The view from this room would have been fairly similar to that of the great hall; framing the northern shoreline of the lough, the sea lanes and the expanse of the lough itself. However the three windows on the south wall would have allowed views of incoming ships, a portion of the harbour area, and the southern shoreline as well. The projective views from this room were far-reaching, and may have been fairly interesting, especially to the south-west.

Photographs showing these windows can be seen in Figures 5.6.43-52. As this room was accessible, it was possible to personally observe the views from all of the windows; this observation definitely supports the projective views suggested by the viewshed.

Lord's private chamber

The lord's private chamber was located on the third- floor of the great tower, directly above the council chamber. Because of an offset, it is slightly larger than the floor below, and is altogether more elegant and commodious than the council chamber (McNeill 1981: 26). As it appears today, it is square, open and airy. A large fireplace sits within the west wall and a mural stair leads off the west end of the south wall to a single latrine. The viewshed from this location was created using an elevation offset of 18m, and because there are openings in each of the four walls, an azimuth of 360° was utilized. These viewsheds can be seen in Figures 5.6.53-55. Percentages of visibility from this location were 16.43% within the 11km radius, 59.4% in the 4km radius and 76.36% in the 1km radius.

Currently there are two windows in each wall of this chamber. Initially the south wall had three windows, but an arch was added to the room during the 16th century, blocking the smaller centre window). The largest windows are set towards the southern half of the room, while those in the northern half are relatively narrow. The affect of the varying size of the windows is that the chamber is bright along the south side, and darker to the north. The windows in the north wall are not centrally placed in the wall, but are off-set slightly to the west (McNeill 1981: 26). They are round-headed loops, externally chamfered in Cultra stone. Each is set within a rectangular embrasure with rear arches, also of Cultra stone. The north window in the east wall is a single light, chamfered externally and splaying to a round-headed rear arch. The southern window in the east wall opens off a rectangular embrasure with a rear arch and is much wider; it is also chamfered externally. Like the window below it in the council chamber, this window has a drawbar hole (albeit in the south splay) and a socket (in the north splay) for a shutter.

The south wall originally had two large double-light windows (although both are currently single lights), with a narrow loop-sized window between them which is gone. The two side windows still dominate the wall and illuminate the room. Each of these windows have drawbar holes for shutters in the splays of their embrasures. The windows in the west wall (like the east wall) differ in size. The southern window is fair sized with a rear arch; its internal and external dressings are of Cultra stone, and the external dressings are chamfered. The northern window is very narrow, essentially loop-sized, its embrasure splaying to a round-headed rear arch.

From the number and position of the windows within this chamber we might expect the projective views to have been splendid, taking in every aspect of the hinterland, and that is indeed the case. However, because the windows vary greatly in size, not all of the views were created equal. The windows of the north wall require the observer to be standing immediately next to the window to get anything more than a sliver of a view. The view looks onto the outer ward of the castle and the rear of the gate house. Some views of the town can be seen beyond the gate house, and to the west, over the curtain wall of the outer ward. During the first two phases of the castle, the later (current) gate house had not yet been built, and urban views from these windows would have been fairly unrestricted. However, the current main gate was lowered in the later part of the 18th century (McNeill 1981: 49), and was originally higher than it is today. After its construction, the gate house would have considerably limited views of the town from this chamber.

From the windows of the east wall an observer can see down into the inner ward, the east curtain and curtain tower. To the north-east the coastline runs to the entrance of the Belfast Lough. Directly to the east and south-east is the lough, with the north tip of the Ards peninsula visible. An observer standing at these windows would have a clear line-of-sight reaching to the Irish Sea. It was interesting to note that during the course of this observation, the weather turned foul and the view of the opposite shore was completely obscured. But while the weather held, the projective view was far-reaching. The windows on the south wall frame a portion of the inner ward, the battlements of the south curtain, and the lough outlined by the opposite shore. These windows afford a clear line-of-sight across to the dark green of the Holywood Hills and Cultra, Co. Down, where the quoin stones for the castle were quarried. The west window in the south wall offers views of Belfast harbour, the Craigantlet Escarpment, Black Mountain, the Cave and Carnmoney Hills; and in good weather... the Mourne Mountains as well. The windows on the western wall afford a view of the mountains west of the castle, and also a view of the entrance to the Six Mile Water river valley. Perhaps the size and position of the largest windows in the south half of the chamber reflects nothing more than that the builder chose to look toward the directions which had the furthest view. But most impressive from the south-west windows is the bird's eye view of Carrickfergus harbour. Photographs of these windows can be seen in Figures 5.6.56-73, and personal observation ascertained that the projective views certainly support the viewshed.

Comparison of these projective views with each other and the views from other castles

Because the great tower at Carrickfergus Castle has seen so little change (in the position of its windows at least) and is still standing and open to the public it afforded a rare opportunity to peer out of the same openings from which the builders themselves had once enjoyed a projective view of their holdings. Only Dundrum Castle and (to a limited extent Nenagh Castle) provided views from a great tower from any room above first-floor height. (Although Trim Castle has also been restored to the point that it is open to the public, the present access format prevents a visitor from standing at a window and taking in the projective view.)

Overall the projective views from Carrickfergus were exceptional. Indeed, they were the second highest in the far-reaching general views of the castles within this study, only Nenagh had higher percentages (Nenagh had 22% visibility within the 11km radius compared to Carrickfergus at 17.5%). However, Carrickfergus *did* have the highest percentages of visibility within the 4km and 1km radii (Carrickfergus had 62% at 4km while Nenagh had 44%, and 80.88% within the 1km radius compared to Nenagh's 79%). Nenagh Castle is actually a good comparison with Carrickfergus; both were approximately the same height (24m) at battlement level, and the excellent general views from both castles are certainly related to the elevation of their great towers. Built within 25 years of each other, both castles had lord's council as well as private chambers within the great tower (although Nenagh's great tower was round). Both were built as the *caput* of their lordships, had massive gate houses and attendant medieval towns. However, there are contrasts as well, for while Carrickfergus was a coastal castle, built on flat land, Nenagh was inland and built on the side of a hill. Another castle which compares well to Carrickfergus is Dundrum Castle, Co. Down (also a coastal castle, with interestingly, a round tower), which was actually begun by John de Courcy as part of his lordship.

The priority behind the siting of John de Courcy's castle on the dolorite promontory at Carrickfergus has traditionally been assumed to be defensive (as is his choice of site at Dundrum). But the structural evidence of the original castle makes this difficult to argue; primarily the siting and type of the original entrance was a vulnerable opening with only a basic wooden door on the east side of the castle. It was reached by a rocky narrow edge of headland not enclosed within the curtain, and positioned where it could not be observed by anyone within the castle's great tower. The obvious and most militarily sound spot to position the entrance to the original castle would have been in the north curtain wall, adjacent to the great tower; facing the land to the north from which any serious threat would have to come. A possible reason for being sited this way might have been to enable the great hall and tower to be side-by-side. This could provide De Courcy with an impressive access from the tower to the great hall; a passage which would be visible to everyone within the hall. Had the castle entrance been placed in the north curtain, between the two buildings, such a grand procession from the first story of the keep - down the steps, across the ward in front of the hall windows and then into the great hall itself - could have been marred by traffic coming and going through the castle gate. This possibility suggests deliberate manipulation of the inner castle landscape with social/display and economic/comfort priorities in mind.

A second structural argument against a defensive priority at Carrickfergus is De Courcy's decision to pierce the curtain wall adjacent to the main gate with two large windows for his low first-floor great hall. Akin to placing the entrance of the castle in a position where it could not be seen from the great tower, penetrating his main defensive barrier was a major military blunder, and for a warrior to do so would have been foolish. But De Courcy was not a military fool. The fact that he disregarded good military practice suggests that defensive concerns were simply not as high a priority for him as comfort or social display (McNeill 1997: 52). This offers us an insight into the degree of De Courcy's self confidence, or at least an indication of how un-threatened he felt by his Irish neighbours. It is interesting to note that history shows that the threat to De Courcy's holdings, when it did come, was not from the Irish but from his Anglo-Norman peers.

During the second and third building phases at Carrickfergus (conducted by the crown and De Courcy's Anglo-Norman peers respectively), the defensive deficiencies of the castle were addressed. Eventually the substantial twin-towered gate house was built, jutting out into the town, with room for accommodation for a constable and most likely a chapel. This tells us that as time went on, the needs of the occupants changed, as did their priorities. The focus became more military and defence based. Certainly there was a desire for more room within the castle but beyond this, the push into, and the increased ability to observe the attendant town suggests a growing interest in the mechanics of the estate itself. The fact that the hinterland to the north was now being addressed suggests that it was more important to the lord. The addition of accommodation for a constable suggests that there was a *need* for one, which indicates that both port and town were well established and a viable concern. It also suggests a willingness on the part of the lord to turn over the peace keeping aspect of the town to someone else.

The projective views from the three gate houses at Carrickfergus are worthy of discussion. The percentages from the three gates were very similar within the 11km radius: 14.24% from the early gate, 13.5% from the intermediate gate and 14.54% from the later gate. It was interesting that the change of locations between the early gate (east of the great tower) and the intermediate gate (north of the great tower) actually caused a *reduction* in the possible far-reaching projective views. However, the percentages of visibility rise steadily in the 4km views: 44.3% from the early gate, 55.8% from the intermediate gate and 58% from the later gate. The rise is even more dramatic from the later gate; near-distance (1km): 55.1% from the early gate, 67.3% from the intermediate gate and 74%. The decrease in the far-reaching projective views between the early and intermediate gates was offset by the increases within the nearer distances. It is also interesting that the increase in the elevation between the first and final gate towers (from a 10m early gate to a 15m later gate) made such a *slight* difference in the far-reaching views (scarcely 0.3%), while increasing the immediate views by nearly 19%.

The priorities for the initial siting of Carrickfergus appear to have been convenience; immediate access to a source of fresh water, a firm foundation upon which to build a stone castle, and a close source of building stone. We might not want to dismiss the idea of the elevation of the great tower being a conscious choice towards a commodious habitation. The roof of the great tower could be accessed privately from the lord's chambers making Carrickfergus an excellent example of social viewing (McNeill 2006: 122-125), perhaps for entertainment. Along this vein, the great tower is very well placed to command both maritime views and scenic landscapes. The scene in the port would have been constantly changing, interesting and even entertaining. The view of the lough would have been attractive and variable as ships came and went along the sea lanes.

Certainly economics played a role in the castle site choice, as evidenced by the location next to best natural harbour on the lough. De Courcy was apparently interested in keeping an eye on the means of his economic solvency, for his views were skewed almost completely to that angle. His choice of positioning of windows in the castle suggests that it was important to him to view the transportation life-line of his small empire and that was what he did, almost to the exclusion of the northern hinterland of the estate. This would suggest that at least in the first generation at Carrickfergus, economics (with projective views geared almost completely towards the source of that economic solvency), and comfort priorities were uppermost. In the succeeding generations, the priorities seem to have shifted towards control, display, and military defence.

Reflective views: viewshed and personal observation Approaches

As has been mentioned above, the modern roads which would have approached Carrickfergus during the subject time frame would have come from Larne (the undifferentiated third class road coming almost directly south through the centre of the Taylor and Skinner map seen in Figure 5.6.74); and from Antrim (the B59 coming directly east through the Carnmoney Pass which is *not* shown on the Taylor and Skinner map). The Taylor and Skinner map shows three other approach roads which were most likely not contemporary to the study time frame. These are the shoreline road coming south-west from Whitehead, the road coming south-east from Strade (B58), and the Marine Highway (A2) coming north-east from Belfast. However, since they are shown on the Taylor and Skinner, and the road from Strade intersects with the B59 from Antrim currently; first-sightings were undertaken on all these approaches. The coordinates of these first-sightings are illustrated in Figures 5.6.6 and 5.6.7.

Surprisingly the road which had the least reflective views of the castle was the Marine Highway (A2) coming east from Belfast City. This road travels along a flat stretch of land, which the viewshed in Figures 5.6.5-7 suggest could be visible throughout the entire 11km radius. However, modern built environment and vegetation make a reflective view along this approach impossible until the traveller is at the round-about approximately only 300m west of the castle. The B59 (Antrim Road) follows the route which is used for the rail line. It comes through the Carnmoney Pass, then intersects with the modern B58. The castle is visible just after this junction, as the B58 descends into the suburbs of Carrickfergus about 2km away from the castle. Until that point topography, built environment and vegetation make viewing the castle difficult. Another factor which lessens the reflective view, even when the castle *is* within line-of-sight, is the colour of stone with which it was built. The North Road from Larne approaches Carrickfergus from (not surprisingly) the north. Because of topography, this road does not enter line-of-sight with the castle until just outside the 4km radius, but vegetation hampered a reflective view of the castle until 2.5km away. The A2 shoreline road coming from Larne through Whitehead (travelling south-west) allowed a reflective view of Carrickfergus as far away as 2.9km. Like the Marine Highway to the west, it is on flat land, but the area is not yet as built up as the road to Belfast. Lastly, it was expected that the reflective view of the castle from the lough would be far-reaching, just as the opposite (projective views from the castle) would be expected. This indeed proved to be the case. When leaving Belfast Harbour by ferry bound for Scotland, the castle was visible (though not particularly *noticeable*) from as far away as approximately 5km. Interestingly however, on the return trip it was raining and the castle was all but invisible within its setting. The interference in this case was two-fold: the effect of the castle's dark stone against the built environment of the town (which made it difficult to recognize), and precipitation. Photographs of the first-sightings from these approaches can be seen in Figures 5.6.75 - 79 and a table of the approaches can be seen in Figure 5.6.80.

Community

Carrickfergus Castle had a successful medieval town (Figure 5.6.2) located directly north of the later main gate. As the old town has been well fossilized within the modern town, it was possible to discern both the location of the parish church and market place from which to obtain reflective views of the castle.

In the present urban environment the castle is only just visible from the parish church of St Nicholas (located approximately 150m from the main gate). This reflective view is illustrated in Figure 5.6.81 which shows that with the present built environment and vegetation, the castle is almost impossible to see. However, looking at this photograph with just a bit of imagination (removing the buildings and the vegetative interference) suggests that had it been physically possible to see the castle, it could certainly have been noticeable from this position.

Compared to the present reflective view of the castle from the parish church, the view from the market place (seen in Figure 5.6.82) allows no doubt about whether the castle was meant to be seen or not. This photo was taken from the position of the approximate centre of the market; within Carrickfergus town this was the key site where port, market and main approaches all met in front of the massive gate of Carrickfergus Castle. This would have undoubtedly been the nexus of the town during the middle-ages. The photo shows how the later gate house on the northern point of the headland extended outward into the urban area, positioning itself firmly into the core of the town in an imposing way. This gate and the enlarged ward not only altered the castle's size, it also pushed the castle right into the growing village. The gate towers would have seemed even more substantial during the subject time frame as McNeill suggests the towers originally contained three floors (1981: 38). It would have been very difficult to escape the silhouette of the massive gate towers, and the rising great tower behind that. The castle as seen from the market place was extremely dominating, both physically and psychologically.

Comparison of these reflective views with each other and the views from other castles

As suggested in Figure 5.6.82 the castle imposed itself *deeply* into its community and would have been a daily presence in the lives of its community. It compares well in this way with Trim Castle whose West Gate reached almost into the centre of its town's market with a visually and physically commanding civic gate. Although the castle can be seen quite well within its environment, there are interesting factors that play into its *noticeablity* within that environment.

After examining the viewshed in Figures 5.6.5–7 (and *before* a site visit to Carrickfergus Castle), this researcher had a somewhat skewed perspective of how well Carrickfergus Castle could be seen in its surroundings, and especially from the three main approaches to Carrickfergus during the subject time period; the maritime approach using the Lough, the western road coming from inland Antrim, and the northern road leading from Larne. It was expected that the reflective view of a traveller coming from the interior of the lordship towards Carrickfergus would be of the castle rising up from the rocks along the coast; the dark stone fortress with its lighter quoin stones profiled against the water. Likewise, it was assumed that travellers approaching across the lough would have had an excellent reflective view for many kilometres. However, after having made a site visit, in fact *several* site visits, reality began to surface and the expectations changed. For instance the castle did not 'rise up' majestically in the reflective view (with the exceptions of along the A2 when it *did* suddenly appear at a distance of 300m at the round-about on the Marine Highway). Instead, the castle was there, battling its environment to be noticed...and generally losing.

The problems with reflective views at Carrickfergus seem to be two-fold, first is the fact that the headland the castle was built on is basically just above sea level; it is not elevated above its surroundings. In fact, the land surrounding it on three sides *rises*, placing the castle in a semi-bowl-like setting. For example, the viewshed from the battlements of the great tower suggests that at least periodically Carrickfergus should be visible from the top of the escarpment to the north-west of the castle (approximately four kilometres from the great tower). In reality, against the backdrop of the water, the dark grey stone of the castle is nearly invisible from the top of the escarpment. It is not until much closer to the castle that one is able to distinguish it from the murkiness of the water. Even at the one kilometre point on the northern approach, the modern built environment and current vegetation coverage obscure the castle quite effectively.

This same scenario was repeated on the approach from Antrim (B58 and 59). And to an extent was true for the maritime approach on the lough. Because the castle is at sea level, it is not silhouetted against a skyline but is instead superimposed upon an urban environment, and much of it the same dull grey as the castle. Adverse marine conditions such as high waves (or even a simple choppy sea) could have made spotting the castle very difficult from a medieval watercraft on the lough. Add to this rain or other inclement weather, and visibility decreases quickly. The second problem is the choice of stone De Courcy used; although the quoins of the castle are of Cultra stone, a soft cream or beige colour, the mass of the castle's fabric was slate-grey basalt. This is not a colour that can stand out strikingly against the dark green hills of Antrim (when approaching by water), or against the slate-grey colour of the lough when approaching from land. The NIE&HS has apparently attempted to make the castle 'stand-out' more by sand-blasting the granite masonry. This has softened the hard grey somewhat, and allows the castle to seem more cream-coloured in certain lighting situations, but this is a recent event. Unless the tower had been treated with some type of wash, the contemporaries would never have experienced a soft cream-coloured castle against the green hills.

If we were to take the GIS viewshed at face value, without ground-truthing it, we could easily come to the conclusion that Carrickfergus was indeed positioned with the intent of being highly visible in its surroundings. We could assume that it stood out boldly against the coastline as seen by ships coming into the port, and imposing and lordly to those approaching from land. The fact is that Carrickfergus fairly blends into its surroundings, whether from water or from land, it simply *does not stand out* in the landscape, and unless an observer knows what they were looking for, it might be missed altogether. This would suggest that for Carrickfergus castle, visibility of the castle was *not* a motive in siting. That said, it should be stressed that De Courcy certainly built the castle to look impressive when it *was* seen, but not necessarily sited it specifically to be visible within its environment. Faced with a low-land site, he chose to build upwards, giving his castle the elevation that the site lacked. This enabled Carrickfergus to have the most impressive possible visibility (with the exception of Nenagh within the 11km radius) of any of the castles in the study.

5.7 *Castleroche, Co. Louth*

Note: At Castleroche, OSi orthophotographic coordinates were used for the gate house, lord's chamber, great hall, and for the original De Verdon castle site of Dún Dealgan. These were found online at: http://ims0.osiemaps.ie/website/publicviewer/main.aspx#V1,699085,811839,7.

Introduction

Castleroche was the second De Verdon castle built in Ireland, but the first constructed in stone. It is sited upon a limestone plateau in an area of rolling countryside, where mountains sweep south-east towards Dundalk Harbour (see topographical map in Figure 5.7.1. It is situated approximately 90km northwest of Dublin, 7.5km north-west of Dundalk, and perhaps most importantly, 5.2km from Dún Dealgan, the mythical birthplace of Cú Chulainn (Killanin and Duignan 1967: 261, 262), where the first De Verdon castle in Ireland was located. Bertram de Verdon came to Ireland with Prince John in 1185 and was granted lands in Louth. He established his *caput* in what is now Castletown, on the outskirts of Dundalk, erecting an earth and timber castle. Although there is some uncertainty about the initial builder and the construction dates at Castleroche (Buckley and Sweetman 1991: 335; Leask 1936: 183; McNeill 1997: 85; Sweetman 1999: 56), this study accepts The Calendar of Documents relating to Ireland dated 6 July, 1236 which gives credit to Rohesia de Verdon, granddaughter of Bertram:

> "Rohesia de Verdun having fortified a castle in her own land against the Irish, which none of her predecessors was able to do..." (CDI I, 2334 p. 347).

Rohesia chose to build her stone castle on a rocky promontory which rises 30m above the plain to the south. The castle's construction fully exploits this precipice; three sides of the castle-end of the plateau are edged by cliffs; the fourth side (to the east) separates the castle from the bailey-end of the outcrop by a rock-cut ditch. Not surprisingly this distinguishingly rugged setting is reflected in the castle's name 'le roche' (the rock). Castleroche has been and is still considered a prime example of the 'frontier' fortification, geared for defence and poised to protect County Louth's fledgling Anglo-Norman settlement from hostile Irish neighbours (Leask 1977: 63; Smith 1999: 46; Sweetman 1999: 58-80).

Castleroche appears to have been a rural castle. A reference to a town of "Roch" is made in a murage charter dated 1376 (CPI 73). However a physical and geophysical survey of the once enclosed limestone plateau directly east of the castle suggests it functioned only as an outer ward or bailey to the castle, with no hint of planned streets, associated plots and houses, market or nucleated population (Oliver 1999: 37 and 51). There do not appear to be any evident seigneurial markers at Castleroche. This may be due to the fact that this was not the initial De Verdon castle in the area. There is a townland known as 'Deer park' located equidistant between Dún Dealgan and Castleroche, and it may be that the two castles utilized the same park simultaneously. This is taken here as documentary evidence of a seigneurial aspect of Castleroche. There does not appear to be any other documentary or circumstantial evidence of a manipulated approach or managed landscape surrounding the castle. The OSi road map of the area surrounding Castleroche can be seen in Figure 5.7.2.

Location and discussion of the primary components

The primary components of gates, great hall and lord's (or in this case, lady's) chambers are found quite straightforwardly in two principal features at Castleroche: gate house and great hall (the lady's chambers occupy the first and second-floors of the gate house). These features were built within the same initial building period, and can be seen in the castle plan shown in Figure 5.7.3.

General views

The best overall projective views from Castleroche would have been seen from the wall-walk of the gate house. This is illustrated by the viewshed in Figure 5.7.4 which was generated with an elevation offset of 12m (approximate height of gate house), and using a full view of 360°. The height of the limestone promontory ensured that the castle rose considerably above the undulating land around it. The rolling hills create an interesting linear pattern in the viewshed, which like the land itself sweeps to the south-east. There are some areas of visibility quite far from the castle, indeed stretching south-east 11km and more to the Irish Sea. Something that is not clearly evident from so large a viewshed is the fact that the top of the motte at Dún Dealgan (the site of the first De Verdon castle, and actually Rohesia's ancestral home) is visible from Castleroche. The most far-reaching views are to the south-east and are patchy, reflecting the rise and fall of the low lying ground between the castle and the coast. The projective views within a 4 to 9km radius to the north-west and north-east of Castleroche are scattered but more densely clustered.

In the 4km viewshed seen in Figure 5.7.5 we can see how closely-set these clusters of visibility actually are, but the 1km viewshed in Figure 5.7.6 illustrates that the best projective views from the castle are the immediate views to the south, west and (to some degree) the north. There is actually a stretch towards 8:00 on the clock-face which offers uninterrupted visibility reaching more than a kilometre from the castle. Problems with the projective views to the east stem from the eastern portion of the Castleroche plateau and the hills beyond which sharply curtail views beyond 150m.

Projective views: viewshed and personal observation

Gate

Because the gate house is in ruins, with only the east wall extant to approximately 12m in height, it was not possible to obtain photographs from the gate house wall-walk, and the viewsheds from this elevation seen in Figures 5.7.4-6 must stand as the only representation of the projective views from this element. The gate house at Castleroche was a double-towered block with a central gate passage, built all of one piece. The towers were nearly semi-circular; being slightly flattened to the front (east) which projects out beyond the curtain (see Figures 5.7.7 and 5.7.8). There were two floors above the gate passage which are discussed below in connection with the lord's chamber component.

Of concern to the look-out stationed atop the gate house at Castleroche would be two types of traffic: official or prestigious travellers coming from Dublin through Dún Dealgan/Dundalk; and *any* traffic which might threaten the security of the castle (though the two might not have been mutually exclusive). The Anglo-Norman settlement's Irish neighbours (the Mac Mathghamhna clan) controlled the lands to the north and west, and if Castleroche was indeed a regional base for defence as has been suggested by Smith (1999: 46), it would certainly have been concerned with any movement into the core of the community from these directions, specifically movement through the two passes leading into the county from those directions; 'The Gap of the North' and the western route to Armagh.

The castle can be approached by four directions along a pair of third-class roads. These roads are undifferentiated on the OSi map but for ease of discussion the road running south-east to north-west will be designated herein as the Castletown Road, and that which runs south-west to north-east as the Forkhill Road. The two roads intersect approximately 400m to the north-west of Castleroche at what is known today as the Castleroche Crossroads. Official/prestigious travellers coming from Dublin via Dún Dealgan would have approached from the south-east presumably along the Castletown Road. It is interesting to note that according to the viewshed in Figure 5.7.5, none of the entire length of the road between Dún Dealgan and Castleroche is visible. In fact, Figure 5.7.6 suggests that traffic from the south-east could not have been visible from the castle until within 250m of its gates. This could have been a huge defensive liability for the castle. The reason for the poor projective view is due entirely to topography (the extent of the bailey's plateau to the east and south-east, and the fact that there are hills rising at least 30m higher than the castle's plateau to the east).

In contrast, the projective view of the Forkhill Road running south-west from the crossroads is excellent and allows almost blanket coverage within a 1km radius of the castle. The views of the northern extents of both the Forkhill and Castletown Roads (at least within the first km), although patchy are still quite good up to 700m from the castle. What is very interesting and quite illustrative is the fact that the pass to the north and the route west (which fall within the 11km radius) are *not* visible from Castleroche. 'The Gap of the North' is located between 12:00 and 1:00 on the clock face in Figure 5.7.4. The route west ran westerly, more or less along the green line of N53 from Dún Dealgan (also best seen in Figure 5.7.4). *Neither* route falls within line-of-sight from Castleroche. Although no photos from the gate to the west were possible, the limitation of visibility to the east is illustrated to some extent by the photo in Figure 5.7.9 which shows the bailey area. Taken at ground level, the rising hills in the far-ground (seen between the trees) and the masking effect of the plateau itself can clearly be seen. In this case personal observation certainly supported the viewshed.

Great hall

The great hall was located to the south of the gate house, nestled in the south-east angle of the curtain wall. It was a ground-level hall with a basement, and measured approximately 18m in length by 14m in width. The long axis ran east to west with gabled ends, the eastern end of which is extant. The hall had three grand windows along the south wall, measuring approximately 3.6m in height by 2m in width. Each window had window seats to encourage active projective viewing across the low rolling hills to the south south-west. The hall had an exterior door in the west end of its north wall which opened onto a porch in the centre of the castle ward. Neither the door nor the porch is extant (with the exception of the west porch wall), so measurements were not possible. Two internal doorways entered the hall, one from the west (low end) which leads to the basement below the hall by way of a mural stair, and one from the east (high end) which led to the lord's chambers through a mural passageway running through the east curtain wall. These openings are illustrated in the ground-floor portion of the castle plan in Figure 5.7.3.

The viewshed created from the windows of the hall can be seen in Figures 5.7.10-12. It was generated using an elevation offset of 6m and an azimuth of between 130° and 240°. (Note 6m was used as an elevation offset even though the hall was at ground-floor to compensate for the basement). In the viewshed we can see that patches of projective views are possible out to at least 8km towards 5:00 on the clock-face. Between 6:00 and 7:00 on the clock-face long swathes of hillside are visible within 3km from the castle. The best projective view from the hall is seen in the 1km radius around the castle (Figure 5.7.13). This localized view encompassed 13.4% of the immediate hinterland surrounding the castle, and totally focuses on the rich resource of the river valley. The area framed by the hall windows was the excellent farmland which housed and fed much of Rohesia's Anglo-Norman community. These were her tenants, and this was her land spreading out, literally, as far as the eye could see.

The projected view from this hall would have been picturesque (even sylvan), though not extensive. The hall windows were inaccessible, but photographs taken of them from the porch area can be seen in Figures 5.7.13-17. These photos provide a measure of the projective views available from the windows, as does Figure 5.7.18. This photograph was taken from the vantage of the south-east corner of the castle wall, though slightly lower than the ground-floor level of the hall windows; still, it offers a projective view fairly similar to that which would have been available from the great hall windows. What we see in these photos certainly supports the viewshed from the hall, and illustrates the topography stretching to the south of the castle.

Lady's chambers

The lady's chambers at Castleroche were located within the gate house on the east side of the castle. The gate house as mentioned had two floors above the gate passage, with the towers and the space between them forming the east wall of the lord's chambers. The gate house was a sturdy stone building, roughly rectangular with the long axis of the upper rooms running north to south. Although the west and north walls of the building are now gone, from the remaining south and east walls we are able to gain rough dimensions of 18m in length by at least 8m in width (external measurements). The first and second-floor seem to have had only one room each, and these two rooms collectively formed the lady's chambers.

At Castleroche we have a chance to look at both the 'public' or council chamber (similar to the third-floor of the great tower at Trim) and the 'personal' chamber (similar to the second-floor of the great tower at Trim). The first-floor of the gate house at Castleroche would have been the business floor, with access restricted to members of the lady's council. The floor above it would have been the more private chamber, dedicated to the lady and members of her family. In order to see if, or how the amount or quality of projective views offered by the two spaces may have differed, it is necessary to address both of these views: that intended for the intimate public (those persons Rohesia would have most wanted to impress); and that reserved for the personal domestic use of Rohesia.

Entrance to the lady's chambers was through the door at the high end of the great hall, through a mural corridor within the east curtain wall lit by four arrow loops, then up a short flight of steps to the first-floor council chamber. Presumably another set of stairs led from this chamber to the second-floor private chamber above, however, evidence of this was lost with the collapse of the west and north walls of the gate house. Although photographs illustrating basic views of the exterior and interior of the windows of these chambers can be seen in Figures 5.7.19 and 5.7.20, these openings could not safely be accessed, so there are no photographs to illustrate the views available from them and we must rely solely on GIS viewsheds for projective views.

Lady's council chamber; first-floor of gate house

Along the east wall of the gate towers on the first-floor were a total of five openings (arrow loops approximately 20 cm in width by 100 cm in height). These openings were positioned to specific directions; one facing north, one facing north-east, two facing east and one facing the south-east. There were probably windows in the west wall of this room, looking out onto the inner ward of the castle, but it is impossible to state for certain. Viewsheds from this space were created using an azimuth between 10° and 150°. Though it should be stressed this projective view is a 'best case scenario' and would not have been possible from any one opening, such a projective scope would have been possible over-all by taking views from all of the extant openings into account. The viewshed seen in Figures 5.7.21-23 were generated using an elevation offset of 6m, and give an indication that although patchy and extremely limited (taking in only 0.59% of the hinterland), the projective views from this chamber were quite far-reaching between 4:00 and 5:00 on the clock-face, stretching to the edge of the 11km radius. Although difficult to see distinctly, this viewshed suggests that a projective view of Dún Dealgan was a possibility from this chamber (Figure 5.7.21). Another interesting thing about this viewshed is that beyond 200m from the castle, (between 1:00 and 3:00 on the clock-face) projective views are cut by the promontory itself. This reduction of the far-reaching view is most noticeable in Figure 5.7.23.

Views from this location looking east would mostly have been of the bailey on the east of the castle and of Dún Dealgan, the initial power-base of the lordship. The view of the bailey may have been interesting, focused as it was on the day-to-day workings of the castle in general. If we keep in mind that Castleroche was a rural castle, lacking in overt seigneurial signs such as cathedral, bridge and market, it is possible to imagine that the glimpse of the crowded and busy bailey may have reinforced the authority of the lady by exhibiting to the most prestigious guests (from the vantage of a slight elevation) the extent of the her immediate command of people and goods. But the not-so-subtle reminder of power Rohesia built into her council chamber by assuring that Dún Dealgan was visible may have been most important projective view this room had to offer.

Lady's private chamber; second-floor of gate house

Along the east wall of the gate towers, on the second-floor of the gate house are a total of four openings (all the size of arrow loops or smaller). Three of these openings face directly east, and one small opening (most likely a simple garderobe light) in the south gate tower faces south. As with the floor below, these openings could not safely be accessed, and there are no photographs to illustrate the views available from them. Once again, there were probably windows in the west wall of this room, opening onto the inner ward of the castle. The viewshed seen in Figures 5.7.24 -26 was generated using an elevation offset of 9m and an azimuth between 30° and 150° (20° narrower than the room below). Again it should be stressed that this projective view is a 'best case scenario', taking views from all of the openings into account. As with the room below there are patchy projective views that are quite far-reaching between 4:00 and 5:00 on the clock-face, stretching to the south-east edge of the 11km radius. A projective view of Dún Dealgan was also a possibility from this chamber (Figure 5.7.24). What we see is a viewshed very similar to the council chamber, but with the elevation rise there is a slight increase in the percentage visible. In the 11km radius it increases from a very small 0.59% to an only slightly better 0.87%. However when the two rooms are compared within the 1km radius view, the added 3m elevation more than *doubles* the amount of ground visible, from 1.29% to 2.92%.

Comparison of these projective views with each other and the views from other castles

The classic 'castle debate' stems around the purpose of castles. The tendency in the past has been for archaeologists and historians to suggest there was *one* all-encompassing *raison d'être* for these enigmatic structures. Of course each castle is unique, created for (possibly many) reasons specific to its individual builder. And only a careful examination of the history, extant fabric and position of the castle within its environment can illuminate the priorities of that builder. There is no 'one-size-fits-all' analysis for castles, and Castleroche is almost the perfect example of why this is true. On the surface it appears obvious that it was built as a defensive fortification, having all the right elements. But a deeper examination of this one castle can illustrate how important it is to take each castle separately, to consider as many of the different processes and choices that went into its construction as possible. What can such an examination of Castleroche reveal about its purpose?

Rohesia de Verdon is the only Anglo-Norman woman documented to have built a castle in Ireland (O'Keeffe 2001; 79). Unlike many of the other castle builders in this study, she was not the first member of her family to arrive in Ireland. In fact Castleroche was conceived and erected by a third-generation Anglo-Norman. Many of the castles in this study built by the first group of Anglo-Norman colonists seem to have been constructed on land that already had a significant claim to authority. Some examples of the latter are: Trim and Clonmacnoise which had been built on church lands; Dunamase which was built on an important Irish *dún*; and Limerick which may have been built on either church land or atop a previous Irish fortress (as may have Adare). Like these, the *first* De Verdon castle at Dún Dealgan was built on a motte that may have been an adaptation of an earlier esker ring fort (Killanin and Duignan 1967: 261, 262), and appropriated if not the exact site, at least the *name* of an important Gaelic mythological location. Castleroche, the second De Verdon castle was, in this sense, built on a true green-field site; claiming no ties to any pre-Anglo-Norman power base. By the time Rohesia was born, the De Verdon name and power had been thoroughly established in Louth, and she may have felt there was no need to assume a site of earlier Irish significance to give her residence authority.

Although it is impossible to know exactly why Rohesia moved the De Verdon *caput* 5km inland, away from a highly prestigious and well placed site near the growing economic centre at Dundalk Harbour, it would certainly be helpful to try to understand her motives (and hence her priorities). This study examines the three most obvious possible motives: that Castleroche was built for military expansion and subsequent frontier defence; that it was built as a replacement of an out-dated centre; and that it was a vehicle to enhance prestige through social and administrative display.

Military expansion and frontier defence

Rohesia may have felt it was time to expand her holdings in Ireland, and there is the possibility that Castleroche was intended as part of that expansion, pushing the De Verdon centre north-west towards the hills and off the plains. Much has been made of the military nature of Castleroche and its location; its position on the promontory, its importance along the frontier of Anglo-Norman Louth, even the excellent defences of the castle itself. This can be illustrated in the following three references: first, Leask wrote "Commanding a pass in the hills of south Armagh, some miles north-west from Dundalk is Castleroche" (Leask 1977: 63), unfortunately we are not told which pass/route Leask was referring to. Secondly, Smith suggested that Castleroche was a frontier castle, set to strengthen the defences around Dundalk (1999: 46); and lastly, Sweetman stressed the warlike character of Castleroche, and pushed the castle's apparent defensiveness by repeatedly mentioning the wooden hoarding around the curtain walls (Sweetman 1999: 58, 59, 63, 76, 80).

The "conquest, colonisation and defence of English Louth" is a term used by Smith (1999: 46-7) that conjures an image of an initial 'blitzkrieg' subjugation and colonisation manoeuvre against the Irish by the Anglo-Normans. Although this is true enough, the term also implies that the two societies were afterward locked in constant battle for Louth. However, during the time period of this study the colony in Louth was prosperous, doing well, and actually expanding (Duffy 1997: 112). If the fear was of the Irish, danger would have come most likely from the north or the west. Though the reality was that the De Verdon's had more to fear from their *Anglo-Norman* peers than from their Irish neighbours (ibid. 115), and an Anglo-Norman enemy could conceivably have approached from any direction. (Of the three attacks between 1196 and 1224 mentioned by Smith, only the first involved the Irish, who were actually working in tandem with John de Courcy (the Anglo-Norman lord of Ulster), the other two were the work of Hugh de Lacy (1999: 33).

We have already seen in the viewshed from the gate at Castleroche (Figure 5.7.4) that the castle was not able to observe *either* of the major routes to the north and west, nor was it able to monitor the road coming directly from the east. What was visible was the last kilometre of the roads leading from the *general directions* of the major routes. The inference here is that the builder felt it necessary to observe only the last kilometre of approach from these vital passes (an immediate threat to the castle itself). There is basically nothing to suggest the passes (on their own) were considered critical to monitor. To put it simply, the castle was not in a position to observe movement into the core of the community from the frontier of the county, and would have been of little use in an over-all defence of the district.

It might be helpful at this point to compare the projective views of the two De Verdon castles in Louth to determine if the elevated rugged setting of Castleroche offered an improvement in terms of visibility. The viewshed from the first De Verdon castle, seen provided in Figure 5.7.27 illustrates just how well Dún Dealgan was placed to provide projective views. It shows that approximately 17% of the 11km radius surrounding the castle was visible, 32% of the 4km radius and 70% of the immediate 1km. This is in contrast to the comparatively meagre projective views allowed by the 11km and 4km viewsheds at Castleroche: 4.5% and 12.9% respectively. Although the 1km radius at Castleroche was fairly significant at 45.7%, it is clear that increasing over-all percentages of visibility from her centre of power could not have been a motive for the relocation.

What is most interesting here is that Dún Dealgan was actually better placed to observe the routes to the north and west than Castleroche. In Figure 5.7.27 we can see that the entrance to the Gap to the North (located between 11:00 and 1:00 on the clock face) is amply covered by line-of-sight from Dún Dealgan. Although the visibility is patchy, it is fairly comprehensive. The route to the west shows much the same story, with dense patches of visibility between 9:00 and 10:00 on the clock face. GIS suggests that Rohesia, in moving her centre to Castleroche in a sense 'traded down' in terms of projective views.

As for the strength of the location chosen, there is no doubt that for the mode of warfare at the time, the promontory that held Castleroche and its bailey were defendable (cliffs around three sides of the castle and the robust ditch across the fourth side make that quite obvious). And there are technological aspects of Rohesia's castle which suggest that security was a high priority: the strength of the gate house, a mural tower to provide flanking fire, battlements, arrow loops placed along the battlements within the curtain walls, and evidence of wooden hoarding. Castleroche certainly appeared avant-garde defensively. But a close look at the *manner* in which the technology was utilized suggests that either the builder did not understand the dynamics of the various devices, or simply did not care whether they were functional or not.

This is evident in several examples; firstly the single mural tower at the north angle of the curtain wall did not project out far enough to allow for cover of the gate, the most vulnerable position of any castle. Secondly, the long stretch of wall-walk between the mural tower and the gate house was fairly narrow and would have made the usage of the arrow loops by bow-men along that wall difficult. More importantly the arrow loops here were so regularly spaced that they could not have provided an efficient field of fire, creating an area of dead-ground (a spot not covered by fire) at the castle's entrance (Oliver 1997: 28-9). Oliver's study highlighted a third example of the poor use of cutting-edge defensive technology at Castleroche; the hoarding. He determined that hoarding (a passive defensive system), as applied at Castleroche would have actually decreased the defensiveness of the whole by preventing use of the arrow loops (an active defensive system) (1997: 23). The fourth and last example was offered by Oliver when he examined the arrow loops in the east curtain wall south of the gate, within the corridor leading to the lord's chambers. This corridor measures approximately 13m in length by 91cm wide and is lit by four arrow loops. While these four arrow loops were actually well spaced to provide a tactically effective field of fire, the passage was simply too *narrow* to accommodate the stance of an archer operating either a longbow or a crossbow (Oliver 1997: 31). Oliver determined that although the castle had been equipped with the most up-to-date defensive technology, it had been so poorly placed as to be impractical, making any tactical defence at Castleroche accidental (Oliver 1997: 27-32).

Replacement of an out-dated centre

By the 1230s Dún Dealgan may have been perceived as too small (it was roughly two thirds the size of Castleroche), or perhaps the rising standards of lordly accommodation had left the timber and earth castle seemingly uncomfortable and outdated. Castleroche certainly made use of its space to provide good accommodation (McNeill 1997: 86; Sweetman 1999: 56, 76). Advances in castle architecture as a whole seen in England and on the continent may also have been a factor; Rohesia (like Walter de Lacy at Trim) might have wanted an up-to-the-minute citadel to show that she was aware of, and could afford the most current trends. McNeill suggests that the integration of the hall with the lodgings in the gate house at Castleroche was a sophisticated improvement in castle design (1997: 166). And though Castleroche was more domestic than Beeston Castle in Cheshire, McNeill proposes that Rohesia may have used Beeston as the proto-type for her castle in Louth. The two castles do have much in common: both were sited on inland rocky promontories with inner wards of similar size and shape, and both inner wards were cut off by rock-cut ditches, both castles relied on their curtain walls, mural towers and massive gate houses for strength (see McNeill 1997: 86 for a full discussion of the similarities). Construction on Beeston started *c.* 1225 by Ranulf, Earl of Chester, and as mentioned above, the Calendar of Documents suggests a date before 1236 for Castleroche (CDI I, 2334), so it is possible that Rohesia did copy Beeston's bold style for her castle in stone. However, Beeston was a genuinely martial castle, built using cutting-edge defensive technology, so placed as to be effective.

Social and administrative display

The third possibility for the creation of Castleroche is that the castle was intended as a vehicle to enhance prestige through social and administrative display. This seems likely, supported as it is by Rohesia's use of fashionable architecture and her inclusion of sophisticated military advances in non-functional ways. We have seen that for all its trappings, Castleroche *appeared* secure but was basically no more than a defensive show-piece (Oliver 1997: 27-32), which offered spectacular projective views of the most productive De Verdon holdings. We also have some evidence in the design of the castle that status, and the presentation of that status was important to

Rohesia. This is seen in the lord's chambers, which were often one of the most difficult components within any castle to access. Using access analysis as a framework to understand the route to the lord's chambers at Castleroche we see that a guest had to pass no fewer than five points where access could be denied before reaching the lady's council chamber. These include the castle gate itself, the outer door of the porch of the great hall, the inner door to the great hall, the doorway to the corridor at the high end of the hall, and finally the door at the top of the steps into the council chamber. An extra two access points were necessary to pass before reaching the private chamber on the second-floor: the door to the mural stair and the door at the second-floor. In fact, Castleroche is a beautiful example of the idea of the "honourial route" suggested by Dixon and Marshall at Castle Hedingham (2003: 303-4), and is all the more interesting because it occurs so early in the 13th century.

In comparing and contrasting Castleroche with other castles within this study, we might look to two castles built by the lords who most hounded De Verdon lands in Louth: Hugh de Lacy and John de Courcy. In many respects De Courcy's castle at Carrickfergus and (the elder) Hugh de Lacy's castle at Trim have much in common with each other. This includes great towers which offered spacious accommodation for the lords and their families; adjacent towns with successful and busy markets; and the fact that both were built next to waterways (Belfast Lough for Carrickfergus and the River Boyne for Trim). At first glance it would not seem that these castles (which were erected at least 40 years earlier than Castleroche), had many similarities with Rohesia's stone castle. But there are similarities, and the similarities are important: all three castles had massive gate houses; all three of these castles exhibit display of prestige as a high priority, and all have clear emphasis on commodious accommodations for their aristocrat and his family.

Another important similarity between the three is that all are held up today as obviously martial structures. Trim and Carrickfergus were first-generation fortifications, built at a time which traditional thought suggests defence and security should have been key issues. Castleroche, located near the edge of Anglo-Norman held lands, has traditionally been considered a frontier castle poised to rebuff attack and protect the fledgling colonial community. But although each of them exhibits some true martial aspects, they all were built initially (and refurbished later) with hints of ostentation which actually weakened their defensive capabilities as a whole.

It is tempting to see in Rohesia's castle, a combination of the more successful pieces of the castles of her peers, and difficult not to imagine that she had visited them and taken away the best bits to use in her own new administrative centre. Both Trim and Carrickfergus had built their impressive gate houses by the 1230s, both had utilized their great towers to good effect by combining within them what, for the times, were sophisticated and almost lavish public and private chambers. (Trim cleverly allowing for true privacy by restricting access to the family space on the second-floor while making the more public council chamber on the third-floor relatively easy to access. With Carrickfergus doing the same only in reverse; the council chamber at second-floor level, and the family chamber at the top of the tower.) There is a striking similarity in the way in which the very practical council chambers at both Trim and Castleroche provided necessary light without offering visual distraction, effectively keeping the attention of the attendees focused inward. Along this same vein, the early castles had separate great halls in stone (though Carrickfergus' initial hall was very small), with grandiose windows framing spacious and imposing vistas, set with window seats to invite guests to sit and admire the power and resources at the disposal of their host. What Rohesia seems to have done was to artfully place the privacy of the lord's chambers within close proximity of the great hall, by utilizing the massive gate house for accommodation and integrating the two features. And all of it was tastefully done within the relatively restricted space of the castle's inner ward.

In concluding this section, we can point out the ways in which the use of GIS in an analysis of the projective views has contributed to the resolution of the debate on the purpose of Castleroche. The viewsheds from Castleroche have suggested that the views we might expect to see from the various features of the castle coincide well with the probable interests and agendas of the communities that might have gazed at them. Although the castle may not have been strategically placed to defend the community at large, it did look to its own safety; the top of the gate house is oriented to view that area from which threat was most likely to come, the west. The hall was oriented to dramatically exhibit to the guests and inhabitants of the castle the richness of the fertile river valley to the south-east, Rohesia's greatest resource. And each of the lady's chambers tell a different story; in the council chamber where business was conducted, the number of openings allowed for good lighting, but the narrowness of the openings did not allow for expansive views. Instead these constricted 'windows' kept the focus of those in attendance firmly in-turned on the matters at hand. The projective views that *were* possible from this chamber were split between the bailey where the daily business of the castle was taking place; and on the ancestral home at Dún Dealgan, a vivid reminder of Rohesia's source of power. The views within the private chamber were focused mainly on the bailey to the east and hence day-to-day life.

The viewshed and the castle fabric suggests that the builder had four main priorities: the comfort of herself and family; psychological dominance shown by a re-affirmation of authority (harking back to the initial De Verdon power-base at Dún Dealgan within the council chamber); security (although *not* for the Anglo-Norman community as a whole, but specifically for the castle itself); finally and certainly most significant is a social display of power and wealth (seen in the way the hall capitalizes on views of the prosperous south-east quadrant).

Reflective views: viewshed and personal observation
A look at Figures 5.7.4 – 6 from the point of view of an observer looking at the castle suggests that the castle was quite visible, especially to the south-east within the 4km radius between 7:00 and 9:00 on the clock-face (towards Dún Dealgan and Dundalk). Although the percentage of pixels visible in the 11km viewshed is quite low, 4.5%, the percentage increases steadily the closer the observer comes towards Castleroche, 12.9% within 4km and 45.7% within 1km. Problems with reflective views are generally related to the topography to the north and east, but vegetation plays a large part in obscuring the reflective view from all other directions.

Approaches
The Taylor and Skinner map (seen in Figure 5.7.28), illustrates the approach routes to Castleroche in 1778, and although we cannot be certain that the roads in use in 1778 were used initially from Castletown to Castleroche, by comparing the OSi map in Figure 5.7.2 to the Taylor and Skinner map it is obvious that the country roads used today are close (if not exactly the same) as those from 1778. Travelling along the Castletown Road towards Castleroche, an observer today is unable to glimpse the castle until within 250m, practically at the base of the promontory. The problem here is (as the viewshed suggests) entirely related to topography. Unfortunately there is not a photograph to represent this reflective view.

Coming from the north-east and the north-west, both vegetation and topography compete with the reflective views until the observer is at the Castleroche Crossroads, 387m from the castle, an only slightly better reflective view than when coming from Castletown. The first-sighting from this spot can be seen in the photograph in Figure 5.7.29. And although the viewshed suggests that the castle *could* be visible throughout the entire 1km radius along the road coming from the south-west, current vegetation does not allow reflective views. The farthest, and indeed the most impressive reflective view of Castleroche was actually seen from one of the few patches of inter-visibility between the castle and the N53 (Figure 5.7.30. This modern road basically represents the path of the western route from Armagh into Louth. Although reflective views of the castle along this route are few and far between, this photo gives a good indication of how stunning the castle might have looked when it was visible from the south, looming atop its plateau. It was interesting (and confidence building) to find the GIS viewshed and personal observation were supportive, in-that a reflective view of the castle was possible, but only from one of the few spots along the N53 that the viewshed suggested it could be visible. In general, personal observation from the approaches to Castleroche suggests that, with the exception of the reflective view from the N53, the castle may not have been easily visible to travellers in the past. Even today it does not announce its presence to strangers, except perhaps accidentally. It is difficult to see beyond casual vegetation, and when first glimpsed along the country roads which pass closest to it, the castle does not stand-out. A summary table of the approaches to Castleroche can be seen in Figure 5.7.31.

Community
As a rural castle, Castleroche did not have a large immediate community sitting at its gate, though it certainly had a good sized bailey just beyond the gate house. Its closest neighbours (those living within a 1km radius of the castle) could have had relatively good reflective views (45.7%). The 1km viewshed from the great hall (Figure 5.7.6) suggests that those living in the hinterland directly south-west of the castle would have been hard pressed *not* to have had a reflective view of Castleroche. And the photograph taken from the N53 illustrates that when visible from its hinterland (even from as far away as 4km), Castleroche was an impressive and dramatic sight. This ostentatious cliff-top setting seems impregnable, dominant and imposing (Figure 5.7.30). Snatches of *projective* views available from the photographs of the great hall windows (Figures 5.7.14-18), hint at the scope of the area from which excellent *reflective* views could have been available to the south-east.

The closest urban areas would have been Castletown and Dundalk, and it may be significant that the De Verdon heirs continued to be active members of those communities long after Castleroche was built (Killanin and Duignan 1967: 261; Smith 1993: 561-588). With this in mind it might be correct to include these near-by towns as members of a rather extended-immediate community for Castleroche; and it is certainly worth the effort to determine if the stone castle may have played any part in the community consciousness from those towns. The 11km viewshed from the gate at Castleroche (Figure 5.7.4) implies that given the right conditions, Castleroche might have been visible all the way from Dundalk, especially from the area in-and-around Dún Dealgan. Unfortunately there are no photographs to represent the possible far-reaching reflective views from these communities; however the 11km viewshed created from the motte at Dún Dealgan seen in previously in Figure 5.7.27 suggests that the initial castle certainly *could* have had reflective views of its successor.

Comparison of these reflective views with each other and the views from other castles
It is interesting to examine the audience Rohesia built to impress. Bertram de Verdon and his descendants underpinned the development of subinfeudation in Louth by encouraging large numbers of English peasants to settle on their lands in Ireland. Eventually the community ties at Louth became so strong that the county itself rebelled against the crown (in the 1312 rebellion led by Rohesia's great-grandsons Nicholas and Milo de Verdon). This community spirit within Louth became so cohesive that Smith suggests it was akin to the "squirearchy" of the Welsh Marches (1993: 563-587). We can be fairly sure that those living in the area with the most impressive reflective views of Castleroche would have been those socialized to appreciate the sight.

When viewed from the south-east or the south-west, it seems obvious why this site was chosen for the De Verdon stone castle. It is in a stunning location which appears unassailable, and from afar the castle looks completely impenetrable. Similar to Dunamase Castle (which is also known as 'the rock'), reflective views are not available wholesale. But in those areas where the castle *is* visible, it imposes visually, commanding attention and inspiring awe. Interestingly, the strong gates of Dunamase Castle faced away from its town (actually hidden behind the 'rock' itself). The gate house at Castleroche faces east, and would have angled slightly away from the Anglo-Norman community of Louth to the south-east. This might have taken the edge off its daunting appearance, making it slightly more benign to its Anglo-Norman observers. It could in fact only be seen head-on from the bailey. However that hardly mattered, for the gate house towered so far above the curtain wall that even in ruins it can still be clearly seen from the *rear* of the castle. And although this massive gate house was an integral part of the castle, it was *not* intrusively looming down on its own society. Still such a structure, visible from three sides, must have been a presence in the lives of the people who would have gazed up at it daily; a constant reminder that its lady had the power, knowledge and the resources to erect such a fortification.

However, we must also contrast Castleroche to Dunamase. Dunamase was obviously a martial castle, built by a man, for men; as an outpost first and fore-most, and certainly not as an accommodating habitation site. It has no extant lodgings other than evidence of a narrow first-floor apartment attached to the ground-floor hall. Unlike Castleroche, the high-tech defensive devices at Dunamase were built to function in the capacity for which they had been designed. They were not simply for show; from the arrow loops to the substantial gates, anyone approaching Dunamase with mischief on the mind would have found its fortifications were genuine physical barriers, not simply psychological barriers. With this in mind, the priority of defence and security suggested by the reflective views can be relegated to a pseudo-military 'show' of security rather than true functional and structural defence. The views suggest that appearance was Rohesia's primary priority when she built Castleroche. This supports the priorities of social and psychological dominance, suggested by the projective views.

5.8 Clonmacnoise Castle, Co. Offaly

Note: OSi orthophotographic coordinates were used for gate, great hall and Temple MacDermot. These were found online at: http://ims0.osiemaps.ie/website/publicviewer/main.aspx#V1,600821,730623,8.

Introduction

Clonmacnoise Castle, Co. Offaly is an example of a problematic castle as discussed in the methodology section of this work. At the heart of the problem is the fact that there are only two reliable documentary sources regarding the construction of the castle, upon which all castle researchers have based their chronologies. These are the 1214 *Annals of Loch Cé* which records a defeat of the "...Foreigners of the castle" at Clonmacnoise, and the reference in the *Calendar of Documents relating to Ireland* in which the bishop of Clonmacnoise was compensated by the crown for the loss of his property due to the construction of a castle in 1216:

> "for his land occupied in fortifying the castles of Clonmacnoise, for his fruit trees cut down, his cows, horses, oxen and household utensils taken away." (CDI, vol. I, no 694):

Although we know a castle was built *c.* 1216, we cannot be certain the extant stone and mortar inner-ward and great hall represent the castle referred to in the above reference. Several authors suggest that there are at least two scenarios possible at Clonmacnoise; the first being that the stone castle seen today was a later build upon an earlier earth and timber foundation (Sweetman 1999:5-7, 36; McNeill 1997: 129). The second scenario is that the original castle was erected on a low esker which was scarped and cut by massive ditches; the great hall upon the esker's summit, the stone inner ward and the massive earthworks which form an outer ward and barbican all being part of one initial design (O'Conor and Manning 2003: 138, 157).

The situation is not helped by the quite poor state of preservation of the castle, or the fact that much of the castle was, in either instance, largely of earth and timber. With this controversy and the state of preservation of the castle in mind, including Clonmacnoise as a subject castle proved a challenge to the study methodology. Since the choice of site for the current stone castle may have been pre-determined by the presence of an earlier Anglo-Norman motte, we must determine if visibility may have been a consideration in the *initial* choice of site. In effect, at Clonmacnoise we must allow for both of the above scenarios by not only comparing and contrasting *known* entities (the extant stone gate and hall), but an *unknown* entity (a possible earlier structure on the same site) as well.

Since we cannot analyze the latter using the extant fabric, some consideration must be given as to what the type, location and height of such a structure might have been. In doing so, this researcher is not asserting that this structure did in fact exist, but simply suggesting that if it had existed, examining the views it could have offered may help us determine the factors leading to the choice of this particular site for its location. For purpose of comparison, we will suppose that another structure (very likely a timber tower) would have stood where the ruined stone great hall is today (here the earthen bank is at its widest and highest point), at approximately the same elevation. Finally, as will become clear, the lord's chamber component was not treated within this analysis.

Clonmacnoise Castle is located 12km south-west of Athlone, 70km east of Galway and roughly 116km west of Dublin. It was built on the flood-plain along the east bank of the River Shannon, in the relatively low lying central Irish lowlands. Approximately 150m to the south-east of the castle is the *Eiscir Riada* or *Slí Mhór* (important route way). This is an east-west esker which was used as an ancient transportation route (Tubridy 1987: 2) across the region's expansive raised bogs (Ruffel *et al.* 2004: 251; Moloney 2003: 7). The esker can be seen in the topographical map shown in Figure 5.8.1 as the most northerly line of high elevation running west towards the Shannon. Just before the river the esker curves 150° and heads south-westerly until it intersects with the Shannon at Shannonbridge, and then continues to the west. The monastery of Clonmacnoise, which was established *c.* 545 sits directly atop the *Eiscir Riada,* (Geissel 2006: 81-86; Mytum 2003: 38), just 250m north-east of the castle.

Clonmacnoise Castle, (either an earlier entirely earth and timber version or the present stone, mortar and earth hybrid) was built *c.*1213-16 (McNeill 1997: 129; Sweetman 1999: 77; O'Conor and Manning 2003: 163). It was built as a royal castle on the orders of John de Grey, justiciar. It has been suggested that the ground used for the castle was the location of the abbot's fort itself, set at a distance from the rest of the monastery (MacAlister 1909: 152). The stone portion of the castle appears to be the result of a single building episode, with the exception of the construction of a later fore-work leading to the great hall (McNeill 1997: 129; O'Conor and Manning 2003: 147). It was an urban castle, set on the outskirts of an already existing 'monastic town'; however the market location and the street patterns remain unknown, and could only be revealed by excavation (Bradley 2003: 50). A plan of the castle in relation to the monastery and the river can be seen in Figure 5.8.2. There do not appear to have been any seigneurial features or evidence of a managed landscape in relation to the castle, although Bradley has suggested that the monastery itself contained seigneurial aspects (2003: 46). The OSi map of the immediate area surrounding the castle can be seen in Figure 5.8.3. Note that there are two modern roads coming from the east, both following eskers; an undifferentiated road (to the north) which is the Pilgrim's Road along the *Sli Mhor* or *Esker Riada*, and the modern R 444.

Location and discussion of the primary components of the stone castle

Clonmacnoise Castle consists of an inner ward which measures approximately 18m by 18m internally, and a great hall. The corners of the castle are oriented to the four points of the compass. Only the principal components of gate house and great hall can be securely identified and as mentioned, both appear to have been built during the same construction phase. In this examination both the first-floor (hall level), and the battlement level have been treated. The *tops* of the great towers and great halls are not normally considered as viewing platforms within this study, but as the great hall's battlement position and elevation may very likely represent the views provided by an earlier timber tower at this same location, it was considered in this study. The lord's chamber component, as with other royal castles in this study is not straightforward. As there was not a specific lord associated with the castle (but instead a steward, constable or garrison commander), it is possible that there were no actual lord's chambers erected. If there had been these rooms would most likely have been within a wooden building in the inner ward and cannot be identified by any extant fabric. For this reason, the lord's chamber component was not treated here. The castle plan can be seen in Figures 5.8.4-6.

General views

The viewshed which best illustrates the overall views at Clonmacnoise can be seen in Figures 5.8.7-9. It was created from the position of the extant great hall, using an elevation offset of 10m with an azimuth of 360°. This position quite adequately serves to help us visualize the projective and reflective views from a possible earlier timber tower surmounting a motte, but it also offers a look at the views available from the battlement level of the later stone hall.

The most far-reaching views from this location are between 8:00 - 3:00 on the clock face. At these points views as far as 11km can be had. There is a general smattering of visibility between the castle and the 9km radius between 7:00 - 12:00 on the clock face, with some pockets of visibility between the castle and about the 6km radius. The best over-all views are between 7:00 and 11:00 on the clock face within the 1km radius; this is the area immediately to the south-west, west and north-west of the castle. There is little uninterrupted visibility from this position, with the exception of within a 500m radius around the castle. This is best illustrated in Figure 5.8.9.

There are several problems with the viewshed, for instance the interesting pattern of no-visibility which looks like a 'torch-light' pointing towards 11:00 on the clock face best seen in Figures 5.8.8 and 5.8.9; this is caused by a small peak to the north-west, on the opposite banks of the Shannon. Though only a few metres higher than the great hall, it effectively blocks visibility from the great hall for 1.5km. Also, some anomalies relating to the DEM are seen as sharp lines of 'visibility' to the north and north-west of the castle (Figures 5.8.7 and 5.8.8). These artifacts are caused by the over-lapping of satellite images; which can be clearly seen on the topographic map and are apparent in each viewshed relating to Clonmacnoise.

Projective views: viewsheds and personal observation Gate

The viewshed created from the gate house is seen in Figures 5.8.10-12. It was created using an elevation offset of 10m and an azimuth of 140° - 90° to simulate how the projective view from the gate house would have been interrupted by the great hall to the south-east. This small gate house would not have been the only gate at Clonmacnoise Castle; presumably there was a more substantial timber gate at the northern angle of the outer ward. It separated the inner and outer wards at the west angle of the curtain wall with the gate facing north-west. The extant stone gate had two floors above the gate passage and measured roughly 5m by 6m; the longer axis running south-west to north-east. It had a gabled roof which is still seen in the extant south-west wall. This building currently only has portions of its north-west and south-west walls remaining.

During the study time frame, travellers could have approached Clonmacnoise from at least five different directions: on water from the north and west along the Shannon; by land *and* water from the north-west (over a togher across the Coolumber bog and ultimately crossing the Shannon presumably by ferry) (O'Sullivan and Boland 2000: 2), from the north-east along the *Sli Mhor,* or Pilgrim's Road, and from the continuation of the esker to the south-west. A lookout stationed atop this gate house would have had good projective views to approximately 700m up the Shannon, and about 1km downriver (although the viewshed suggests there are patches of visibility for 2.5km along the river). The viewshed suggests that the togher across the Coolumber bog would only have been visible for 700m at best. This seems surprisingly scanty coverage considering the apparent flatness of the surrounding flood plain. Yet, when we look at traffic approaching by *land* across the esker to the east, the story is quite different. According to the viewshed it could have been possible to monitor the *elevated portions* of this ancient highway for at least 4km to the north-east, and up to 2km to the south-west. This is best seen in the 4km viewshed in Figure 5.8.11.

As with the viewshed seen in Figure 5.8.8 the slight hill to the north-west of the Shannon creates a 'torch-light' zone of non-visibility. However the lower elevation of the gate house means that this 'beam' extends a further 3.5km. Photos of the gate and the extant walls of the gate house can be seen in Figures 5.8.13-17. Access to the level of the wall-walk or roof was not possible, however photographs from the approximate height of the gate house were taken from the position of the great hall, and these can be seen in Figures 5.8.18-20. Personal observation supports the projective view seen in the viewshed.

Great hall battlement level or possible earlier tower

The great hall was located at the south-east side of the inner ward, which was 3m higher in elevation than the north-west side. The south-west, south-east and north-east walls of the hall form the curtain wall at this end of the inner ward, with the hall's north-west wall facing the inner ward. The building was two storeys high (O'Conor and Manning 2003: 142), with the hall at first-floor level. It was rectangular in shape with the long axis running south-west to north-east, measuring approximately 18m by 12m (external dimensions). There was a cap-house at the top of an internal set of stairs at roof level leading, onto a wall-walk. There is evidence of a turret on the south corner of the building, as well as wooden hoarding along the south-west wall of the hall (ibid. 146). The viewshed from this elevation has already been discussed (Figures 5.8.7–9 above) and the lack of modern access to this height means that no photographs are available from this position.

Great hall

The main door to the great hall was on the western end of the hall's north-west wall. It was probably originally entered by means of a wooden stair (O'Conor and Manning 2003: 147), but at some point a stone and mortar fore-building was erected which was highly defensive in nature, controlling access to the hall by means of a drawbridge. Access to the fore-building was via a set of stairs attached to the south-west wall of the inner ward. Only small portions of the north-west and south-west walls of the great hall are extant. The south-east wall of the fore-building has dislodged from the foundation, and now leans precariously against the remnants of the hall. With so much damage to the fabric, we cannot be sure of the total number of windows the hall originally had. There is currently evidence of at least three windows; north of the main entrance in the north-west wall, in the centre of the south-west wall, and in the displaced portion of the south-east wall. This last window appears to have had seats, and the one surviving jamb-stone is carved (O'Conor and Manning 2003: 146). Judging from the placement of this opening, it is probable that there was at least one more window along this wall. There would most likely not have been a window in the north-east wall as this would have been the high end of the hall.

The viewshed from the great hall component can be seen in Figures 5.8.21-23. This was generated using an elevation offset of 6m to estimate a first-floor height, and an azimuth of 100°-40° to reflect the lack of a window to the north-east. We can see in the 11km radius view that there are no extensive views to the south-east or east, though there are some very truncated projective views towards those directions within the first 1km. The most far-reaching view is close to the 11km radius in the south-west, towards 7:00 on the clock face, but the spot that is visible is very small and patchy. There simply is not enough elevation to provide extensive far-reaching projective views. Like the views from the gate house, there are smatterings of visibility within 6km of the castle to the south-west, west and north-west. The 4km and 1km views in Figures 5.8.22 and 5.8.23 illustrate the scattered pattern of visibility and the extent of coverage of the Shannon.

The lack of a north-east hall window accounts for an almost 'torch-like' slice of non-visibility between 1:30 and 3:00 on the clock face, but it is the esker ridge to the east and south-east of the castle that block extensive views of the area between 3:00 and 7:00 on the clock face. As with the viewshed from the gate house, the artifacts caused by overlapping of satellite images is apparent in the 11 and 4km viewsheds. The viewshed suggests the view was exactly what we might expect to see from the hall; excellent projective views up close, especially of the river to the north and west. The projective view of the gently rolling low-lands to the west and south could have been bucolic; while the prospect of the river traffic on the Shannon could have been both interesting and informative. What would *not* have been visible from this hall were the bustling monastery and the monastic town. Photographs of the openings in the hall can be seen in Figures 5.8.24-29. The collapse of the walls within the hall allowed relatively safe access to the first-floor level at the position of the window in the north-west wall. A series of photographs were taken, and these are shown in Figures 5.8.30 – 32. These very clearly illustrate the quality of the projective views to the immediate north and west; a view of the Shannon is unavoidable. They also show how *short* a distance can be seen to the east and south-east which would have had regions of impassable bog during the study time frame. These panoramic shots support the GIS viewsheds well.

Lord's chamber

As mentioned above, there may not have been a lord's chamber at Clonmacnoise. If this component had been present at this castle, it may have been in timber as there is no extant evidence of it. This being the case, the lord's chamber component was not treated in this exercise.

Comparison of these projective views with each other and the views from other castles

Of the three viewsheds generated within Clonmacnoise Castle, the most far-reaching was definitely from the battlement level of either a previous timber tower or the great stone hall (due to the higher elevation and wider viewing azimuth) (Figures 5.8.7-9). It is interesting to note that although the great hall was built on land which is at least 3m higher in elevation than the gate house (and the elevation offset of both views was 10m), the percentage of pixels visible from the great hall battlements was only *slightly* higher than those visible from the gate house. In fact they were very nearly equal: only 1% more pixels were visible at the 11km radius from the great hall battlements, only 3% at the 4km radius and only 5% at the 1km radius.

If the stone hall was the primary structure on the rise, we would assume that it was oriented and built to the specifications of the builder (John de Grey). Visible from the hall would have been the Shannon to the north and west (which the castle could control to some extent), the gently rolling hills to the south-west and *Slí Mhór* to the south (which it could certainly control). But perhaps more telling is what the castle did *not* deem important to look at; the monastery (which it did not at this point control). By placing the high end of the hall to the north-east, the castle simply did not have to address that well-established neighbour or its accompanying monastic town. It is more than likely church guests would have regularly been entertained in this hall; by preventing such guests a glimpse of their power-base perhaps De Gray may have been asserting a little of the crown's own clout. (It is possible that there was more than a slight amount of animosity between the church and the crown in regards to this castle. De Grey, after being forced to compensate the bishop for the wealthy, prestigious church lands and valuables he had (as had been done so often before) appropriated, may have felt no obligation to highlight any portion of the monastery by framing it in the hall's windows.)

The location of the gate house may have specifically been chosen as the best possible position within the confines of the castle to monitor the approach routes. But the reality is that the land around the *Sli Mhor/Eiscir Riada* is relatively flat, and the elevation of the esker would have allowed for even better projective views. The most elevated points in the immediate area are atop the eskers: the highest is 600m south-east and the second highest is along the second esker, 500m directly east. John de Grey might have sited Clonmacnoise Castle at one of these points if visibility was a high priority. Although both these spots would have been in the middle of what were major thoroughfares at the time, it could have been done.

Interestingly however, the castle was not built on the esker, but was sandwiched between the esker and the River Shannon. The Shannon was not only a major north-south transportation network, it offered a ready supply of water to fill the moat and serve the castle, as well as an excellent natural harbour only 30m north-west of the castle's moat. De Grey chose to take advantage of the close proximity of water rather than the superior elevation of the esker. This suggests that access to and partial control of that extremely valuable resource outweighed the strategic advantages of building on the high-ground, including the advantage of larger projective views.

The intent of Clonmacnoise Castle was to establish the presence of the crown, and provide an Anglo-Norman military centre at a strategic crossing of the Shannon. The priority exhibited at Clonmacnoise seems to have been the physical control of the harbour and a portion of the Shannon. Projective views suggest that although the approach routes could be monitored quite well from the gate, visibility does not seem to have played a role in the choice of site of the castle itself.

Reflective views: viewshed and personal observation

Approaches

Although Clonmacnoise had been an important destination during the study time frame and for at least four centuries before that, by the time Ireland was being extensively mapped it had ceased to be a vital religious centre or even an English power-base. Because of this it was not included in the Taylor and Skinner *Maps of the Roads of Ireland* printed in1778 which has been used to help determine the approaches of most of the other castles in this study. However, most of the approaches mentioned above are depicted on the Townsend 1810 'Bog Commissioner's Map' illustrated in Figure 5.8.33. Not shown in this map is the *togher* on the west side of the Shannon, or the north-east portion of the *Slí Mhór*, the Pilgrim's Road; this researcher was unable to find any illustration of the *togher*, but the approximate position of the Pilgrim's Road has been manually highlighted on the map.

Although the Shannon itself would have been a major transportation route-way, reflective views from it were not analysed for this study. However, the viewshed in Figure 5.8.8 does suggest that an individual coming from the north-east on the Shannon could have seen the castle from a distance of about 700 m; coming from the south-west *occasional* glimpses of the castle were possible as far as 4km away, with uninterrupted views starting within 700 m.

The land based routes seen on the Bog Commissioner's Map remain virtually unchanged today and are both designated as R 444. The modern R 444 coming from Athlone runs atop the esker ridge just south of the *Sli Mhor/Eiscir Riada,* dipping and rising with this second esker. Headed west the road crests a final rise and the castle ruins come into view, approximately 250m away. Interestingly, the 1km radius view of the viewshed from the gate house (Figure 5.8.12) suggests that a reflective view of the castle should not be possible for at least another 10m. Although this may seem to suggest that personal observation and the viewshed were not mutually supportive in this case, it is important to remember that a viewshed will only be as accurate as the digital elevation model upon which it is based, and the data available for this study is only accurate to within 25m. The photo of the initial sighting from this road can be seen in Figure 5.8.34, and (within the limitations of the data) does support the viewshed.

The Pilgrim's Road runs between and atop the intermittent ridges of the *Sli Mhor/Eiscir Riada*, downgrading to a mere track as it nears the monastery. The viewshed in Figure 5.8.8 suggests that when walking the ridges, a traveller along this path could have had occasional reflective views of the castle from as far away as 4km. Closer to hand, Figure 5.8.9 suggests that a reflective view may have been possible near the 1km radius. Unfortunately, not having realized that the small undifferentiated county road which now roughly follows the esker was actually an important ancient route-way, this approach was not driven to any great length (it *was* however walked for the final 500m of the approach). The experience of this researcher was that the castle was not visible at any point between the Nun's Chapel and Temple Mac Dermot. The reasons were vegetation, built environment and for the last 40m simple topography. It was not until actually stepping *off* the Pilgrim's Road and leaving the enclosure at Temple Finghin (to the north of the cemetery) as seen in Figure 5.8.2 that the castle came into view, 290m away. Although the photo seen in Figure 5.8.35 was taken 70m to the north-east of the Pilgrim's Road, it has been included to illustrate the reflective views from this direction (if not the actual road), and the spot is designated as the Pilgrim's Road sighting in Figure 5.8.11 and 5.8.12.

The most impressive reflective view from an approach was found heading north-east on R 444 from Shannonbridge. Here the castle was visible from 2km, the viewshed in Figure 5.8.11 illustrates that this point is actually the first spot along this modern road from which a reflective view should have been possible. Before this (and after it to an extent) topography blocked any views. In this particular case, personal observation and the viewshed were completely in agreement. The photograph of this initial sighting is seen in Figure 5.8.36. In each of the above photographs we can see that although Clonmacnoise Castle does not look dominating or intimidating in its setting today, it could have been impressive upon approach. In general it can be said that personal observation of the modern approaches to Clonmacnoise support the viewsheds. A summary table of the approaches can be seen in Figure 5.8.37.

Community

Bradley has argued that Clonmacnoise was almost an ideal monastic town; analogous to Armagh, Kells and Tuam, and that it had been in existence for several centuries before the arrival of the Anglo-Normans (2003: 50). A surface and geophysical survey at Clonmacnoise by Mytum suggests that the areas immediately to the north and east of the monastery showed the 'densest' usage. And although the town seems to have surrounded the monastery on all sides, the area Mytum refers to as showing 'dense' usage is most extensive along the route of the *Sli Mhor/Eiscir Riada*. There was what he termed 'significant' usage of the area directly to the west of the monastery (just east of the castle's earth works), but the land on which the castle is sited proved to have seen the sparsest amount of use (2003:55). This paints a fairly clear picture of the possible shape of the monastic town, spreading outwards from the monastery, with its epicentre to the north and east. Bradley states that monasteries were a normal place for buying and selling, and that a wide enough range of artifacts have been discovered at Clonmacnoise to suggest this was true here as well. The position of the monastery along two major trade routes made Clonmacnoise the perfect site for commerce, and he suggests that there was certainly a fair held here before the arrival of the Anglo-Normans (2003: 47). If there was a weekly market, it is not certain where this was held, though the area to the north of the monastery which Mytum considers as seeing the densest usage would be a reasonable guess. If this is the case, perhaps the view from Temple Finghin in Figure 5.8.35 might be at least *somewhat* similar to the reflective view of the castle from the monastic town's market place.

A viewshed was generated from the vantage point of Temple Mac Dermot, (Clonmacnoise Cathedral) to determine what kind of reflective view of the castle the cathedral might have had. This can be seen in Figure 5.8.38. An elevation offset of 1.7m was used to represent the eye-level of a standing observer with an azimuth of 360°. This viewshed suggests that at least the ground beneath the castle is not visible from outside the cathedral. Unfortunately built environment (in the form of the visitor's centre to the west of the cathedral) made it difficult to determine whether a reflective view from that vantage point was possible or not, but if modern buildings could block views, it is feasible that medieval buildings could have as well. The castle could certainly be seen from the western side of the visitor's centre. It was in fact, unmistakable as seen in Figure 5.8.39. Only 200m away, unless obstructed by built environment, the castle could have loomed somewhat over the core of the religious life of the community. However, even single-storey buildings in this area, which Mytum listed as having seen 'significant' usage, could have blocked reflective views of the castle.

Currently the 'feel' of the castle is ineffectual. It is so ruinous as to seem completely harmless. Whether this would have been the case during the subject time frame is far from certain. The pile of stone and mortar we see today was only a *portion* of a once rather expansive earth, timber and stone fortification. In general however, the GIS viewshed offers a fairly realistic portrayal of the reflective views from the monastery.

Comparison of these reflective views with each other and the views from other castles

A castle which has several similarities to Clonmacnoise is Athlone, located just 12km to the north; both were royal castles, both built by John de Grey as justiciar at approximately the same period (1210-1220), both monitored crossings on the Shannon and acted as staging points for forays into Connaught. Most importantly however, both were built in locations that had already been established as important crossroads: places of power. These were locations where Anglo-Norman presence was expected to be contested, and often was. Although Athlone was preserved and used through the centuries while Clonmacnoise fell out of use relatively quickly, both castles have little of their original fabric remaining. But a benefit of visibility studies using GIS viewsheds is that, regardless of the amount or condition of the standing fabric, we are able to get an indication of how visible the castle was in its environment. How visible it was may tell us something of how it was *perceived* by those who saw it.

The viewsheds seen in Figures 5.8.7-9 gives a good indication of how much of Clonmacnoise Castle's hinterland would have seen it. In the broadest view (the 11km radius) only 5.7% of the land surrounding the castle could have had some reflective views of the castle, 15.75% of the area within a 4km radius could have had views, and 41.64% of the immediate 1km area may have been able to see it. These figures compare well with the reflective views offered by Athlone which were 6.27%, 20.4% and 39.29% respectively. Barring blockage by past environment, Clonmacnoise Castle (like Athlone), could have been visible from *most* places within the adjacent town, and on both sides of the River Shannon (the most local environment). But neither castle figures largely in their wider environment, and we can assume that at Clonmacnoise Castle reflective views were simply not a high priority.

For all their similarities, it is also possible to contrast Clonmacnoise Castle with Athlone. For instance, at Athlone, John de Grey took advantage of an esker (or of Geoffrey de Costentin's motte) to position the castle slightly *above* (6 – 7m) its environment. After an initial disastrous stone tower collapse in 1211, De Grey had the entire esker or motte (an area measuring roughly 50m x 75m) revetted with masonry. Part of that work (the base of the polygonal tower, two ¾ round towers along the frontage of the Shannon, and the substantial wall between those towers) still stands today. The effect of this dedication of stone and mortar was two-fold; it created a castle with an aggressive, even belligerent aspect that dominated its immediate surroundings; and it illustrated the crown's commitment to permanency in the area.

We do not see the same degree of commitment at Clonmacnoise. While the stone castle and its outer earth works were part and parcel of what might have been a somewhat daunting, defensive whole, we cannot ignore the fact that John de Grey used stone only as the central 'nugget' of the royal castle at Clonmacnoise (permanently enclosing an area of only approximately 22m x 32m). His construction of a motte or use of an already existing esker as a foundation for the stone hall did little more than place that hall on a *level* with the west base of the monastery knoll. It certainly did not elevate the castle above its neighbours. In addition, there was no great tower built at Clonmacnoise, (as there was at Athlone), no strong stone angle towers in the curtain, no extensive stone walls. Nor, in fact did the great hall (the only significant stone building) rise very high. It was probably tall enough for its battlements to peer over the timber palisade of the castle's outer ward, but it surely couldn't be described as having 'towered' over, or dominated the monastic town. Compared with Athlone, Clonmacnoise was certainly a respectable royal outpost, and probably a very practical administration centre; but it was not threatening, controlling or even intimidating.

In general, an analysis of the reflective views of Clonmacnoise Castle as seen by GIS viewsheds and personal observation seem to rule out visibility as a priority for constructing a castle at this particular site. In addition, although the castle was obviously military in nature, intimidation does not seem to have been a high priority at Clonmacnoise Castle; instead control of the harbour on the Shannon and basic administration of the crown's presence in the region appear to have been the priorities of the builder of Clonmacnoise Castle.

5.9 Dunamase Castle, Co. Laois

Note: OSi orthophotographic coordinates were used for the village of Dunamase, and were found online at: http://ims0.osiemaps.ie/website/publicviewer/main.aspx#V1,653234,697930,5). Also, as a point of comparison (which is discussed in detail in Chapter 6: section 6.2), Dunamase Castle was actually analysed using two different pixel sizes; the Landmap DEM, accurate to 25m, and the OSi DEM, accurate to 10m. The 10m DEM gave a slightly better representation of the visibility, i.e. the site of the village of Dunamase did not fall within the LoS in the 25m DEM, but fell just within the LoS using the tighter pixels. This small difference can have a larger effect on the story offered by the viewshed; suggesting on one hand that visibility of the town from the castle might have been important, and on the other that it might not. We can do little about this, except to be aware that both scenarios are a possibility. However, it does point out the fact that data precision can certainly alter results. As the use of the tighter 10m DEM on Dunamase was meant only to be a comparison of the two pixel sizes and not to replace the original data, the information obtained from the Landmap 25m DEM is used in this study.

Introduction

Dunamase Castle, Co. Laois is built upon a massive rock outcrop which rises out of an otherwise relatively flat plane. The topographic map can be seen in Figure 5.9.1. The 'Rock of Dunamase', on which the castle was built had been the site of an Irish *dún* but was appropriated by the Anglo-Normans very soon after their arrival in Ireland. It appears to have changed hands between prominent Anglo-Normans quite a few times early on within the study time frame.

Exactly who built the castle as we see it today is somewhat controversial, and the fact is it may have had several builders. O'Conor suggests that the hill was first fortified by Meiler fitz Henry, with William Marshall (the elder) re-modelling it to increase its defensive capability (O'Conor, 1996; 111-114). Recent work by Hodkinson who excavated the castle between 1993 and 1997 suggests that Geoffrey de Costentin may have begun the castle, with the re-model being the work of Meiler fitz Henry, and William Marshal (the son) adding the stone barbican at some point later (Hodkinson, 2003: 48). While the exact builders are unknown, at least the approximate dates of construction are more or less agreed upon by both archaeologists: initial work probably began quite early, possibly between 1181 and 1189; with the re-modelling occurring between 1208 and 1210. Although O'Conor feels that the stone barbican and the Barbican Gate were part of the re-modelling in 1208 (1996; 114), and Hodkinson proposes that they were added as late as the middle of the 13th century (2003: 35-43), the fact is that building on the rock progressed in stages, unfolding steadily until the mid 13th century. Because of this, the construction of Dunamase is treated here as a one period site.

Dunamase is now a rural castle, not associated with a modern urban centre. There was a town (or borough) located approximately 600m west of the Rock during the time frame of the study (Glasscock 1970: 172: Hodkinson http://www.clanomore.com/dunarch.htm), which may not have existed past the 1340s (Glasscock 1970: 172). But there are historical references to several seigneurial items; there had been a mill (Davies and Quinn, 1941: 13, 17), a garden, a warren and possibly a deerpark (or at least a park of some type) (CDI 2, 466-469). Murphy and O'Conor suggest that the fact Dunamase Castle is located within a townland known as Park (which has a boundary which curves "in a deerpark-like fashion"), and that there are two other townlands (Park Upper and Park Lower) just to the east of the Rock implies there may have been at least one deerpark near the castle during the Anglo-Norman period (2006: 64). There is no other documentary or circumstantial evidence pointing to a managed landscape at Dunamase. Figure 5.9.2 is an OSi map of the network of modern-day roads surrounding the castle.

Location and discussion of the primary components

The primary components of gates, great hall and lord's chambers are to be found in five principal features at Dunamase; the Gate Tower, the Gate House, the Barbican Gate, the Postern Gate and the Donjon (Figure 5.9.3). The Gate Tower was the first to have been built and faced south-east. It was blocked after only a few decades when the inner Gate House (which also faces south-east) was constructed. The Postern Gate, which was built at the same time as the Gate House (Hodkinson, 2003; 41), was located on the extremely steep south-western side of the hill, this was a private entrance and, it would seem, quite difficult to use. The final gate built at Dunamase is the Barbican Gate, which sits lowest on the Rock, facing south-south-east. The Barbican gate and Gate House are sequential; an individual would enter the Barbican first, and then the Gate House in order to reach the castle compound.

This castle was straightforward in respect to determining the location of the great hall and lord's chambers, as both were located within the great tower. The great tower has a ground-floor hall, with a two-storey solar block at the northern end, and is unique in Ireland (Hodkinson, 2003: 36).

General views

As illustrated by the topographic map, the Rock of Dunamase sits near the tip of a range of hills reaching northward into an otherwise low region. There are several large hills directly south and east of the castle, and three somewhat smaller hills to the north-east. The general viewshed seen in Figures 5.9.4-6 were taken from the approximate elevation of the wall-walk atop the great tower (10m). These figures show that the hills obscure much of the far-reaching projective views except to the north and west. Using the analogy of a clock-face there is a slice of visibility towards 2:00 in the far-distance (which begins at approximately 7km and stretches to the 11km radius). The slope of the hill to the south-east of the castle (at 4:00 on the clock-face) can be seen as an 'island' of visibility between 5 and 8km. From 8:00 to 1:00 there is a patchy visibility. One obvious decrease in visibility within the near and middle-distance beginning at about 9:00 is the result of an outcrop at the west edge of the castle hill. The best over-all projective views from this position are to the north and south within 1km of the castle, especially to the south where there are uninterrupted views. Figure 5.9.5 illustrates how, from this position, the entire Rock of Dunamase is observable, as is the village of Dunamase to the west.

Projective views: viewshed and personal observation
Gates

As Dunamase Castle lies in ruins, no access to the tops of buildings (gates or great tower) was possible; viewsheds using an elevation offset of 10m were relied upon to represent the maximum views from the gates. The Rock of Dunamase is surrounded on three sides by steep, craggy inclines, and easy access to the summit of the hill can only be had from the east-south-east where the hill is less precipitous. This access was closely guarded by three of the gates mentioned above, the Gate Tower, the Gate House and the Barbican Gate. However, the Gate Tower was in use as an actual gate for only a brief period, probably not more than three decades.

Gate Tower

The viewshed from the battlements of the Gate Tower can be seen in Figures 5.9.7-9. It was generated using an elevation offset of 10m with an azimuth of 360°. Figure 5.9.8 shows that the road to the north, which would have been the route taken by traffic coming directly from Dublin, was almost constantly observable from approximately 1.7km until within 500m of the Gate Tower. The road from Stradbally to the east is uninterrupted through the last 1km before the castle, and the road from Port Laois to the west was uninterrupted for approximately 1.5km, and from that point is still observable in patches almost until the 3km radius. The photograph seen in Figure 5.9.10 was taken at ground-level from the Gate Tower; it illustrates how the hill to the east obscures visibility beyond that point, but the cross road south-east of the castle is clearly visible. The roads leading to the junction are hidden by modern vegetation, but can be traced by the lines of darker vegetation branching off the cross road.

Gate House

The viewshed from the battlements of the Gate House can be seen in Figures 5.9.11-13. It was generated using an elevation offset of 10m with an azimuth of 360°. Figure 5.9.12 illustrates that this gate had almost identical overage of the approach routes as did the Gate Tower, however the opening of the gate was placed facing the Stradbally Road and not the cross road. This is shown in Figure 5.9.14.

Barbican Gate

The viewshed from the battlements of the Barbican Gate can be seen in Figures 5.9.15-17. It was generated using an elevation offset of 10m with an azimuth of 360°. The two southern approaches are covered more or less to the same degree as the Gate Tower and Gate House; however, the road to the north was *not* visible through most of the immediate 1km radius. A photograph from this gate at ground-level can be seen in Figure 5.9.18. The percentages of visibility from these three south-east facing gates do not vary much; the far and middle-distances are all within a few tenths of a percentage point. The near-distance is slightly better covered by the earliest gate, the Gate Tower (55.6% compared to 44.4% and 52% at the Gate House and Barbican respectively).

Postern Gate

The Postern Gate is located on the cliff along the south-west side of the hill. It would have had an entirely different view from the other gate houses, as is illustrated by its viewshed in Figures 5.9.19-21. The views from the battlements of this gate were hampered by the fact that the battlements were at approximately ground-level on the Rock of Dunamase, just south of the great hall. Because of this location, the rise in slope of the hill itself curtailed all projective views to the north-half of the gate. The viewshed was generated using an elevation offset of 0m (to represent the position at ground-level). While the great hall would have interfered with any views from the battlements of this gate, it was obvious that the contours of the hill were severe enough to restrict any visibility to the north, and an azimuth of 360° was used. The Postern appears to have opened facing north-west, but from the battlements it had fairly dense views to the south within the 1km radius. While it could not directly view the cross roads west of the castle, it could observe the approach of the roads from both Port Laois and Stradbally through most of the 1km radius. Outside 1km however, the views to the east are non-existent, and though the road from Port Laois is visible for another 250m, views to the south-west are patchy beyond 1.5km. Looking due west (towards 9:00 on the clock face), a hill removes the far-reaching view. Personal observation at this point suggests this gate had a near 'bird's eye view' of this side of the Rock of Dunamase; although current vegetation limits an extended view of the approach from Port Laois. Unfortunately there are no photographs from this viewpoint to illustrate the projective views.

Great hall

The Great Hall is located in the great tower, a rectangular building with its long axis running north to south, built on the highest point on the Rock (196.5m). Excavation has shown that there were three windows and two doors in the hall, one window in each of the exterior walls (east, south and west), and a door in both the east and the west wall. The doors were approximately opposite each another, and after the addition of fore-buildings during the 1208-1210 re-modelling, both opened onto porches. The western door was the main entrance into the hall and was quite large (current outside measurements being 2.27m in width by 3.65m in height), though the door to the east is no longer extant, excavation showed it to be slightly narrower, and Hodkinson interpreted it as a service entrance (Hodkinson 2003: 36).

Projected views from the hall should have been excellent, though the choice of building a ground-floor hall instead of a first-floor hall decreased the over-all view. The actual size of the windows may not have been substantial. Although Hodkinson refers to the west window as a 'light in an embrasure', the walls surrounding the west and south windows are ruined to such a degree that the openings could not be measured. There are no measurements in the excavation reports or subsequent articles, so it is difficult to determine the size of the original openings. The window in the east wall *is* extant, but was remodelled into a doorway during the 18th century and it is impossible to know its original dimensions. The windows on the west and south walls were fitted with intriguing chutes for water cisterns within the embrasures (Hodkinson 2003: 37), presumably to deal with the fact that the castle had no fresh water supply.

The viewshed from this component can be seen in Figures 5.9.22-24. As the hall was at ground-floor level, the elevation offset used was 1.7m (the approximate height of a standing man). The viewshed demonstrates how the ground-floor elevation, even built as it was on the highest point of the hill would have limited the close-range views (nearly one-half of the first 1km radius is blocked by its position on the Rock itself). The hills due east and south of the castle severely truncate the view past the 1km ring. However, views beyond 2km can be had to the west (though not due west). The site of the village of Dunamase does not appear to be visible. In all, this view would have been interesting, but probably not terribly informative in regards to observation of the hinterland. Photographs taken from the locations of the south window were relatively unhelpful in illustrating views as fallen masonry almost completely restricts observation. However, it was possible to get photographs from the approximate position of the hall's east window (now a doorway) (Figure 5.9.25), the west window (Figure 5.9.26), and from the main (west) entrance (Figures 5.9.27 and 5.9.28).

Lord's chamber

The lord's chamber was located at the northern end of the hall at first-floor level. This one room arrangement apparently had only two windows, one facing north which measured approximately 204cm in width and 407cm in height, and one facing east which measured approximately 80cm in width and 340cm in height (a window seems to have been added to the west wall during a post-medieval rebuild). The viewshed in Figures 5.9.29-31 (taken from the approximate elevation of the first-floor of the great tower or 6m) suggests that the projective views to the north-west were quite far-reaching, though they are blocked to the north-north-east by hills. Photographs of this view from ground-level (1.7m) can be seen in Figures 5.9.32-34; although somewhat misty due to weather conditions, they closely support the viewshed.

Comparison of these projective views with each other and the views from other castles

The elevation afforded by the location of the castle, and the position of the great tower in particular gives Dunamase a fairly far-reaching projective view. The height of the Rock takes the castle above visibility limiting hedges and trees, and allows viewers to see beyond the immediate vegetation. We can contrast Dunamase Castle with Carrickfergus Castle, which is located basically in a bowl-shaped bay, only barely above sea level and either at, or below the elevation of the surrounding town and hinterland. At Carrickfergus the height of the great tower (which contained the lord's chambers) takes the viewer above immediate obstructions and allows for rather far-reaching views, but the projective views from the other principal components of gate and hall are curtailed by the low elevation of the building site, built environment and foliage. At Dunamase, not only does the great tower (which contains the hall and lord's chambers combined) take the viewer above immediate obstructions, but simply *standing* upon the hill, almost anywhere in the castle ward commands projective views of at least a kilometre, while from the castle ward at Carrickfergus all that is visible are the insides of the curtain walls. As for the gates at Dunamase, although they have somewhat limited projective views (compared to those available from the top of the hill), they are still substantial in scope. Even the slight elevation afforded by their location on the side of the Rock raises them above the level of much of the immediate vegetative cover.

We can contrast the hall at Dunamase with the great hall at Trim, both of which were basically ground-floor venues (Trim's great hall was built on sloping ground, and though the west entrance was at ground-floor, the east wall over-looked the River Boyne at a first-floor level). Dunamase does not appear to place as much emphasis upon projective views as does the hall at Trim. Trim's great hall had four substantial windows opening onto the castle ward on its west side, and four more massive windows trained on the abbey and deerpark on the east side, offering impressive displays of power, wealth and influence. At Dunamase, the exact proportions of the hall windows are not known, but there were only three total, one in the east, south and west exterior walls. These windows seem to have been functional rather than decorative with the emphasis on water catchment and light, not on projective views. Poised as it was at the apex of the hill, the stage was set to provide spectacular viewing, *if* viewing had been the priority. But at ground-floor level, the projective view is not expansive. The choice of building the hall at ground-floor level, rather than on the first-floor (which could not have been much more expensive as the stone was local to the hill), effectively curtailed extensive projective views. It can only be assumed that spectacular views from the great hall were *not* a priority for the initial builders at Dunamase, or for their successors. The great hall at Dunamase does not appear to have been a show-piece, or a forum for the display of wealth, status or power. It seems to have been simply a functional space, designed to serve as a gathering place for a working garrison or a place for relaxation after hunting.

Judging from the aggressive and functional gates, the over-riding priority of the builders of Dunamase Castle seems to have been security and defence. Excavations conducted by Hodkinson from 1993 to 1997 discovered few artifacts to suggest the Rock was used as a primary residence for a lordly household. Most of the items uncovered suggest that a largely male population inhabited the castle throughout most of its history (2003: 45), and the castle appears to have been first and foremost a military establishment, a rather 'rough and ready' outpost, with no independent water source apart from the precipitation captured and stored in hall cisterns.

Although there was certainly room in the castle ward for excellent accommodation in the form of timber buildings, there is no proof of them remaining today. The accommodation we do have in the solar portion of the donjon would have been very limited, amounting to an area measuring only 9m north to south and 14.4m east to west on the north end of the first-floor of the hall (the ground-floor of the north end of the tower would almost certainly have contained the service rooms needed for the great hall) (Hodkinson 2003: 36). Such a small space would have been a far cry from the comforts of Pembroke or Chepstow castles (Marshall's castles in Wales), but may have been adequate (on a temporary basis) for Marshall or his constable or captain. It is certainly difficult to imagine that Isabel de Clare (Marshall the elder's wife) would ever have chosen to stay at Dunamase whilst in Ireland when she had the more commodious castles of Carlow and Ferns available instead (O'Conor 1996; 101: Sweetman 1979: 218, 240, and 1992: 326).

Lastly, we can contrast the lord's chambers at Dunamase with those found both at Trim and at Carrickfergus. The lord's chambers at Trim and Carrickfergus appear to place a priority on observing the functional, wealth generating machinery of the lordships; at Trim it is the marketplace, at Carrickfergus it is the harbour. The lord's chambers at Dunamase on the other hand did not apparently contain a west facing window, a fact which ensured that the market, the parish church, nor even the town could be observed by the lord while he sat at leisure or in office in his personal chambers. This would seem to strengthen the hypotheses that the driving function of Dunamase Castle was not to act as an administrative or economic centre of the lordship. From the projective views and the remaining castle fabric the main priority of the castle seems to have simply been defence.

Reflective views: viewshed and personal observation

Approaches

The 1778 map drawn by Taylor and Skinner (1969: 157) shown in Figure 5.9.35 illustrates that the basic layout of the roads leading to or around Dunamase have not changed greatly in over 200 years (compare with Figure 5.9.2). Today, built environment and vegetation very effectively conceal the castle from the view of travellers on N80 headed east from Port Laois, and the hill just west of the rock continues to block reflective views until a traveller is within 1km of the castle (Figure 5.9.36). The most impressive reflective view from an approach was found headed south-west along the N7 from Dublin to Port Laois, where the castle is visible from a distance of over 2km (Figure 5.9.37). The castle is still striking from the flat of the plane, silhouetted as it is against the skyline of the hill. Figure 5.9.5 clearly illustrates just how extensive a reflective view from the plane to the north and north-west would have been. The N80 going west-bound from Stradbally winds between two tall hills (each over 300m) just east of the Rock of Dunamase, and it is not until the road enters the pass between these hills (approximately 1km away) that the castle is visible along this approach (Figure 5.9.38). In general, personal observation suggests that the castle, when *not* hidden by vegetation or topographic elements features large in its landscape and is highly visible. A summary table of the approaches can be seen in Figure 5.9.39.

Community

The medieval town of Dunamase was located west of the castle, nestled between two hills. The exact locations of the market and parish church were not available (if indeed they are even known). Because of this it was not possible to assess the impact of the castle from those external components. The viewshed in Figure 5.9.40, which was generated from the position of the village (with an elevation offset of 1.7m) suggests that the hill west of the castle may have interfered with reflective views of the castle. The great tower, which is still the most arresting feature in the reflective view, would have been taller yet in the 13th century. The solar end of the great tower was a two-storey structure, and Hodkinson suggests that the hall would have appeared much larger and even more austere than it does today (1996: http://www.excavations.ie/Pages/Details.php?Year=&County=Laois&id=2048). Figure 5.9.41 taken just north of where the town may have been located, illustrates how even today the castle hovers over the countryside.

Comparison of these reflective views with each other and the views from other castles

Since all the public gates are located to the south-east of the castle, and the town sat to the west, the gates would not have imposed into the life of the town the way that the main gates of Trim and Carrickfergus castles did. In fact the viewsheds generated from the various gates suggest that views of the village would have been spotty; not one gate was located in a position to visually intimidate the local town citizens through an over-bearing presence in their daily lives.

Those most impacted by the heavy martial nature of the gates would have been individuals coming from the south-east, headed either specifically to the castle, to the town on the other side of the Rock of Dunamase, or to some destination beyond. But the very nature of the castle's hill assured that, though these massive gates may have been necessary to guard the entrance leading onto the hill, they were not needed to convince external communities of the obvious strength and power of the castle.

It is possible to contrast Dunamase with Carrickfergus and Trim on another reflective point. At both Carrickfergus and Trim, ground-level vegetation and single-storey built structures are able to completely obscure reflective views of the castle even within the town. But Dunamase, built on one of the highest hills (though not admittedly the highest point in the area), can be seen silhouetted against the skyline from almost every direction within a 1km radius, and from much farther distances ranging from the south-west to the north. But perhaps most importantly, Dunamase was very visible to the lands north-west (the main Irish region north of the Lordship of Leinster), overlooking, and attempting to dominate and control the inhabitants of the "tribe-land" of the O'More's (Orpen 1911: 382).

Visibility (at least reflectively) certainly played a role in the choice for the castle site at Dunamase. The castle is so apparent, so observable in its environment that even today it is a potent symbol of power, poised to loom over local inhabitants and travellers alike. The reflective views of the castle as seen in the viewsheds and from personal observation suggest that psychological dominance (and certainly the intimidation and displays of power associated with that type of dominance) were a high priority of the builders at Dunamase. However there was nothing pseudo-military, as at Castleroche; Dunamase was truly a secure and functional castle.

5.10 Dundrum Castle, Co. Down

Note: OSi orthophotographic coordinates were used for the medieval village of Maghera. These coordinates were found online at: http://www.nimap.net/Mapviewer/Default.aspx.

Introduction

Dundrum Castle, Co. Down was built on a ridge of shale and grits, carved out by glacial ice overlooking the bay of Dundrum to the south-east (Hamlin 1977: 6). It is located 39km south of Belfast, 12km south-west of Downpatrick, approximately 50km north-east of Dundalk, and 135km north-east of Dublin. It was constructed out of local Silurian slates with calcareous tufa and Scrabo sandstone for the dressings (Waterman 1964: 137; McNeill 1980: 43), and built on hilly land which extends to the north-east, but which sweeps dramatically upwards towards Slieve Croob (to the north-west) and the Mourne Mountains (to the south-west). The topographical map can be seen in Figure 5.10.1.

It is in an excellent location to monitor (if not control) the land-based transportation routes into east Co. Down. Two routes; from Newry to Clough which passes between the Mourne Mountains and Slieve Croob, as well as along the coast from Carlingford by way of the ferry, both passed within 2.5km of the castle (to the north and south respectively). The castle overlooks a picturesque inner bay. The OSi map illustrating the modern surrounding roads can be seen in Figure 5.10.2.

Dundrum Castle was originally known as the castle of *Rath* from the parish in which it stands, *Rathmurbuilg* (now Maghera), and the name Dundrum (*Dundroma* or fort of the ridge) is a relatively recent application dating from the early sixteenth century (Jope 1966: 207). Most authors agree that the *dún* upon which the castle was built may have been used by the Irish for quite some time before the coming of the Anglo-Normans (Phillips 1883: 4; Hamlin 1977: 1; Sweetman 1999: 40; McNeill 1997: 27). There have been at least four stages of building at this site, and probably more: 1. a perimeter bank of unknown date, 2. the present stone curtain wall (with an early, rather small gate), 3. the round great tower, 4. the later gate house (McNeill 1997: 27). The castle had at least five owners during the subject time frame: John de Courcy from 1177-1205; Hugh de Lacy (the son) during two periods, from 1205-1210 and 1227-1243; the Crown from 1210-1227 and 1243-1254; Prince Edward 1254-1264; and the De Burgh earls, Walter from 1264-1271 and his son Richard from 1271 on.

As with so many castles in this study, the *exact* builders of the different components are not conclusively documented, but can be surmised. The original Anglo-Norman fortress on this site was probably built by John de Courcy sometime after the early months of 1177 and included; the perimeter bank (which is certainly older than the wall and possibly older than any castle), the stone curtain, timber buildings and the early gate (McNeill 1980: 7). At the expulsion of De Courcy from Ireland in 1205 the castle fell into the hands of Hugh de Lacy III who held it until 1210 when he was himself expelled. A reference in the Pipe Roll for Ireland from 1211-12 shows that shortly after acquiring the castle, the Crown had building expenses at Dundrum for a great tower, a hall, a granary and a stable within the castle. Davis and Quinn originally suggested these costs were related to Carrickfergus (1941: 59), but in his corrigenda Quinn attributed them to Rath or Dundrum (1943: 36). The amounts mentioned for these structures (£4-15-2) would have been too small a sum to cover the total cost of their construction in stone, which might imply that between 1205 and 1210 De Lacy had already commenced erecting these buildings, and that the Crown had merely to finish the effort (McNeill 1997: 27).

There is controversy over the date of some work done to the top floor of the round tower. Jope suggests that the use of two types of ceilings in the mural chambers (wickerwork centring and corbelled roofs) in the second-floor suggests that entire floor was added later, *c.* the 15th century (1966: 210). But while the round tower at Dundrum *did* experience restructuring in the later medieval period (Sweetman 1999: 86-7; McNeill 1997: 195), McNeill points out that the continuity seen in the mural stair, chimney flue and the masonry in general would indicate the tower was of one period (1997: 53). He also noted that there is a change in the types of flooring used between the first and second-floors, but these methods of flooring were in use concurrently and cannot be used to indicate chronology. The radial arrangement of the joists for the second-floor indicates that the first-floor room was one of some prestige, as found at Longtown and Pembroke towers (2003: 105). This study therefore assumes that the windows and loops of the great tower seen today are original. Finally, a later asymmetrical gate house was grafted into the curtain wall in the mid 13th century (Jope 1966: 208), possibly around 1260, for a document of that date records repairs to the "doors" at Dundrum (*Pipe Roll 46 Henry III*, I, 2: 56-57).

Today Dundrum Castle is an urban castle, overlooking the modern Dundrum town. And while there is the natural port 500m to the south, and a mention in the 1212 Pipe Roll for a salt pan (Davis and Quinn 1941: 59; Quinn 1943: 36), the only evidence for a town during the study time frame comes from De Courcy who gave the Bishop of Down license to found one at "at Kilclief or Rath....." (Reeves 1992: 166). As the license granted was to the Bishop (i.e. upon religious land), there is a strong chance that the town referred to in this grant is Rathmurbuilg (modern Maghera four kilometres down the road), where the parish church is located. There are no other references to burgage rents, the existence of a mayor or of a corporate body.

It appears that the castle's only seigneurial feature is the little port, and there does not appear to be evidence of a managed landscape surrounding Dundrum Castle. However John de Courcy did contribute generously to the religious aspects of his new lands in that he was an avid founder of monastic houses in Ulster. He spread evidence of his power and benevolence across a wide range (see Flanagan 1999: 54-178 for a complete account of De Courcy's interaction with the church in Ulster).

Location and discussion of the primary components

The primary components of gate and lord's chambers are to be found in three principal features at Dundrum: an early gate, a later gate house and the round tower (lord's chambers). The roughly oval curtain wall which encloses an area approximately 40m by 60m was built in short, straight sections (the castle plan can be seen in Figure 5.10.3. The first of these components to be erected was the very small early gate positioned on the east side of this perimeter (it was built during the second building stage). The position of this gate was discovered during excavation, and although only a drawbridge pit remains extant from this component, as its exact location is known it was considered in this exercise. The next component built (Figure 5.10.4) was the round tower in the lower south-west portion of the ward (erected during the third building stage), which is the accepted site of the lord's chambers (Jope 1966: 209-10). The final component built was the large later gate house along the castle's south-east perimeter (erected in the fourth building stage).

Although there was most certainly an early great hall during De Courcy's time at the castle, the exact location is uncertain. A gable-roofed double latrine was built into the inside of the south-west curtain (the outline of which is still seen in the masonry of the wall). It shows no signs of attachment to other structures, and suggests that between the time of the construction of the curtain wall and the great round tower there may have been timber buildings of some importance within the ward (McNeill 1997: 27). The spot next to these latrines may have been the location of John de Courcy's earliest great hall (ibid 1980: 7).

The location of Hugh de Lacy's great hall is also unknown. Although a secondary skin of masonry at the north-east side of the ward (which has an offset for a first-floor) may have been the location of the De Lacy hall (McNeill 1980: 9), the fact is that we cannot say with confidence where that component was positioned within the ward. Because of this uncertainty the great hall component at Dundrum Castle was not treated within this exercise.

General views

An interesting by-product of running the GIS viewsheds for Dundrum Castle was a reminder of how much difference the precision of the digital elevation model (DEM) can make in the final product of a visibility study. Initially data accurate to 25m (information available for no cost at the Queen's University GIS lab) were used to generate the viewshed at Dundrum. When completed it was striking how much these viewshed did *not* reflect reality; it simply did not illustrate the visibility seen by personal observation in the field. Amplified examination of the viewshed showed that the coordinates obtained by hand (from the castle at the crest of the hill) were displaced on the DEM. Instead of plotting the castle at the crest of the hill, the elevation model positioned the coordinates halfway down the east side of the hill; the result being a viewshed which was completely unrealistic.

Additionally some of the hand-taken field data coordinates taken at Dundrum appeared to be incorrect, reflecting coordinates that did not match the castle plan which had been used as a guide to pace off the perimeter. This had also happened at Adare and seems to be related to the large concentration of stone (both standing and tumbled) which can 'scramble' the satellite representation of data from a hand-held device. While it was possible to double-check and correct poor field coordinates taken within the Republic of Ireland using the OSi website, such satellite data is not available for Northern Ireland. Fortunately however, in Northern Ireland digital elevation models accurate to within 10m (the same precision information which has to be purchased from the Republic) are available for no-charge through the Queen's University GIS lab. (Clearly, dealing with two separate nations within one study has its challenges.)

The viewsheds for Dundrum were run again, using the more precise 10m digital elevation model. This time the result was exactly what should have been expected. Although it had been observed that 10m DEMs resulted in a generally more refined (and sometimes larger) viewshed than the 25m DEMs (Dunamase and Trim), to this point the 25m DEM had not offered a blatantly *incorrect* picture. This drove home the value of working with 'tighter' (more precise) data whenever possible. Figures 5.10.5 and 5.10.6, the 11km viewsheds from the battlements of the round tower, illustrate the degree of difference in the viewsheds between the 25m and 10m DEM's respectively. Obviously the 10m DEM was used to create all other viewsheds for this castle.

The general views seen in Figures 5.10.6-8 represents the most extensive (the maximum) views available of and from Dundrum. This viewshed was generated from the location of the round tower using an azimuth of 360°, with an elevation offset of 13m to simulate the height of the battlements of the round tower, which would have been the highest viewing platform at the castle.

In these views there are scatterings of far-reaching views which (in several directions) extend all the way to the 11km radius; these are seen from 12:00 to 3:00 on the clock face, from 6:00 to 8:00, again at 9:00 and between 10:00 and 11:00. In the middle-distance (9km to 4km radius) there are good views from 1:00 to 3:00 and towards 7:00. There are two fairly large patches of visibility between 9:00 and 10:00, and a rather continuous series of patches between 10:00 and 11:00. Figures 5.10.7 and 5.10.8 clearly show that there are views covering the full 360° surrounding the castle from this position. There are long swathes of uninterrupted views at 1:30, 3:00, 7:30 and 10:00 on the clock. Indeed this viewshed represents the best uninterrupted visibility of any subject castle within this study. The best over-all view is toward the Mourne Mountains in the south-west, where the coastal approach road from Dublin can be seen (with the exception of a 1km stretch at Newcastle) for the full 11km. However, the majority of the ocean stretching out below the castle could also have seen (and be seen) from Dundrum Castle; this represents a massive general view.

Projective views: Viewshed and personal observation
Gates
Early gate

While there is nothing left today of the original simple gate at Dundrum, McNeill suggests that there was very little to begin with; not a gate house or even a tower (1997: 194). The gate faced east (the most easily accessible slope of the hill), and would have been approached from the area which became the outer ward of the later medieval castle. The viewshed seen in Figures 5.10.9-11 was generated using an elevation offset of 6m (to simulate the height of the curtain wall above the gate). Because the round tower would have been in existence for most of the working life of this gate, an azimuth of 280° - 260° was used. The illustration of the gate viewshed is hampered somewhat by the absence of road designations on the OSNI portrayal of the road systems. Names of the dominant routes have been added to Figure 5.10.10 for ease of analysis. During the subject time frame, a guard at this gate could have been able to see traffic approaching from Clough to the north (along what is now the A2) when they were about 1km from the castle. However there is a portion of this road approximately 500m long which stretches from the east to the south of the castle (taking in much of what is now the town of Dundrum) which is not within line-of-sightfrom this gate. From the south-west (Dublin), along the coastal road via Carlingford and Greencastle (also now the modern A2), approaching traffic could have been visible throughout almost the entire 11km radius, until line-of-sightis lost, as mentioned above, just south of the castle (lacking only a 700m stretch through what is now Newcastle and a 300m span just south-west of Keel Point on Dundrum Bay). The old road from Newry (now B180) could have been visible approximately 2.5km outside of Dundrum. Perhaps the most observable approach to the castle would have been from the open sea which could have been visible (subject to weather conditions), to the horizon. The viewshed in Figure 5.10.10 clearly suggests that observation of the route from Newry to Clough was a possibility; a lengthy portion of this road (approximately 2km) falls within the castle's line-of-sightto the north-west. Photographs of the location of this gate can be seen in Figures 5.10.12-14.

Gate House

The later gate house, built *c.* 1260 was more or less superimposed over the south-east portion of the existing curtain wall. This gate type is unique in Ireland, having one half-round tower outside the curtain wall but two square-shaped towers within the curtain (Sweetman 1999: 58). The oddly asymmetrical shape was dictated by the contour of the hill-top, and though appearing awkward, was quite a clever tailoring of contemporary design to fit the needs of this specific castle. It has been suggested that the design may be related to the great gate house of Pembroke Castle built by William Marshal I (McNeill 1997: 92).

The gate was approached by way of a narrow path from the south-west base of the hill. At the top of this path, the one projecting half-round tower on the east side of the gate house forced a 90° left turn into the gate passage. The gate house had two storeys; the ground-floor having two rooms (one on each side of the gate passage), while the first-floor had three rooms (two on the east side and one on the west which was reached by a separate outside stair). There are extant dressings of sandstone which may have come from Scrabo hill at the head of the Strangford Lough (*Pipe Roll 46 Henry III*, I, 2: 56-57; McNeill 1980: 43).

The viewshed from this location can be seen in Figures 5.10.15-17. It was generated using an elevation offset of 10m. As with the earlier gate, the round tower would have created an obstruction in the gate's projective view, so an azimuth of 320° - 280° was factored in to reflect this. All of the approaches were well covered by this gate's projective views, and in fact, as seen in Figure 5.10.16 this gate corrects the problem of no-visibility along the 500m length of the A2 from the south to the east of the castle. There were trade-offs however, for the 2km portion of the road between Clough and Newry is not visible from this gate. Either this was deemed no longer necessary to watch, or it was so easily observed from the early gate and the battlements of the great tower that monitoring this road was simply not the responsibility of this gate.

This gate house can be seen in Figures 5.10.18 and 5.10.19, while the projective view from the actual gate is seen in Figure 5.10.20. There were two problems with field observation of the projective views from the gate house at Dundrum Castle; vegetation and weather conditions. Access to the exact elevation of the battlements of the gate house was not possible, but fortunately the stairway which led to the door of the first-floor room in the west side of the gate house is still extant, ending in a viewing platform with a guard rail. The photographs seen in Figures 5.10.21-29 were taken from this position. These photographs clearly illustrate the problems with visibility caused by vegetation. Assumedly, a *cordon sanitaire* would have kept surrounding vegetation to a minimum, and if this was so, such views on a clear day would have been excellent from this castle.

Great hall

As mentioned, the great hall component of this castle is not included due to the uncertainty of its location.

Lord's chambers

The round tower in the south-west portion of the ward is the likely location of the lord's chambers. This structure is approximately 13m in diameter externally, and the final storey would most likely have been the battlement level, which is also roughly at 13m. The walls were about 3m in thickness with a pronounced base batter. There were three interior floors with the main entrance at first-floor level. A spiral mural stair just west of the entrance gave access to the ground-floor where a well (fed by seepage) had been created by digging a pit roughly 7m into the foundation rock beneath the castle (Waterman 1964: 138). This basement area had two narrow openings for light. The first-floor had a hooded fireplace and two fairly narrow windows with seats, a door nearly opposite the entrance gave access to the double latrine built into the south-west curtain wall. The second-floor had several rooms; a central main chamber and at least three mural chambers as well as a garderobe. The dating of this floor (as mentioned above) is controversial. This study accepts the evidence of the continuity of masonry in the fireplace flue, stairwell and outer walls which suggests it was present from the conception of the structure. Although it may certainly have undergone cosmetic changes in the fifteenth century, the openings (around which this study is based) seem to be original.

As we have at Dundrum two levels of chambers for the lord: public and private, a first-floor business (public or great) chamber, with second-floor (private) chambers, these floors were addressed separately to see if, and how the views from the two levels may have differed.

Lord's council chamber; first-floor of great round tower
When dealing with the round tower it will be helpful to utilize the concept of the clock to describe the locations of the openings (north being 12:00). The first-floor of the round tower had four openings: the main entrance roughly at 4:00, the door leading to the latrines at 8:00, and two windows, one each at 5:30 and 11:00. The possible projective views from this room (counting the view from the main entrance) would have been to the south-east, south and north-west. The viewshed from this chamber can be seen in Figures 5.10.30-32. It was generated using an elevation offset of 7m. Although the gate house would have interfered with projective views from this chamber in the later 13th century, it would not have been a factor when the tower was built, and a full azimuth of 360° was used.

The first thing to note when looking at this set of illustrations is that the viewshed seems quite comprehensive: 22.7% within the farthest view (11km radius), 46% within the middle distance (4km radius) and 71.8% within the 1km radius in. However, this viewshed may not be an accurate representation of the projective views from this chamber and requires some explanation. There are several factors that confuse the issue; firstly, even though the windows had seats, there were only two and they were quite narrow. The second problem lies in the height of the battlements of the curtain wall which seem to have been at eye-level when seen from the first-floor windows. The battlements may easily have interfered with any far-reaching views; by intermittently blocking visibility as well as obstructing views of areas lower than the wall-walk. In short, although it seems that the view from this chamber might have been quite extensive, it is more likely that the majority of the projective views were of the castle ward; internal to the castle and limited in scope. Photographs of these openings can be seen in Figures 5.10.33-40.

Lord's private chamber; second-floor of great round tower
Within the extensive set of apartments that made up the private chambers on the second-floor of the round tower there were many windows of varying dimensions. Roughly a fifth of the wall at this floor level (from 12:00 to 2:30 on the clock) has no external openings (this is where the mural staircase and fireplace flue pass through the masonry). Another fifth of the wall (2:30 to 4:30) has tumbled and we cannot know the arrangements of the windows. But within the remaining 3/5^{ths} of the tower there are seven fairly evenly spaced openings (occurring at 5:00, 6:30, 7:00, 8:30, 9:30, 10:30 and at almost 12:00). If this pattern of even window spacing continued through the missing wall it could have allowed an overlapping of projective views between the north and north-east, compensating for the length of wall that lacked openings. Views from this floor might have been quite comprehensive; though each window would have focused to a specific direction, an individual on this floor might have had access to an elevated 360° projective view almost as comprehensive as the view from the battlements above them. This extensive panorama of visibility has been assumed here, and the viewsheds created from this viewing platform assume this full azimuth, using an elevation offset of 10m. They are seen in Figures 5.10.41-43 and suggest that the projective views could have been extensive; 30.7% visibility within the 11km radius, 47.45% within the 4km radius and 77% in the 1km radius. Photographs of these windows can be seen in Figures 5.10.46-59. Panoramic views from the battlements can be seen in Figures 5.10.60-64.

Comparison of these projective views with each other and the views from other castles
The first item of discussion regarding projective views at Dundrum actually relates to all of the subject castles. This is the issue of vegetative interference. While it was mentioned in Chapter 3: section 3.2 that the obstruction to views from vegetation was expected, Dundrum highlights the degree of obstruction possible. The photographs seen in Figures 5.10.21 – 29 illustrate the projective views from the gate house in both winter and summer conditions, and clearly show how the extent of visibility can be altered by seasonal vegetation changes. This underscores the benefit of a *cordon sanitaire* to keep vegetation from impeding projective views.

Although Dundrum was mostly likely *not* an urban castle it has much in common with the urban castle at Carrickfergus; both were originally built by John de Courcy, located on the shoreline (or near it) and constructed in stone. Both were later inhabited by Hugh de Lacy III (the son), who also lost them at the displeasure of King John. The castles spent time under the crown and profited by it through royal works; Carrickfergus was enlarged and a substantial gate erected while Dundrum saw work on a tower, hall, granary and stable. When Hugh de Lacy was again in control of these castles he expanded Carrickfergus to the full extent the rocky promontory would allow, and built a massive gate over-looking the town. At Dundrum he built a much stronger gate of unusual asymmetrical proportions, tailoring it to the crest of the castle site. Each of these castles had a great tower with fine lordly accommodations and a separate great hall (although Carrickfergus' great hall was of stone and Dundrum's may have been in timber).

Hugh de Lacy III who succeeded De Courcy at Dundrum (as the first earl of the newly created earldom of Ulster) was a tenant-in-chief of the English crown, the younger son of a man who had been able to draw on significant resources, and to attract tenants who were themselves substantial landholders (Flanagan 1999: 154-6). We can see a change in perspective between the two builders in the types of castles these men erected. De Courcy's *caput* at Carrickfergus was a relatively cramped affair, though there was ample room on the peninsula to build a much larger castle. The gate at Carrickfergus was small and poorly defended (although the smallish gate he installed at Dundrum was slightly more secure). His great tower at Carrickfergus was well built however, with (what for the time) were almost opulent personal accommodations. It was a substantial rectangular structure, tall, certainly impressive, but perhaps unimaginative.

By comparison the round tower which De Lacy might have started before his expulsion from Ulster was altogether more adventurous and stylish (Hamlin 1977: 3; McNeill 2003: 105). This suggests he was familiar with the wider European trends. The fact that either his father Hugh II or his brother Walter had erected a great round tower at Longtown in Wales illustrates that the De Lacy family was well aware of the modes of the day.

The projective views from the lord's private chamber represent the best views available indoors from the castle. Unfortunately it is difficult to compare the differences between the projective views (as represented by the viewshed) from the lord's council and private chambers because the viewshed generated does not take the curtain battlements into account. (We are left again with the concept of GIS offering the maximum amount of visibility, while the *reality* of the projective view from this venue is impossible to know.) However, if we compare the rather meagre exposure to the outside world offered by the external openings on the first-floor (the main door, a door to the garderobe and two narrow windows) with the seven (plus) windows of more generous proportions on the second-floor, it becomes obvious that the *attitude* towards viewing from the upper floor was altogether more open. This is something which Dundrum has in common with at least three other castles in this study which have council chambers: Carrickfergus, Trim and Castleroche. All three of these 'business' chambers tend to limit the amount of distractions possible by being scanty on the amount, placement and/or direction of the windows. Window openings in council chambers seem totally functional; providing a light source while keeping the inner-circle focused on matters occurring within the chamber itself.

Generally the over-all projective views at Dundrum are excellent, especially the castle's ability to monitor its approaches. Although the road from Newry to Clough between the Mournes and Slieve Croob (north-east bound on modern A25) was too far away to actually be controlled by the castle (2.5km), it could certainly have been *monitored* by a watchman placed at the early gate. From the later gate house a view of this pass was blocked by the round tower. Interestingly however, the loss of the ability to monitor this route did not decrease the percentages of the *total* projective views by much, and actually increased the area visible in the closest 1km radius: the early gate could see 22% of the 11km radius, 43.5% of the 4km radius and 66% of the 1km radius, while the later gate house had visibility of 21%, 40.6 and 67.8% respectively. We should note that this length of the Newry to Clough route could have been visible from the battlements of the round tower even before the erection of the gate house, so the loss of sighting from the main gate would not have decreased the exceptional over-all ability of the castle to monitor transport within its environment. Almost *all* of the approaches (the exception being from the immediate north which was occupied interestingly enough by the Irish) fall within line-of-sight from the castle.

Judging from the projective views as well as the remaining fabric of the castle, Dundrum was intended foremost to be a commodious habitation. It would be difficult to argue that the location John de Courcy chose for his castle at Dundrum was chosen for any reason other than its excellent projective views. The amount and quality of these views strongly suggest that visibility played a strong role in its siting.

Reflective views: viewshed and personal observation

A look at Figures 5.10.6-8 from the point of view of an observer looking at the castle suggests that the castle could have been quite visible from its hinterland. In the farthest instance, 25.5 % of the hinterland (not taking into account any views from the ocean or the bay) could have seen the castle. Prominently placed as it was, if vegetation on the hill had been controlled the castle would have been difficult to miss from the east and the south-west (3:00 – 7:00 on the clock face) within the 11km. Within the middle-distance, 50 % of the surrounding land had reflective views of the castle, and in the 1km radius, 80.8 % had views.

Approaches

Judging by the Taylor and Skinner map seen in Figure 5.10.65, Dundrum was an important hub on the south-east coast of Ulster. This 1778 map illustrates at least eight approaches to Dundrum in 1778. For this study, three of these approaches were chosen and then travelled in order to obtain first-sighting information. The amount of vegetation atop the castle's hill very nearly engulfs the castle, making it difficult to pick out from a distance. The least visibility was experienced on the Newcastle Road north-east bound (the modern A2). Along this route the castle was visible from approximately 1km away; the issue here being vegetation. (Unfortunately the GPS coordinate for this sighting was lost due to either computer or user error, and while a photograph of this sighting is included, the position is not highlighted on the viewsheds seen of the approaches to the gates.) The road from Clough (the modern A2) south bound allowed a reflective view of the castle starting from 1.1km distance. This is very close to the point at which the viewshed suggested a reflective view was possible. The farthest reflective view was seen along the Old Newry Road (modern B180), north-east bound where the castle was visible from 2.3km away. The road between Newry and Clough (to the north-west of the castle), while not being an actual approach, was also analysed for a reflective view, but vegetation and built environment does not allow views of the castle from this route. Photographs of the first-sightings can be seen in Figures 5.10.66-68 and a summary table of the approaches can be seen in Figure 5.10.69.

Community

As it is difficult to prove that a medieval town lay between the castle and the port, this study takes the approach that any community located directly below the castle was limited, and that Dundrum was in essence a rural castle. A viewshed seen in Figure 5.10.70 which was created using an elevation offset of 1.7m and an azimuth of 360° illustrates that from the position of the parish church (over 4km to the south-west) topography would have completely blocked a reflective view of the castle.

We cannot be certain where the market place within this town was located, but if it had been close to the church, as is often the case, reflective views from it could also have been blocked by topography. We might assume then that visually the castle could not have factored greatly in the daily lives of community members from any principal components located at Maghera. The round tower (as seen in Figures 5.10.6-8 can be seen from almost every point within the inner bay, and from great distances out to sea as well. From the bay the castle could have been striking (as it still is today), and more than a little pompous. The inclusion of the gate house would have only increased the pretentiousness of the castle. Unfortunately no photograph of the castle from the bay is included in the study.

Comparison of these reflective views with each other and the views from other castles

Although the gate at Dundrum Castle did not impose into the everyday public life of the community at Maghera, the castle itself was positioned perfectly to be persistently viewed hovering above the port beneath it. Barring vegetation and built environment, it could have been seen from great distances, *especially* along the approach roads. In a theatrical sense, Dundrum is similar to Castleroche, which when viewed from the south and south-west seems almost suspended above its environment. However, as dramatic as Castleroche was, it was not visible to as large a percentage of its hinterland as Dundrum.

Not surprisingly, the castles whose percentages of visibility compare well in reflectively with Dundrum are also hill-top fortresses: Carlingford (which was another De Lacy holding), and Dunamase. Carlingford was an urban castle also sited along the coast; it could have been visible from 27% of its 11km radius, 53% within 4km and from 90.75% within the 1km radius. These two castles seem to have been located specifically to either see, or to be seen; their positions and designs ensuring that they possessed their landscapes, if not literally, then certainly figuratively and symbolically.

The differences between these castles lie in their arrangement within the Anglo-Norman settlement as well as their usages. Carlingford and Dundrum are located along the strip of Anglo-Norman settlement between Dublin and Larne. Here Anglo-Norman influence was very strong, and settlers were thick on the ground. These two castles were obviously intended as habitation sites. And although the position of the lord's chamber component at Carlingford was not obvious due to later works, there was ample room within the perimeter of the castle to provide accommodation (indeed the later works themselves suggest commodious lodgings were important to its inhabitants).

Another hill-top castle, Dunamase, Co. Laois was analysed using DEMs accurate to both 25m and 10m. The percentages from the 10m general viewshed at Dunamase (not shown here), corresponds quiet well to both Carlingford and Dundrum (both analysed with 10m data). Dunamase was a land-locked castle, which had a medieval town located at the base of its hill. Although the 10m DEM suggested that the town and castle of Dunamase fell only just within line-of-sight of each other, over-all the castle was highly visible within its hinterland: 39% of the 11km, 45.43% of the 4km and 92% of the 1km radius. Space was somewhat limited atop the hill at Dundrum, and the lord's accommodations went 'up' within a stylish circular tower perhaps reminiscent of William Marshall I's Pembroke Castle (McNeill 1997: 92; 2003: 96). The design of the round tower has been suggested to be a social statement of display originating in France. The use of the round tower as primarily the show-piece feature of a castle was used by great, as well as lesser lords in England, Wales and Ireland (McNeill 2003: 96-98).

Dunamase however, located towards the boundaries of the Lordship of Leinster and overlooking the "tribe-land" of the O'Mores (Orpen 1911: 382), seems to have been a warrior's outpost. Although it has several more documented seigneurial aspects than either Carlingford or Dundrum can claim (Davies and Quinn 1941: 13, 17; *CDI* 2, 466-469), it never appears to have been intended to provide comfortable housing for its inhabitants. McNeill has suggested it may have served as a hunting lodge or men's retreat (2009: pers.com). Archaeological evidence uncovered during excavation suggested that the castle's population was largely male (Hodkinson 2003: 45), and the fact that the townland it sits in is known as Park, while two other townlands (Park Lower and Park Upper) are just to the castle's east make this quite plausible (Murphy and O'Conor 2006: 64). Dunamase seems first and foremost to have been a militarily secure fortress whose blatantly martial style was not simply for display. When looking at Dunamase is clear how the martial model portrays power, which may explain why another lord (in this case De Lacy) might duplicate it even when the castle's circumstances may not have called for true military strength.

This might be some of what prompted the building of the gate house on the south-west curtain wall at Dundrum. This gate has been seen as an attempt at increasing the defensibility of the castle by strengthening its weakest point; the obviously sub-standard (for the times) and easily accessible early gate. The later gate was placed in a location that would have been difficult to approach by either friend or foe. It certainly seems to have met the requirements of a serious defensive gate, even if compared to Carlingford, Carrickfergus or Trim (other De Lacy castles) it may have seemed rather tentative (McNeill 1997: 91). What is interesting however is that the later gate was also (perhaps conveniently) *very* visible to the port, the town and the *entire* southern aspect of the castle. It provided more than increased security; it endowed the castle with a soldierly icon of strength. Paired with the great round tower which rose above the curtain wall just to its west, the new gate certainly amplified the pomposity of the castle.

Today the 'feel' of Dundrum is not necessarily aggressive or menacing, but it is very potent and the castle is undoubtedly a powerful symbol of blatant, lordly consumption. The examination of the reflective views of the castle supports the idea that visibility played a major role in John de Courcy's choice of site for the castle, and continued to be a priority of subsequent builders throughout the subject time frame.

5.11 Ferns Castle, Co. Wexford

Note: OSi orthophotographic coordinates were used for the gate house, south-east and south-west towers, south and east curtain walls, centre point of the castle, the 13th century cathedral, St Augustine's Monastery and the parish church. These were found online at: (http://ims0.osiemaps.ie/website/publicviewer/main.aspx #V1,701684,649889,8).

Introduction

Ferns Castle, Co. Wexford is located 20km west of the Irish Sea, on terrain that is relatively flat though interspersed with gentle hills. Within 8km to the west of the castle, the Blackstairs Mountain range rises to almost 800m, and less than 2km to the north the foothills of the massive Wicklow Mountain range begin. Ferns is situated 9km north-east of Enniscorthy, almost equidistant between Wexford to the south (40km) and Arklow to the north-east (35km), Dublin is 85km to the north-east. The River Bann flows approximately 1km to the east. The castle was built on a bedrock of vertically stratified shale, and it is this same shale that was used to construct most of the castle, with building material coming directly from the rock-cut fosse surrounding the castle (Sweetman 1979: 221; O'Keeffe and Coughlan 2003: 135). The name Ferns comes from the Irish *Ferna* meaning alders, or "a place abounding in alders" (Adams 1904: 13). The topographical map for this site can be seen in Figure 5.11.1.

Ferns had been important for centuries before the arrival of the Anglo-Normans. Here St M'Aodhóg (alias Aidan or Eda) founded a monastery on the flatlands approximately 500m east of the castle site, on land given to him by a newly converted Hy Kinsella ruler (Hore 1910: 300; Killanin and Duignan 1967: 279). By the 12th century Ferns had become the ecclesiastical and political centre of Leinster, and was the home of the ruling Mac Murchada family. About 1160 Diarmait Mac Murchada founded St Mary's Abbey for Canons Regular of St Augustine on the site of St M'Aodhóg's monastery (ibid). It was from Ferns that Mac Murchada sought the help of the English in regaining his kingship of Leinster in 1168 (Furlong 2006: 105-111). The site and Ferns manor eventually went to his daughter Aoife and her husband, Richard de Clare (Strongbow) who answered Diarmait's request in 1170. After Strongbow's death, Henry II bestowed the manor and castle of Ferns upon William FitzAdelm de Burgo (Adams 1904: 178), next Giraldus Cambrensis says that the manor spent some time in the hands of the sons of Maurice Fitzgerald, who built a castle at Ferns (*Expugnatio Hibernicia, The Conquest of Ireland*: 171).

It is not entirely clear who owned or occupied the castle between 1177 and 1192, but when Isabella (daughter of Strongbow and Aoife) married William Marshal the elder, 1st Earl of Pembroke, he became joint owner of her considerable properties, lands and offices in Leinster, Wales, England and France. At the death of William Marshal the elder in 1224, the castle and lands passed to his son William (Adams 1904: 178; Furlong 2006: 172). A reference in the *Calendar of Documents relating to Ireland* places the castle as part of the dower awarded to the widow of William Marshal the son in 1232 (CDI I 1950). In 1247 William de Valence (Henry III's half brother) became lord of part of Leinster, a position he held until his death in 1296 (McNeill 1997: 144; O'Keeffe and Coughlan 2003: 147).

Unlike many of the other castles in this study, Ferns has an impressive amount of historical references relating to its owners. Unfortunately none of these references allows us to assemble a conclusive building chronology for the castle. While we know who the key players were, the exact construction time frame (and therefore the exact builder) of the stone castle we see today is somewhat ambiguous (Sweetman 1979: 240). O'Keeffe and Coughlan have given an excellent report of the various theories of its evolution suggested by historians and archaeologists (2003: 136-147). The full account is not repeated here, but ranges from the idea that the castle was constructed by the Marshal Earls between 1207 and 1225 (Sweetman 1999: 39) to the possibility that William de Valence was the builder sometime in the mid to late 13th century (Leask 1977: 49; McNeill 1997: 144; O'Keeffe and Coughlan 2003: 147). O'Keeffe and Coughlan have even suggested that perhaps De Valence marked his acquisition of Irish lands in 1247 by erecting this very castle (ibid). It is this construction time frame (mid to late 13th century) and builder (William de Valence) that has been assumed in this study.

We cannot know how long it took to build the present stone castle, or if it was built over several periods or all in one effort. The castle offers some hints however; the south-east corner tower has three different forms of masonry, each taking up approximately one-third of the height of the tower. Sweetman suggests that this indicates at least two building phases (Sweetman 1999: 81), while O'Keeffe and Coughlan have proposed that the varying lines of masonry may represent patterns of acquisition of stone from different quarry sources (2003: 145, 147) rather than distinct periods of construction separated by substantial intervals of time. The fact is that we do not know.

Ferns was undoubtedly an urban castle, either imposed onto, or hovering on the outskirts of the town which had developed in relation to St M'Aodhóg's monastery. Today the town and castle occupy the hill, overlooking the monastery and cathedral, but the direction of the development of the town of Ferns is not entirely clear; it is possible that during the subject time frame there were *two* centres of habitation; an ecclesiastical centre on the flatland surrounding the monastery, and a secular/political centre around the castle on the hill.

Seigneurial aspects mentioned in documentary sources include St Mary's Abbey for Canons Regular of St Augustine, which though founded by Diarmait Mac Murchada (Killanin and Duignan 1967: 279), was part of the legacy left to his daughter Aoife, and so to the

Marshal successors and perhaps even subsequent lords. A cathedral and a parish church were built approximately 100m north-west of the monastery between 1220 and 1240, very likely by the younger earl Marshal. Local folklore suggests that the three small hermitages (Cill Dara, Bethany and Cherith), which still stand near this spot were built in the 8th century, so it is possible that this part of the hill was church land even during the study time period. A weekly market in the manor of Ferns was granted to the bishop of Ferns by Henry III in 1226 (O'Keeffe and Coughlan 2003: 146). No documentary evidence was found for the location of the market, and as it was granted to the church, it may have been located near the cathedral and abbey. However, given the importance of the town religiously *and* politically, it is almost certain that some type of formal market was held at Ferns from a much earlier date. It was noted that between the castle and the hermitages (at the intersection of the Carnew Road, the R745 and the main route from the south into the town) there is a wide, triangular shaped area which may have been the market place for the secular community of Ferns. An inquisition taken at the death of Joan de Valence, Countess of Pembroke (Lady Palatine of Wexford) in 1307 suggests that other seigneurial features at Ferns included a water mill, the parish church and literally thousands of acres of land under cultivation (Hore 1910: 306). Although the town was originally episcopal, it might also be considered a seigneurial element as it must have increased considerably in size with the influx of Anglo-Norman settlers. When Aymer, William de Valence's son and heir died in 1324 there were 160 burgages in the town of Ferns (ibid; Sweetman 1979: 218). A plot of land still known as 'Ferns Demesne' is shown north-east of the town on the OSi map seen in Figure 5.11.2. (Note that the OSi map suggests the location of the 'Cathedral and High Crosses' at approximately 1km north-east of the castle, this is a misrepresentation; the cathedral and high crosses referred to are actually much closer to the abbey site, literally on the eastern outskirts of the town.) It is considered here that a structured approach is represented in the way the road from the north passes St Mary's Abbey before climbing the hill to the castle. The manner in which the approaches from the west and north-west brought travellers to the castle by-way of the possible market place may also be evidence of a managed landscape.

Location and discussion of the primary components

Determination of the primary components at Ferns was quite difficult, the reason being that the standing fabric is not only scanty but has been highly restored. Figures 5.11.3 and 5.11.4 illustrate the amount of reconstruction that has taken place since the excavations here in the 1970s. These 'before and after' photographs were taken by McNeill; the first shows the south-west tower and its attached east and south walls as they appeared before work began on the castle in 1975. The second has been taken since completion of the restoration and shows the same tower and walls. The only primary components clearly discernable at Ferns Castle are the lord's chambers. Judging by the numerous and fine window embrasures and fireplace at first-floor level seen in the two walls still attached to the south-east tower, these rooms could be considered part of a suite of lord's chambers. Because we cannot know which room functioned as which type of chamber, these rooms are herein treated in the context of the lord's chambers as a whole, and the views from them (to the east and the south) were analysed.

Not clearly discernable at Ferns are the locations of the main gate and the great hall. Although the windows and fireplace seen in the rooms on the first-floor level are grand enough to suggest the great hall was located within the castle walls we see today, neither space would have been very large, nor would they fit the model of 13th century halls which were open to the rafters (both of these grand rooms had floors above them). Likewise, the area utilized as the modern entrance to the castle does not make a convincing 13th century main gate. This area was excavated by Sweetman who suggested this was the location of what he termed "some type of gate house", but admitted that it would have been a vulnerable entrance way. He also determined there had been a drawbridge but did not explain how such a mechanism might have worked in this instance (Sweetman 1979: 223). Indeed, it was difficult to determine during field examination how such a drawbridge might have been functional.

The castle plan can be seen in Figure 5.11.5 and illustrates how little there is left at Ferns. The paucity of evidence is undoubtedly due to the fact that much has been lost to time and local quarrying of stone. McNeill has suggested that what is seen at Ferns today represents only the internal core of what must have been a much larger castle complex. In this way it represents (in the purest sense) the *donjon* or inner stronghold of the castle at Ferns. Indeed evidence of an outer ward was uncovered during the excavations in the 1970s when a short trench revealed approximately 3m of an outer fosse running north and south, about 12m east of the base of the south-east tower. This ditch was not explored further, but Sweetman suggests that there were outer defences surrounding the castle (Sweetman 1979: 224). Along this line, *Heritage Ireland* has integrated a model of a possible design of the original stone castle into the interpretive centre on the castle grounds. A photograph of it is included in this study and can be seen in Figure 5.11.6. This model and a look at the modern street layout of Ferns is enough to hint that the roads to the north and west of the castle could mark the line of the outer curtain of the castle in those directions.

If this was indeed the case, and the entire outer ward has been overtaken by the encroaching town, it would explain much about the lack of a substantial main gate. This would mean that the main entrance to the complex was located somewhere along an outer curtain which has long since disappeared, and that what we see today as the gate to the castle is little more than a formal gate to the castle's inner core. There are examples of castles with main outer gates which also utilize a ceremonial gate to demarcate the entrance to the inner ward: Conwy Castle in North Wales and within this study Adare Castle, Co. Limerick. This seemed a logical possibility to this researcher, and so the very unconvincing gate excavated in the 1970s is examined in this study within the context of it being an *inner* gate, the ceremonial access to the prestigious nexus of the castle. As the location of the main gate to the castle is then unknown, it was not treated in this study.

Along the same vein, if it is the case that the outer ward is lost to us, there is a high probability that the location of the great hall (which could easily have been in the outer ward, as at Sandel Castle in Yorkshire, as well as at Adare, Co. Limerick) is similarly lost. Of course the great hall may have been placed within the standing perimeter of the *donjon*, perhaps against the west or north walls. But if so, the exact location is still unknown. Because of this uncertainty, the great hall component was not treated in this study.

General views

The angle towers and battlements of the *donjon* at Ferns Castle would have been extremely high. Presently the walls rise to second-floor height and there is evidence (in the form of a garderobe) that the south-east tower had another floor above that. This means that the battlements of the corner towers could have reached 15m, with the walls between them rising to at least 10m if not the full 15m. To illustrate the potential views from such a structure, a composite viewshed was created using the centre point of the *donjon* as the viewing platform. This is obviously a constructed model, meant to deliver the maximum view that could have been available from the battlements at Ferns. This viewshed can be seen in Figures 5.11.7-9. It was generated using an elevation offset of 15m and a viewing azimuth of 360°.

In this viewshed we see that the farthest reaching views are between 2:00 to 5:00, 7:00 to 10:00 and at 11:00 on the clock face. At these positions there are pockets of visibility between the 9 - 11km radius around the castle. Between the 9 - 4km radius it is almost easier to describe which direction *does not* have a view, for although there is no real 'blanket' coverage, visibility extends in every direction except a small slice at 1:00 and another at 6:30 on the clock face. The viewshed within the 4km radius in Figure 5.11.8 shows that between the 4-2km radius, the only area that does not have some kind of visibility is, again, towards 1:00. This is due to the hills north and north-east of the castle within the 1 – 2km radius which effectively block further visibility in that direction. The close-range viewshed (Figure 5.11.9) suggests that within the 1km radius there are actually more points visible than not visible. A photographic approximation of these projective views, taken from the vantage point of the (restored) top of the south-east tower can be seen in Figures 5.11.10-14.

The best over-all views from Ferns are within 300m of the castle, overlooking the town. Beyond this a line of nearly uninterrupted visibility can be seen running from 3:00 to 7:00 on the clock face 2km to the south-east of the castle. It is interesting to note that this stretch has a country road (noted here as the Killabeg Road) running along its entirety. If this road had been viable during the subject time frame, it could have had inter-visibility with the castle (barring built environment and vegetation) for a full 6 km. Problems with the viewshed are related entirely to the topography, which dips and peaks around the castle site. Much of this is caused by the foothills of the Blackstairs and the Wicklow mountain ranges, and the river-beds of the Bann and Slaney rivers. This creates an interesting pattern of almost a 'bull's-eye' of visibility around the closest one-third of the castle's radius, followed by a ring of invisibility which is, in turn, surrounded by another ring of visibility towards the farthest extent of the 1km radius.

Projective views: viewshed and personal observation

Gate

The modern entrance to the castle was analyzed as an inner gate to the *donjon.* As such a structure no longer exists, access to its battlements and therefore photographs from such a vantage point were impossible. A viewshed was created from the position where Sweetman suggests the gate house might have stood (1979: 223). It was generated using an elevation offset of 10m to simulate the battlements of this structure, with an azimuth of 130-250° to reproduce the obstruction to visibility that would have been caused by the walls and towers of the *donjon* itself. Although we cannot know the height of the outer curtain wall, it has been assumed that a projective view would have been possible above the outer curtain. This viewshed can be seen in Figures 5.11.15-17.

The viewshed illustrates the extent of the visibility of approaching traffic from the gate house. Although Killabeg Road, mentioned above, can be observed for at least 3km from this position, the modern route from Enniscorthy is only intermittently visible and not really observable to any degree past the 1km radius. Obviously all routes to the west, north and east would have been unobservable. The best projective views from such a gate would have been the closest 200m of land to the south of the castle. Photographs of the gate area (externally and internally) can be seen in Figures 5.11.18 and 5.11.19.

Great hall

As discussed above, the great hall is not treated within this study.

Lord's chambers

On the first and second-floors of the east and south sides of the *donjon* are windows which clearly denote important rooms. The internal arrangements of the *donjon* are enigmatic and have been discussed by Leask (1936: 171), Sweetman (1999: 78), McNeill (1997: 123) and O'Keeffe and Coughlan (2003: 136-138). Leask and McNeill agree that the internal span of the structure was too large to have been a single space. Sweetman discovered in his excavations that the internal ground-level varied considerably within the *donjon* from the south to the north, the south side being about 2m higher than the north (1979: 218-20).

In 1997 McNeill had suggested that the castle consisted of four ranges surrounding a narrow internal light-well or courtyard (1997: 124), and Sweetman later suggested that as there was no evidence for internal structures, we must assume that interior timber buildings were tied to the curtain wall as at Lea and Carlingford (1999: 78). O'Keefe and Coughlan seem to disagree, interpreting the variability in the levels of the windows seen in Figure 5.11.4 as reflecting the existence of partitions (like Ballymoon Castle in Carlow, which they suggest has several chronologically-significant parallels for details at Ferns (2003: 137). However, they did not provide an answer for *how* these partitions were constructed, and suggested only that that they reflected the organisation and ranking of the household (ibid. 138).

In looking at the various analyses of the *donjon* and after field observation, this researcher decided that the concept of four separate buildings, or ranges around an internal courtyard or light well seemed the most plausible description. There is little standing fabric for the north wall, and none for the west, so the quality of windows and accommodation along these walls is impossible to know. But we do have evidence of at least four rooms attached to the south-east tower, and a chapel and second-floor chamber within the tower itself that were linked to the south and east ranges, all of which suggest sumptuous domestic planning along the east and south walls.

Lord's chambers (east)

On the east wall of the *donjon* at first-floor level there was a fine room having two large two-light trefoil headed windows with seats, which flank (what has been reconstructed as) a massive fireplace. It is this room that has been assessed in the context of a possible lord's chamber. This room appears to have been rectangular, approximately 13m in length (width unknown), with the long axis running north and south. Directly above it on the second-floor is another room which has two large two-light trefoil headed windows (without seats). It is difficult to tell from the photograph in Figure 5.11.3 how much of the fine windows (including the window seats) that we see in this wall today were original, but it is obvious that all were a substantial size.

The projected views from these windows could have been interesting and bucolic, looking out on the town and the River Bann. A viewshed was created from the point of the first-floor room using an elevation offset of 6m, and an azimuth of 60 - 140° to allow for a field of visibility narrowed somewhat by the towers at the north and south corners. This viewshed can be seen in Figures 5.11.20-22. These illustrations show that the projective view from this space would have been fairly far-reaching in several areas; at 2:00, almost at 3:00, 4:00 and at 5:00 the views stretch nearly 11km. Still, even with views extending almost to the edge of the viewshed, only 2 % of the hinterland is visible. The west slopes of several hills are very visible at approximately the 5 – 6km radius, and the most comprehensive views are seen in the 4km radius view in Figure 5.11.21 between 4:00 and 5:00 on the clock face. Here a steady line of visibility stretches across most of the width of the field of view, continuing uninterrupted from 3:00 to almost 5:00 on the clock face. As good as this line of visibility appears however, overall within this mid-range only 4.8% of the hinterland is visible. The projective views within the immediate-range are also rather poor as we can see in Figure 5.11.22 where only 2.8% of the immediate hinterland can be seen.

Photographs of this approximate view can be seen in Figure 5.11.23. These were taken from the altar window of the chapel within the south-east tower. They support the viewshed quite well. Figures 5.11.24-34 are photographs of the east wall (showing internal and external views), as well as internal and external illustrations of each window opening.

Lord's chambers (south)

On the first-floor level of this wall was another fine room which had two windows and a cross loop. The window to the east is a single-light with a trefoil head, the window in the centre is a two-light, also with a trefoil head. The cross loop is in the west end of the remaining wall, but does not appear to have been the last window on the original wall, as the east edge of another embrasure marks the area where the wall now ends in ruin. All of the embrasures, including the ruined one on the west edge currently have windows seats. It is difficult to know if the window seats are original, as we see in Figure 5.11.3 that the two-light and cross loop had been blocked by later additions and refurbishments to the castle, and were only exposed during the restoration work. Because of the grandness of these windows and their embrasures, this room was assessed in relation to a possible lord's chamber. Directly above this room on the second-floor was another large room having two windows and a fireplace. The window to the east, like the window directly below it is a single-light with trefoil head. This does not appear to have a window seat, though the one to the west of it, a two-light with trefoil head does. This two-light window had also been blocked until restoration work, and the seat appears to have been a modern restoration insertion. West of this window is a small fireplace.

Both of these fine rooms along the south wall seem to have been rectangular, measuring at least 13m in length (width unknown) with the long axis running east and west. The extant windows of both of these rooms overlook the entrance of the *donjon.*

Projective views from these rooms could have been interesting, as they face the approach road from the south and look out on good farm land. The viewshed from the first-floor room along this wall can be seen in Figures 5.11.35-37. It was generated using an elevation offset of 6m with an azimuth of 150°-250°. These illustrations show that the projective view from this space, although patchy would have been fairly far-reaching in three areas between 6:00 and 9:00 on the clock face. Here the view stretches nearly 11km. From 4 – 9km there are smatterings of visible areas at approximately 6:00, 7:00 and 8:00 on the clock face. Within the full 11km radius only 2.4% of the hinterland is visible. Within the 4km radius viewshed seen in Figure 5.11.36, we see that the road from Enniscorthy is visible only within 1km of the castle, but Killabeg Road can be observed for at least 2km in the south-east. Overall, only 5.3 % of the hinterland is visible in this view. The projective views within the immediate-range of 1km (Figure 5.11.37) are slightly better at 9.8%, with the best views falling within 200m of the castle.

Photographs of this approximate view can be seen in Figure 5.11.38. These were taken from the south-west facing window on the second-floor of the south-east tower. Although they do not cover the entire field of view that might have been visible from the room on the south wall, in general they support the viewshed well. One thing that is obvious is that vegetation and built environment compete strongly for the current views, and this could certainly have been the case within the study time frame. Figures 5.11.39-56 are photographs of the south wall (showing internal and external views), as well as internal and external illustrations of each window opening.

Comparison of these projective views with each other and the views from other castles

The projective views from the centre point seen in Figures 5.11.7-9 give a good indication of how strong the location of Ferns Castle was visually. The percentages of pixels visible from the castle are 10.57 % within the 11km radius, 22.6 % in the 4km radius and 59.4 % within the 1km radius. The reason for this wide projective view is that the castle was built upon a hill nearly 70m higher in elevation than the riverbed of the Bann only 1km to the south-east. Other castles seen within this study that were built on the top of a hill or prominence of some type are Athlone, Castleroche, Dunamase, Dundrum and Greencastle (though this was a very slight prominence indeed). Of these, the castles whose projective views compare well with Ferns are Dunamase: which had visibility of 17.98%, 21.9% and 42.34% respectively, Athlone: with percentages of 6.27%, 20.4% and 39.29% respectively, and Castleroche: which had 4.5%, 12.9% and 45.78% respectively. What is interesting is that of these three castles, only Castleroche was comparable to Ferns in its use of the styles of the day, ostentation and in the amenities it provided. The other two (which might be considered marcher castles) were defensive in nature, martial in bearing, and merely serviceable in accommodation (Athlone was a royal castle).

Unlike the general view, the projective view from the east chamber however was not strong. The fine windows in this room looked out onto the good farmlands to the east and south, and the River Bann. This was territory which by the mid-13th century would have been (for the most part) under Anglo-Norman control. Figure 5.11.21 illustrates how the shape of the hill on which the castle sat acted to curtail views of the immediate hinterland. Essentially, these grand windows allowed for light, but added little of interest to the view. What was surprising was that the viewshed suggests the monastery (which was only about 500m from the castle) would not have been visible from this chamber. This raised the question of whether the monastery could have been visible from the room on the floor above this. To answer this query an 11km viewshed (not included as an illustration in this study) was generated from an elevation offset of 9m to simulate the second-floor level. The result was that the monastery *did* fall within the projective view of the room above this proposed lord's chamber, but only barely (the increase in projective view gained by raising the elevation 3m amounted to only 0.29%).

From the south chamber the projective views are only slightly better. Figure 5.11.35 illustrates how the view appears to be funnelled down the length of the River Bann to the south. This also would have been good farmland, but as with the east chamber, not a great deal of it could actually be seen.

One thing that has become noticeable within this study is that often the quality of the projective views from the lord's council chambers has been quite low in relation to the other views the castles have offered. This has been due to various reasons such as the size, the number or the placement of the windows within these rooms. While the east chamber at Ferns offered two large two-light windows with seats, and the south chamber had a variety of openings, all with seats, they did in fact have very poor projective views. Indeed the almost dismal quantity of visibility from the chambers at Ferns are matched by only one other castle in the study; Castleroche, where views from the lord's council and private chambers were 0.59%, 0.68% and 1.29%, and 0.87%, 0.97% and 2.92% respectively. Even Adare Castle, built on low-lying land offered better projective views from the later lord's chambers: 5.5%, 10.34% and 28.14%.

Adare would seem to have few similarities to Ferns, especially as far as siting is concerned. But the two castles do share some affinities (though they are not immediately apparent); both these castles had an outer gate as well as a rather ceremonial inner gate, with the great hall located within the outer ward. These two castles were alike in other, more subtle ways; both appear to have been built with habitation in mind, specifically *designed* to be commodious and fashionable. Neither castle appears to have been seriously defensive (Sweetman 1979: 240; McNeill 1997: 124). But perhaps most interesting of all, both castles were utilized as habitation sites for long periods of time, probably due to the fact they *were* so accommodating.

If we were to judge the choice of site solely by the projective views offered by the centre point, it would be easy to suggest that the castle was located *specifically* to take advantage of visibility. If we were to judge the site by the projective views offered by the lord's chambers within the *donjon*, our decision might be quite different. But we must also remember that this castle appears to have been built (as were Dunamase and Dundrum) on land previously important as part of a pre-Anglo-Norman power-base. William de Valence may have built here simply because this was where Diarmait Mac Murchada had located *his* castle. We cannot know Mac Murchada's draw to this site, but overall, Diarmait could have had an impressive projective view, and it is difficult not to draw the conclusion that visibility played a role in the siting of Ferns. As to the priorities of the castle in general, the extant fabric suggests a commodious habitation was the upper-most priority for the builder.

Reflective views: viewshed and personal observation
Approaches

It would appear from the Taylor and Skinner 1778 map shown in Figure 5.11.57 that the modern roads around Ferns are almost fossilizations of the early roads. This map shows five roads converging at, or close to the castle. These correspond quite well with the Carnew Road which approached from the north, the N11 from the north-east, the Killabeg Road (which branches off what Taylor and Skinner label as the Wells Road), the N11 from Enniscorthy to the south and the R745 from Ballycarney to the west.

Of these approaches, the N11 north-bound from Enniscorthy offered the shortest reflective view of 1.2km. The viewshed seen in Figure 5.11.8 suggests that the point at which the castle was first visible on this road is indeed exactly where topography should allow for a sighting. Although the viewshed suggests inter-visibility between the road and the castle for much of the rest of the distance, vegetation and built environment did not allow for good reflective views after this point. A photograph of this first-sighting can be seen in Figure 5.11.58. The R745 from Ballycarney was only slightly better, giving a reflective view at 1.3km. Figure 5.11.8 suggests that there are two or three very small patches of visibility before this, but vegetation did not allow a sighting at these points. Again, although the viewshed suggests a reflective view is possible through much of the remaining distance to the castle, this was not the case. A photograph of this first-sighting can be seen in Figure 5.11.59.

From the north and the south-east approaches, the castle was visible from the same distance; the Carnew Road (north) and the Killabeg Road (south-east) allowed a first-sighting from 3.6km away. Figure 5.11.7 illustrates that these points correspond very closely with where the viewshed suggests first- sightings along these routes might be possible. The Carnew Road then goes through a long stretch where a reflective view of the castle would be impossible, but except for a short distance the Killabeg Road (as mentioned above) more or less retains visible contact with the castle (allowing for vegetation and built environment). Photographs of these sightings can be seen in Figures 5.11.60 and 5.11.61.

The farthest reflective view of the castle was seen coming from the north, travelling through the foothills of the Wicklow Mountains. The N11 passes over two hills before descending to the river valley of the Bann, and it was on the south-western slope of the closest of these that Ferns could be seen, 7.5km away. This is one of the longest reflective views from an approach route seen in this study. After this point however, the elevation of the road drops and topography blocks views of the castle. As the road nears Ferns, although the viewshed suggests there could be reflective views, built environment and vegetation kept the castle hidden until near the top of the castle's hill. The photograph of this first-sighting can be seen in Figure 5.11.62. Although the castle is difficult to pick out in the photograph it was quite clear physically (when an observer knew what to look for).

In general, personal observation suggested that the castle is still highly visible within its environment, however knowing what to look for helped in recognition. It is important to remember that what is available to view today is vastly different from what could have been viewed during the subject time frame. The standing fabric represents only a fraction of what the *donjon* would have been originally. When the castle was whole, the massive towers at each angle, with connecting walls rising at least 10m in height would have been difficult to miss. Ferns Castle would have been not only highly visible, but easily could have dominated its wider landscape. A summary table of the approaches can be seen in Figure 5.11.63.

Community

An Episcopal settlement of some sort had been associated with the monastery of St M'Aodhóg since its foundation in the 7th century. As mentioned, there may have been two centres of habitation during the subject time frame; one ecclesiastical and the other secular/political. Excavation in 1999 within 60m to the south-west of the castle found very few shards of medieval pottery, and the excavation report suggested there were no features of medieval date in that part of the town (Walsh 1999: http://www.excavations.ie/Pages/Details.php?Year=&County=Wexford&id=875); presumably the town extended more towards the north, north-west and east of the castle. A photograph of the castle from 30m to the east of the possible secular market site (where built environment allowed a reflective view) can be seen in Figure 5.11.64. It illustrates how clearly the castle could be seen from this point. Note that the tower visible in this photo is the south-west tower and that the north-west tower would have been 20m *closer* to a viewer standing at this spot. Such a view of the castle (if not blocked by built environment) would have been very impressive.

Comparison of these reflective views with each other and the views from other castles

The castle was not built on the highest point of the hill, and this raised a question; had Diarmait Mac Murchada (or any subsequent builder) built on this higher spot, might Ferns Castle have been more visible than it was? To answer this question, a viewshed was generated from highest point of the hill (not shown), using an elevation offset of 15m, and with an azimuth of 360° (the same elevation and azimuth used for the castle's general centre point viewshed). The answer was not really, the percentages were slightly higher, but only within a few points in the far and middle-distance, and less than 10 points in the near-distance.

Since there is no extant main castle gate, we cannot know how the gate may have appeared from a possible market place between the castle and the hermitages, so we cannot gain a 'feel' for the way it may have impacted the lives of the community members. What we can say is that regardless of the position or design of the gate house, the angle towers and walls of the *donjon* alone would have loomed over a market at this location. In its current ruined state, Ferns does not dominate or intimidate, but it does impress. To community members living in humbler circumstances within the shadow of Ferns castle, the inference suggested by the *donjon* is of massive wealth. The ability to control the resources necessary to erect a structure of such magnitude would have been impressive in and of itself. To see proof of that degree of wealth as a daily backdrop and as a constant presence would certainly have impacted the community.

The castle was highly visible within its immediate environment (59.4% visibility within the 1km radius), and reasonably so within the mid-range landscape (22.6%). We have seen how even in its diminished state it is visible from over 7km away (along the N11 north). For travellers passing through the town or coming just for the market, the *donjon* at Ferns Castle may not have seemed quite as formidable as the great tower at Trim Castle (which still retains its outer curtain and redoubtable gate houses), but it could easily have appeared as unassailable as Dunamase, Athlone or Castleroche, which we have already established as having similar projective views. Even keeping in mind that the site was inherited from Diarmait Mac Murchada, and that its high visibility may have been a matter more of fortune than intent, the reflective views certainly support the idea that the castle was sited to impress visually, and that visibility played a role in its placement. Reflective views also suggest that social dominance (over-awing through ostentation and wealth display) was a major priority at Ferns.

5.12 Greencastle, Co. Down

Note: OSi orthophotographic coordinates were used for the south-east, south-west, north-east and north-west towers, great hall, lord's chambers, parish church and motte; coordinates were found on-line at: http://ims0.osiemaps.ie/website/publicviewer/main.aspx#V1,724500,811787,7, and at http://www.nimap.net/Mapviewer/Default.aspx using the coordinates of
X: 324686.6110 Y: 311853.6110.

Introduction

Greencastle, Co. Down was built upon a slight projection of rock approximately 275 metres north of the east shoreline of Carlingford Lough. Although north of the castle the land begins its climb into the Mourne Mountains, the immediate terrain is relatively flat, as can be seen in the topographical map shown in Figure 5.12.1. Greencastle is located only 6km from Carlingford Castle which sits on the west shore of the lough, and the two castles are inter-visible in clear weather. Greencastle is located approximately 70km south-west of Belfast, 25km south-east of Newry, 45km south-west of Downpatrick, 32km south-west of Dundrum, 60km north-east of Dundalk and 98km north-east of Dublin.

No documentary references were identified to suggest the site was in use by the Irish prior to the arrival of the Anglo-Normans. The first castle in the area may have been of earth and timber, as represented by what appears to be a motte 440m to the west of the present castle site (Jope 1966: 194). While there are no firm references establishing the initial builders of either the possible earth and timber or the extant stone castle, at least two authors suggest that the stone castle was the work of Hugh de Lacy (the son), constructed by or around 1240 (McNeill 1997: 88; Sweetman 1999: 76). If this was the case, at De Lacy's death in 1242 the castle would have reverted to the crown, and indeed the first documentary reference to it is seen in 1252 when the crown orders the justiciar to re-pay one Adam Talebot the amount he had spent as custodian of "Greencastle of the King" (Lynn 1988: 68-9). Lynn suggests construction took place in or about 1250. Excavation of the ditch to the east of the curtain uncovered stratified fragments of glazed pottery jugs dated to the 1250s, and the external bank revealed medieval sherds dated to this period (1988: 68-9). It would appear that the whole of the main layout of the original stone castle occurred within one period (curtain wall, angle towers and great tower, along with several other buildings within the ward) (Jope 1966: 211; McNeill 1973: 112). The castle was damaged by the Irish in 1260 and the Pipe Roll of 1261-1262 lists work done on the hourding as well as the roof of the tower (*Pipe Roll* 45 & 46 Henry III: 56).

Greencastle certainly never had an attendant town the size of that seen at Carlingford Castle, and the settlement has been described as a rural-borough by Glasscock, who suggests the community was located somewhere in the vicinity of the motte, the stone castle, the church, or along the shore below the current village (1970: 172). The only reference to the settlement is from an Inquisition which suggests that by 1333 the borough was worth nothing (though it had formerly been worth £22 6. 8.) (Orpen 1914: 60). Trench excavation prior to the laying of cables revealed a number of medieval burials a few hundred metres to the south-west of the castle site (Ward 2005: http://www.qub.ac.uk/schools/CentreforArchaeologicalFieldworkCAF/Reports/MonitoringReports/Filetoupload,64475,en.pdf). So although Greencastle could not be termed an urban castle, it undoubtedly had at least a small community connected to it.

Greencastle has always been seen as an important site because it controlled the eastern side of the Carlingford ferry, and therefore the route into east Down (the implication being that the castle was sited *specifically* for the ferry). Unfortunately the exact site of the ferry landing at Greencastle is unknown today. There are currently two piers on the spur (one of which is visible in the inset of Figure 5.12.3). As the ferry at Carlingford Lough was part of a main transportation route into and out of the Earldom of Ulster, the ferry would mostly likely have needed to allow for the transport of horses, carts and even cattle. Piers (which require the use of ladders and steps) would not provide the easiest method of off-loading and on-loading live animals or horse-drawn carts. It is reasonable to assume that the ferry would have had a low draft (i.e. flat bottom), allowing it to be pulled up onto a shale beach across which domestic animals could be led. The ferry would have been significant *economically* as an important leg of the journey into the Earldom of Ulster. Control of this revenue-generating means of transport could be deemed important, and the visibility of such an asset could be significant in a seigneurial context. No marine-archaeology has as yet been done at Greencastle to determine the site of the medieval ferry port; however, an examination of the shoreline along the southern shore of the spit of land revealed an area of shale approximately 500m to the west of the castle that has no large rocks (an important concern for watercraft) which *could* have been the site of the ferry port.

There are several seigneurial aspects related to this castle: the borough, the parish church (which is mentioned in a reference dated 1305 (*Calendar of Justiciary Rolls of Ireland*: 6d, 141-2), the ferry-port and a demesne mentioned in the 1333 Inquisition (Orpen 1914: 60). However, the only seigneurial markers evident today are the ruins of the parish church, and the road which led from the ferry port into the earldom of Ulster (suggestive of a managed or structured approach). The road passed south of the motte (believed to be the original site of the castle), then came within a few dozen metres of the parish church before passing the stone castle itself. The OSNI map illustrating the modern roads around Greencastle can be seen in Figure 5.12.2. This map is interesting in that while a road (Greencastle Pier Road) currently leads to the western tip of the spur on which the castle was sited, OSNI does not show *any* roadways within a 1km radius of the castle. This was one indication of the difference in quality between the OSi and OSNI digital road information.

Location and discussion of the primary components
The castle plan can be seen in Figure 5.12.3 and shows that the castle is trapezoid in shape. The curtain walls enclose an area which measures approximately 52m east to west by 40m north to south. Each angle had a D-shaped tower of significant size projecting from the perimeter. Definite primary components can be found in only one principle feature at Greencastle, the first-floor great hall in the centre of the ward. Figure 5.12.4 shows the excavation plan of the north-east tower and a rectangular building attached to it (both now quite destroyed). These areas have been identified by two authors as the private chamber and great chamber respectively (Gaskell-Brown 1979: 65; McNeill 1997: 91). Unfortunately the ruinous state of these structures means that there is no way to know where, or to which direction any window openings on the first or second-floors might have faced. Because of this, the lord's chambers were not treated in this exercise.

Regrettably, positive identification of the location of the castle gate was not possible. It is feasible that the original approach followed the course of the present drive which comes off the Greencastle Pier Road and leads directly to the south-west corner of the castle (Jope 1966: 215); and indeed several authors have suggested that the south-west angle tower was the most likely location for the gate (ibid. Lynn 1988: 66; McNeill 1997: 88). The fact is that there is no extant gate house, and the *position* of the gate is unknown. Because of this, the gate component was not treated in this exercise.

General views
The viewshed seen in Figures 5.12.5-7 illustrates the projective views from the location of the battlements atop the great hall. This viewshed represents the maximum projective and reflective views available from a principle component at Greencastle. It was generated using an elevation offset of 10m to simulate the height of the battlements, with an azimuth of 360°. A portion of the Republic of Ireland has been added to Figure 5.12.5 to provide spatial reference for Carlingford Castle, which, though visible from Greencastle, is not included in the OSNI 10m data used to generate this viewshed. This is due to the fact that the digital elevation material came from two national sources, each interested in portraying only their own territory. Another issue with the viewshed is that the castle is located on the lough, less than 4km from the open sea. This means that only one-third of the 11km radius portrayed in the OSNI data includes land. Since visibility over water is not taken into account with the DEM used for this castle, two-thirds of the actual viewshed is not taken into account, and, beneath the superimposed portion of the Cooley Peninsula appears white. We can rightly assume that the reflective view of the castle would, under good weather conditions have been extensive from the water, but as with the projective view *of* the ocean, it is not represented on the viewshed.

The furthermost projective views over land are to the north, and stretch to the base of the Mourne Mountains from the 5km to the 8km radius (between 10:00 and 12:00 on the clock-face). To this direction (land held during the subject time frame by the Irish), the road coming from the north can be seen very clearly for a distance of at least 7km. Very small patches of visibility can be seen at the 11km radius to 12:00 and again at about 2:00. Although these views to the north seem quite comprehensive, they are interrupted by a drop in elevation when the hill upon which the castle sits reaches down to the lough (best seen in Figures 5.12.6 and 5.12.7). To the south, the projective view goes all the way to the shoreline, and would continue uninterrupted to the extent that the visual acuity of the observer and weather conditions might allow. This suggests a possibly massive span of uninterrupted projective view. The best over-all views on land from the battlements atop the hall would have been to the south and could, conceivably have extended to the full 11km radius over the water. The surprising thing about these views is the odd lack of visibility directly to the south-west and west of the castle, due again to a drop in elevation along the shoreline. Also, to the north-east of the castle a stream flows into the south-east angle of Mill Bay (Figure 5.12.6); the drop in elevation along this stream-bed creates a large V shaped area which would not have been visible from the castle, and likewise would have had no reflective views of the castle.

Projective views: viewshed and personal observation
Gate
As mentioned, the gate component has not been attempted on this castle.

Great hall
The great hall was originally a two-storey building (see Figure 5.12.3), it was remodelled in the 15th and 16th centuries, increasing it by one storey and enlarging the windows. The building measures approximately 15m by 8.5m internally, the long axis running east to west. It had a gabled roof which was hidden behind the parapets and surrounded by a wall-walk. This was a first-floor hall, with the main entrance on the west end of the south wall. The door was reached through a fore-building (now destroyed) on the hall's south side (Lynn 1988: 67). A part of the lower dressings (roll-moulding) of this door's west jamb still remain (Jope 1966: 215). The head and jamb of a second entrance can be found in the west end of the north wall. This led apparently to another structure built within the 8m space between the hall and the north curtain wall (ibid). This door is not directly across from the main entrance which suggests that there was probably not a screens passage; McNeill suggests that this doorway may have accessed the hall's service rooms located in the north building (1973: 113). Had this north building risen higher than ground-level, it could have curtailed the views from the north windows of the great hall. As this structure is now destroyed we cannot be certain how high it went, and analysis of the projective views included views to the north. A third door, located in the north end of the east wall led to a garderobe in the north-east angle

of the building, something unusual in a great hall of this period.

The original windows of the great hall have been greatly altered, but there is evidence of at least three; two along the north wall and one in the south wall. (Though there are currently openings in the east and west walls, these are later insertions). The two windows in the north wall were set towards the east end of the building, but at different heights. The window to the east was positioned quite high above the present stone flag floor (which is a 15th century upgrade from the hall's original wooden flooring); a tufa dressed scontion base began at 1.2m rising to 3m (Jope 1966: 216). This may suggest that the east end was the high-end of the hall and contained a dais (McNeill 1973: 113).

The viewshed from this hall can be seen in Figures 5.12.8-10, and was generated using an elevation offset of 6m (or first-floor level) with an azimuth of 310°-60° to simulate the view to the north and 150°-240° for the south. Although the windows of the great hall were fairly wide lancets and not mere loops (McNeill 1973: 113), the projective view from this hall would have been somewhat circumscribed simply because it would have taken in only two directions: north and south. One of the results was that the parish church (located only 180m south-west of the castle) could not be seen from the great hall. This is illustrated in the viewshed, and was confirmed by field observation. Nevertheless the view to the south could have been far-reaching, for barring vegetation or buildings, there would have been nothing blocking the projective view to the horizon. To the north there would have been excellent views of Mill Bay with the Mourne Mountains in the distance. This may have been quite informative, as the elevation of the first-floor would allow observation of almost 6km along the lough's shoreline (and subsequently the approach from Newry).

Extant features of the windows include the tufa scontion dressing from the original windows (mentioned above) which rises to a height of 2m above the floor on the north wall. On the south wall, close to the south-east angle, the scontion of another window rises to 2m, this was dressed in sandstone below and tufa above (Jope 1966: 216). Photographs of the hall's main entrance and the remains of these 13th century windows can be seen in Figures 5.12.11-26. Figures 5.12.27 and 5.12.28 are photographs of the projective views from these windows. These photographs do reflect the projective views suggested by the viewshed (as far as the weather allowed).

Lord's chamber

As mentioned, the lord's chamber component has not been attempted on this castle.

Comparison of these projective views with each other and the views from other castles

The first item to discuss regarding the projective views is the way in which weather conditions can influence the quality and amount of visibility at any castle. While this was anticipated, the degree to which it was true was not fully appreciated. The dramatic manner in which the projective views were hampered by a summer rain shower is seen in the photographs in Figures 5.12.27 and 5.12.28. These photographs serve to highlight how weather conditions can greatly affect visibility.

Over-all the percentages of the projective views from the great hall across land are fair; 20% of the 11km radius, 39.6% of the 4km radius and 25.8% of the 1km radius could be seen. (To this point in the study, the viewsheds in the near distance (1km radius) have delivered higher percentages of visibility than the far and middle distances. In this case, because there is so little landmass to the south, the percentage in the 1km radius is only slightly higher than the far distance view, and actually *less* than in the middle distance.) The view to the north, specifically of the Mournes would have been aesthetically stunning from this venue. While we do not have the advantage of knowing where the gate of the castle was located, the projective views from the battlements of the centrally placed great hall give a reasonably good indication of what the gate's maximum views *may* have been. This viewshed suggests that the projective views from this vantage point were excellent; 26.3% visibility within the 11km radius, 75% within the 4km radius and 79% within the 1km radius.

While the percentage of pixels visible over water is unknown, an imaginative look at the 4km views from the battlements of the great hall, the hall itself and the lord's council chamber (Figures 5.12.6, 5.12.9 and 5.12.30) suggest that a large proportion of the area represented as white in the viewsheds could in fact have been visible (with the right weather conditions of course). The reasons for the good projective views are the slightly elevated hill upon which the castle was built, the relative flatness of the land immediately around the castle as well as the relative uniformity of the lough and the ocean surrounding the castle on nearly three sides.

The coordinates of the possible ferry port site were added to the viewshed to determine if it was visible from the great hall. The lack of a window to the west precluded a projective view of this area, and even from the *battlements* of the great hall this location could not be observed. This might suggest the ferry port was located elsewhere, or, that a seigneurial projective view of it was not a priority to the castle builder. But overall, based purely on these projective views it is reasonable to suggest that the castle site was chosen with visibility in mind. Certainly Hugh de Lacy could not have chosen a more elevated *natural* piece of ground and still be in a position to control the ferry crossing. This raised a few questions about the motte located 400m to the west of the stone castle, thought to be the site of the (pre-mortar and stone) original earth and timber castle. There is of course the possibility that the motte was not the original site of Greencastle, but was a later construction for which there is no documented account; but if the motte *was* the original castle location, might visibility have played a role in *its* siting? And, along that same line, could a desire to increase visibility have played a role in the abandonment of the motte and the relocation of the castle to the natural hill?

To determine if visibility played a part in a repositioning of Greencastle, a viewshed from the location of the motte was generated using an elevation offset of 10m (tower height) and an azimuth of 360°. Comparison between the motte and the battlement viewsheds gave an interesting result; the location of the motte provided very similar, but actually slightly *better* projective views than the location of the great hall. The motte offered 25.3% visibility within the 11km radius, 75.9% within the 4km radius and 88.6% within the 1km radius (compared to 26.3 %, 75 % and 79 % respectively from the battlements). The differences may seem slight, but as can be seen in Figure 5.12.33, the nearly 10% increase in the motte's visibility within the close-range 1km radius allowed for observation of virtually 100% of the shoreline on the Greencastle spur. What this suggests is that the site of the motte, if it was indeed the original castle site, was almost certainly carefully chosen with visibility (and possibly control) of the immediate hinterland in mind.

It also suggests that an *increase* in over-all projective views could not have been a factor in a decision to relocate the castle; in fact, a comparison of Figure 5.12.33 and Figure 5.12.7 illustrates the loss of visibility (if not control) of the immediate shoreline that such a move would have meant to the inhabitants of the castle. This begs the questions of why then did the builder of the castle not just increase the size of the motte, and build his stone castle atop that spot? There are at least two answers to this; firstly, stone castles take time to construct, and the lord would have needed a base of operations (accommodations) while the stone castle was in construction. Secondly, the location of the motte may have been seen as deficit in some way; too public, or too exposed to the elements on the tip of the spur. It could even have been considered not defensible enough.

It is a legitimate possibility that the choice of the stone castle site included defence as a priority. The commonly held principle that castles were first and foremost defensive architecture is shown at Greencastle in Leask's suggestion that the general effect of the castle is striking and military (1977: 40). Certainly the blanket-like projective views observing the road coming from the Irish held lands to the north (the direction from whence the greatest threat might have been expected) suggests that defence was taken seriously. And indeed Jope rated Greencastle as the "best example of a fully developed military castle" in Co. Down (1966: 213). However, just as cracks in this principle have been realized in the past decades, a few chinks in Greencastle's defensive ability have been pointed out. McNeill noted as early as 1973 that several characteristics about this castle were not consistent with military architecture: the curious slant at which the south-west D-shaped angle tower projected from the curtain either nullified or greatly reduced the defence of the castle at that point; likewise there did not seem to have been a true gate house at Greencastle; nor was the great hall a very defensive building; and the curtain walls (which required serious repair soon after their construction) did not seem to have been particularly challenging to attackers. Indeed, aspects of the castle are "contrary to common defensive sense" (McNeill 1973: 115-117). For these reasons serious defence would not seem to have been a high priority at Greencastle.

Perhaps the motive for siting the castle upon this slight hill was simply that it allowed for a spacious castle perimeter; the current site is over 4 acres or 1.62 hectares (Jope 1966: 213). Judging by the amount of lodgings built within the angle towers, an important priority of the builder may have been comfort and accommodation. The careful inclusion of abundant living space during the initial construction of the stone castle seems to suggest Greencastle was intended to be a rather commodious place right from the outset. For example, there were two garderobes installed in the north-east angle tower alone, and the north-west angle tower appears to have been similar in design (ibid), there was a garderobe just west of the north-east tower in the north curtain, and one in the hall itself. The south-east tower has completely disappeared, but the three storeys of the south-west angle tower appear to have supplied a block of carefully contrived private rooms (McNeill 1973: 116). That the castle continued to be an attractive habitation site throughout the study time frame is evidenced by the fact that Richard de Burgh held the marriages of two of his daughters at Greencastle in 1312 (ibid. 118).

This analysis suggests that projective views almost certainly played a role in the choice of siting of the stone castle at Greencastle. The placement of the castle directly along the only transport route hints that economic control in the form of management of the road was important, while the remaining fabric of the structure and excavation has suggested that the main priorities of the builder were comfort and accommodation,

Reflective views: viewshed and personal observation Approaches

Unfortunately by the time Taylor and Skinner were assembling their map series in 1778, Greencastle must not have figured as an important destination in Ireland, for the road to the castle and the area around it are not included in their itinerary. And although a map of Co. Down by James Kennedy dated 1755 (Figure 5.12.34) clearly shows transportation routes as well as basic topography of the region, it serves to underscore the degree to which the area had declined. While the castle is portrayed on the map, a road *going* to the castle is not shown. Perhaps this road was nothing more than a path by this point, yet it is interesting that even *today* the road map provided by OSNI does not show a road going to the castle. In general however the larger roads shown on the Kennedy map appear to have fossilized very cleanly into the modern roads travellers still use today.

One of the main approaches to this castle would have been from the west, by way of the ferry. Indeed, all travellers coming into or out of Lecale via Carlingford would have to pass within 250m of the castle. For purposes of this study, a water approach across the lough was not undertaken; but an investigation of the area suggested that due to the low elevation of the western end of the spur, the castle is difficult to see when on land. However, from the stretch of beach most likely to be the ferry port site, a glimpse of the castle *was* possible. While this seems contrary to the projective interpretation of the viewshed, which suggests the two sites were not inter-visible, we must take into consideration that this line-of-sight represents a line from the battlements of the great hall to the *ground* on the beach. A human standing on that same ground (1.7m) might easily be visible. However, because GIS recognizes only topographical features, this is not something that can be determined through a basic viewshed. Likewise, while the ground on which the castle is built would not be visible from the possible port site (even at standing height), the *battlements* of the great hall, which rise at least 10m can definitely be seen. The reflective view of the castle from the probable ferry site can be seen in Figure 5.12.35.

An approach, whose use by the Anglo-Normans during the study time frame may not have been heavy, was alongside the eastern shoreline of Carlingford Lough (coming from Newry). This land was held primarily by the Irish, and may not have been part of the Earldom. This is significant primarily because it would probably have been the Irish which would have seen reflective views from this approach, and such reflective views would have shaped their opinion of the strength and power behind the curtain walls. The viewshed seen in Figure 5.12.5 suggest that when approaching Greencastle from Newry (from the north-west) intermittent reflective views of the castle might be possible along a 7km stretch. Although the viewshed suggests the castle might be continuously visible by approximately the 6km radius around the castle, vegetation and built environment prevented reflective views until within the 3km radius (this was 3km from the castle 'as the crow flies', but over 5.5km by road). This first-sighting was at the point in the junction where the Benagh Road intersects with the A2, seen in Figure 5.12.36. The final main approach to Greencastle would have been the route along the coast from east Down (coming from Downpatrick). While the viewshed suggests the castle might be visible along roads coming from the east as far out as 4km, field experience suggests that vegetation and built environment interfere with castle siting, and that reflective visibility of the castle begins only when the traveller turns onto Greencastle Pier Road, approximately 2km from the castle (Figure 5.12.37). Vegetation and built environment again interfere with reflective views from that point until directly at the modern gate to the property, just off the Greencastle Pier Road (Figure 5.12.38). As mentioned, due to the imprecision of the OSNI road information for this area, that final length of road leading up to the castle is not shown herein. A summary table of the approaches to Greencastle can be seen in Figure 5.12.39.

Community

It is impossible without excavation to prove where the rural-borough of Greencastle was located, it may never have been very large, and as mentioned, appears to have already been in decline by 1333 (Glasscock 1970: 172). No references were found of a market at Greencastle, but approximately 180m west of the castle the 13th century parish church stands as evidence of parishioners living within the immediate hinterland. This suggests that some form of community was associated with the castle during the subject time frame. An individual standing to the east of this church could have had a clear reflective view of the castle, but the rise upon which the castle was built is not so high that a reflective view from this point couldn't be blocked by vegetation or even low buildings.

Because we do not know for certain where the castle gate was located, or what form it may have taken, we cannot know the impact that it might have had on an observer at the church, or what role it had to play in the lives of the community members or passing travellers. For instance, we do not know whether the gate was substantial enough to have loomed impressively (which would seem unlikely given the lack of any sign of it today), or if it was just a simple gate marking the boundaries between the castle and the outside world. A photograph of the reflective view of the castle from the church can be seen in Figure 5.12.40, and Figure 5.12.41 shows the castle as it appears from the top of the motte.

Comparison of these reflective views with each other and the views from other castles

There are two castles within this study whose reflective views compare well with Greencastle: Trim and Dundrum. It is interesting that Trim was built by Hugh de Lacy (the father), and Dundrum was owned and refurbished by Hugh de Lacy (the son) who is also the most likely builder of Greencastle. The great tower at Trim can be seen by 18% of its 11km radius, 41.5% of the 4km radius and an amazing 87% of the immediate 1km radius. The great round tower at Dundrum is quite similar, providing possible reflective views to 30.7% of the 11km radius, 47.4% of the 4km radius and 77% of its 1km radius. But, what is exceptional is that while the great tower at Trim is exactly that...a great tower, rising at least 30m above its surrounding hinterland, and the elegant round tower at Dundrum stands atop a prominent hilltop, Greencastle is a rather basic hall-block atop a modest rise within otherwise low lying terrain. There seems little that is remarkable about Greencastle other than its size, and besides the strength of their reflective views, these three castles all seem to share a pomposity which might be nothing more than a De Lacy need for spectacle.

The reflective view of the castle *today* is unfortunately insufficient to suggest what the impact of the reflective views may have been like in the 13th century. And given the amount of decay the castle has seen, gaining the 'feel' of it is difficult. The building that was the great hall is still striking and stands out quite well in its environment visually, and we must remember that it represents only a portion of what was once at this site. Using just a bit of imagination it is possible to conjure up the four tall angle towers inter-connected by the curtain wall. Such a structure could have been a potent symbol of power. However it may not have been terribly *intimidating*; the outer defences of curtain, ditch and bank do not seem to have been well maintained and possibly were never fully repaired after the attack in 1260 (Lynn 1988: 66). So the whole structure, though solidly impressive, may not have been convincingly daunting. Currently the solitary great hall seems enigmatic and somehow tragic. Even in its ruined state, Greencastle has a terrible majesty about it. Like the projective views, the amount and quality of the reflective views suggest that visibility certainly played a role in the siting of the castle. Reflective views also suggest that one of the main priorities in the construction of this castle was social domination.

5.13 Kilbolane Castle, Co. Cork

Note: OSi orthophotographic coordinates were used for the south and north towers, and the centre point of the castle, these were found on-line at: http://ims0.osiemaps.ie/website/publicviewer/main.aspx#V1,542953,621301,8.

Introduction

Kilbolane Castle, Co. Cork was built on the north-facing slope of a bean-shaped hill at the north-east edge of the Mullaghareirk Mountains. There is no indication that this location had been an Irish power-base before the arrival of the Anglo-Normans. The topographical map can be seen in Figure 5.13.1. The castle is located approximately 70km north-west of Cloyne, Co. Cork, and it is of the Bishop of Cloyne that the land was held. It is 50km north-west of Cork, 26km south of Adare, 35km south of Limerick, and approximately 140km south-west of Dublin. The OSi map of the roads surrounding Kilbolane Castle can be seen in Figure 5.13.2.

The exact dates of construction of this castle are unknown, but it is generally accepted that it was built sometime in the late 13th century (Killanin and Duignan 1967: 412; McNeill 1997: 139; MacCotter and Nicholls 1996: 205). As mentioned the lands were owned by the Bishop of Cloyne, and the first reference of Kilbolane in association with an Anglo-Norman lord is in the Pipe Roll of Cloyne for the year 1251. This entry records the death of Gerard de Prendergast who was in possession of the knight's fee for Kilbolane by that time (MacCotter and Nicholls 1996: 205). His younger daughter Maud (age 10) inherited a portion of his lands in Wexford and Limerick, including Kilbolane. She was immediately granted by Henry III in marriage to one of his Poitevin courtiers, Maurice de Rocheford. The descendents of Maud and Maurice would hold Kilbolane for the next century, starting with their son (also Maurice) who was born sometime before 1258 when De Rocheford died. It was probably Maurice the son who built the castle at Kilbolane. The earliest he could have taken possession of his lands and begin construction at Kilbolane would be in the early 1270s, towards the very end of the subject time frame for this study (see MacCotter and Nicholls 1996: 205 - 245 for an excellent account of the history of the manor of Kilbolane). The castle was continuously inhabited until at least 1656 when its owner David Power was transplanted to Connacht (ibid. 245).

It is not known if Kilbolane Castle had an attendant village, the OSi lists a ruined church located approximately 380m south-west of the castle on the southern slope of the hill. It also shows a graveyard approximately 300m to the east of the castle. No excavation reports were found of these sites to suggest their antiquity, and they were not located during the field work for this study. If church and graveyard *were* associated with the earliest phase of the castle, they may suggest a community in the near vicinity, but this is unknown. Currently the closest village is Milford, approximately 700m to the north-west, and Kilbolane was treated here as a rural castle.

Evidence of seigneurial aspects at Kilbolane are few; the name of the nearest town, Milford suggests a mill on the River Deel, and a 1302 reference shows that Maurice de Rocheford (the son) was granted a free warren in all his Irish lands, including Kilbolane (MacCotter and Nicholls 1996: 205). The above mentioned church may also represent a seigneurial aspect, if it was associated with the castle during the subject time frame. There is no current indication of a managed landscape.

Location and discussion of the primary components

The plan of Kilbolane Castle can be seen in Figure 5.13.3. Obviously much of the castle has disappeared, but the standing fabric suggests it was a square enclosure. From the two remaining corner towers to the south and west, it is possible that the north and east angles also contained towers. The main gate was most likely within the south-east curtain wall facing the road coming from the east, but with only a small amount of material still standing it would only be conjecture to suggest either the location of the gate or of the other prime components. With so little to work with, only a basic analysis of the general projective and reflective view was attempted.

General views

The viewshed at Kilbolane (Figures 5.13.4-6) was generated using a central point within the castle ward. The elevation offset used was 10m (the approximate height of the corner tower battlements), and an azimuth of 360° was assumed. The topographical map of Kilbolane in Figure 5.13.1 has an interesting anomaly; the edges of the digital elevation model tiles can be clearly distinguished. These tiles are 12m square, and within the 11km radius of Kilbolane portions of 8 different tiles can be seen. Although this does not create much of a problem with viewing the topography of the area, when this DEM was used to produce a viewshed, the corner of one of the tiles created two linear artifacts in the form of lines of suggested visibility. These are seen to the west and north of the castle. This is most obvious in the 11 and 4km views in Figures 5.13.4-5. The viewshed is still valuable in giving the general scope of the views, but it is necessary to take these artifacts into consideration, especially in comparison of total percentages of visibility within the 11 and 4km radius views.

A problem with the views to the south and south-east is due to topography, namely the obstruction caused by the crest of the hill, which stretches approximately 900m in length south to east of the castle. The crest of the castle's hill is at 148km above sea level, 20m higher in elevation than the site the castle was built on (128m above sea level). The corner towers (using a 10m elevation offset) do not provide enough elevation to clear the crest and offer views to the south. The Seefin Mountains to the south-east rise to 197 km, 69m higher than the castle site, and the viewshed suggests the castle and these crests are inter-visible, but they in turn block any views to the south and east beyond that point.

The most far-reaching views in Figure 5.13.4 can be seen between 11:00 and 1:30 on the clock face, stretching all the way to the 11km radius; there are also patches of visibility at the 11km radius at 2:30, 5:00 and 7:30 to 8:30 on the clock face. Within the 4 - 9km radius, visibility lies to the north between 12:00 and 1:30, to the east between 3:00 to 5:00 and to the south between 6:30 to 9:00 on the clock face. The total visibility within the 11km radius (keeping in mind the artificial line of suggested visibility caused by the edges of the tiles) is only 7.8%. In Figure 5.13.5 there are smatterings of visibility from 7:00 to 3:00. To the west and south-west, the areas of visibility are quite dense within a 2km radius of the castle. The total visibility within the 4km radius (again keeping in mind the artificial line) is 16.5%. In the immediate distance seen in Figure 5.13.6 we can see the best over-all views at Kilbolane are to the north-west towards 11:00 on the clock face. Here there is almost an uninterrupted line of inter-visibility at least 1km in length. Figure 5.13.6 is *not* plagued by the linear artifacts seen in Figures 5.13.4 and 5.13.5, and we can accept the total percentages here at 'face-value'; visibility of 34.67% within the 1km radius. This figure clearly shows the viewing obstruction caused by the bean-shaped crest of the hill.

Projective views: viewsheds and personal observation

Gate / great hall / lord's chamber

As mentioned, none of the prime components are extant at this castle, so an analysis of them cannot be made. However, the viewshed in Figures 5.13.4-6 suggests that the west and north offered the best projective views from Kilbolane Castle. Photographs of the views from first-floor level, taken mostly from the south corner tower and the south-west curtain wall can be seen in Figures 5.13.7-14. These illustrate that the problem with projective views from Kilbolane today is vegetation. Luckily photographs taken from ground-level just to the north-west of the castle (seen in Figures 5.13.15 and 5.13.16) were able to get under some of the surrounding canopy of foliage and offer a hint at the quality of the view to the west (the direction of the town of Milford) and the north-west. Because of the amount of vegetation blocking projective views in every other direction, it is difficult to determine if the viewshed is accurate or not.

Kilbolane had no approach roads from the north, but Figure 5.13.5 suggests that the traffic approaching from the west could have been visible intermittently for at least 2km. Because of topography, observation of the roads approaching from the east and south were nowhere near as comprehensive; the road from the east was visible for approximately 500m while the road to the south was only visible for approximately 200km from the castle. This might suggest that views to the north-west were important to the builders, but that observation to the east and south were not considered vital.

Comparison of the projective views

Overall the projective views from Kilbolane are not high, with totals of 7.8% visibility within the 11km radius, 16.5% within the 4km radius and 34.67% within the 1km radius. We can be somewhat sceptical about the totals in the 11 and 4km views (Figures 5.13.4 and 5.13.5), and the removal of the artifacts we see in these views would decrease the percentage of visibility within those radiuses. If we do take them at face value however, these totals compare well with several castles within this study; Adare, Co. Limerick, Athlone, Co. Westmeath; Castleroche, Co. Louth, and Clonmacnoise, Co. Offaly. Of these, Athlone and Clonmacnoise were royal castles, built for administration and defence, meant to be garrisons for men. Adare and Castleroche on the other hand were genuine family habitation sites; built to be accommodating and commodious (Kilbolane like Adare was continuously used as a family home for centuries). Athlone and Castleroche were built on natural promontories, and Clonmacnoise utilized either an esker, or a man-made motte. Of these five castles, only Adare was built on flat land.

It is interesting that the castle site chosen was not the highest point on the hill. This crest is only 200m to the south of the castle and siting the castle here could have increased the possible projective views considerably. To see just how much the projective views would be improved by this higher location, a viewshed was generated from the crest using a 10m elevation offset and a 360° azimuth. The differences were impressive, as can be seen in Figure 5.13.17. The crest of the hill gave totals of 18.4% within the 11km radius, 47.2% within the 4km radius and 72% within the 1km radius; the amount of visibility more than doubles in every view. Building on the crest of the hill would have allowed excellent observation of all of the approaches; however, the castle would have been farther away from each of them. This suggests that perhaps gaining the maximum projective views possible may not have been as high of a priority for Maurice de Rocheford as was ease of access; the chosen site at the crossroads is almost three times closer to both the east-west and the southern road than the crest of the hill would have been.

A castle which was built in a similar location is Ferns, Co. Wexford. Like Kilbolane it was built adjacent to a major transportation route, on a hill but not on the highest point of that hill (which was approximately the same distance from the road as the crest at Kilbolane hill). While the total percentages of visibility from Ferns were not substantial, they exceeded the visibility possible from Kilbolane in each of the views and could not be considered truly comparable view-wise. An interesting thing about Ferns is that building at the crest of the hill would not have increased the available views to any great degree (less than 0.5% in each of the views), while the extent of the increase the crest of Kilbolane hill would have offered is significant. We have seen this at Ballylahan as well, where if the builder had chosen to build on the hill less than 200m to the west; the projective views would have more than doubled in each of the views. This alternate site would have allowed an almost uninterrupted view of the River Moy stretching 1500m (the chosen site offered a *very* patchy view of approximately 1300m of the river). But interestingly, the site that was chosen was less than half the distance of the alternate site to the main north-south transportation route, with the added benefit of being directly adjacent to the closest water source.

Judging from the length of time it was used as the heart of its manor and as a family home (approximately 380 years), Kilbolane Castle appears to have been primarily an administrative centre and habitation site. From the general viewshed and field research, which suggested that vegetation could easily prove a barrier to projective views from Kilbolane Castle, it is tempting to say that projective views were not a high priority of the builders. However when it *was* possible to breach the vegetation barrier, the views to the west and north were bucolic and impressive as Figures 5.13.15 and 5.13.16 suggest. It is important to remember that during the subject time frame, had a *cordon sanitaire* been kept about the castle's hill, views to the north and north-west might have been very pleasing. As we do not know where exactly the lands of the manor of Kilbolane lay (and it could easily have been to the west and north), we cannot rule out the possibility that visibility played a role in the siting of this castle.

Reflective views: viewshed and personal observation
Approaches

Taylor and Skinner's 1778 map included the ruined Kilbolane Church but only alluded to the location of the castle (their depiction of the main road stops just east of the ruined church). The Ordnance Survey Ireland 6 Inch County Map engraved in 1844 gives a clear indication of the road layout (which can be seen in Figure 5.13.18) and a comparison of this map with Figure 5.13.2 shows that the roads have not changed in the past century and a half. Figure 5.13.5 suggests that the western approach (from Milford and Drumcolliher) may have had reflective views of Kilbolane from approximately 2.5km away. Figure 5.13.6 suggests that the castle is visible only within the last 500m of the approach and the road coming up the south slope of the hill could not have had a reflective view until within 200m of the castle. The reality was that vegetation now makes reflective views of Kilbolane almost impossible from the east or the south until the traveller was almost at the castle's south tower; 60m away when coming from the east (515 from Charleville), and 80m from the south (country road from Liscarroll). The road from the west did allow a bit better reflective view from 1.4km away, but only when the researcher was familiar with the castle, and knew what to watch for. Photographs of the first-sightings along these roads can be seen in Figures 5.13.19-21 and a summary of the initial sightings is found in Figure 5.13.22.

Community

As mentioned, it is unknown whether Kilbolane Castle had an attendant village. For this reason the castle is being treated as a rural castle.

Comparison of these reflective views with each other and the views from other castles

If the ruined church south-west of the castle belonged to the subject time frame, the viewshed seen in Figure 5.13.6 suggests that there could not have been a reflective view of the castle from it. This inability to observe the castle from its associated religious building would be similar to Ballylahan, which had no projective views of its friary in Strade; and to Dundrum, which could not be seen from its parish church in Maghera. All three of these rural castles appear to have placed a high priority on their functions as habitation sites.

As tricky to spot as the ruins of Kilbolane are today, it may seem difficult to imagine the reflective view that might have greeted local inhabitants or travellers when they saw the castle during the subject time frame. However when standing on the roadside facing the barest hint of a moat and the broken south-west curtain wall, the effect of the remaining fabric is almost story-tellingly romantic. This actually makes it quite easy to visualize the effect such a building in its original condition may have had on an observer. Kilbolane is a modest castle compared to Trim, Carrickfergus or Castleroche; more along the lines of Swords, Adare and Athenry in its compactness and benign nature. It is not today an intimidating structure, and though we cannot guess at the impact its main gate would have had on an approaching traveller, the fact that nothing remains of that gate might suggest it was not substantial, and its stones perhaps too easy to haul away.

The reflective view, hampered as it was by vegetation, does not seem to support the idea that visibility played an important role in the choice of site for the castle. But again, a *cordon sanitaire* could have made a great deal of difference. Also, the possibility that the intended audience may have been living to the north-west where the reflective views would have been greatest (an area from which reflective views were not observable due to the fact that there were no roads to this direction) makes it difficult to reject the role of visibility in site choice. The priorities of the builders would appear to have been to provide a commodious habitation combined with administration and economics (judging by the importance placed on easy access to the main transportation route).

5.14 *Kiltartan Castle, Co. Galway*

Note: OSi orthophotographic coordinates were used for the great tower, gate house and parish church; these were found online at: http://ims0.osiemaps.ie/website/publicviewer/main.aspx#V1,545877,704763,7.

Introduction

Kiltartan Castle, Co. Galway was built of limestone upon Carboniferous limestone in low laying terrain (below 60m). Although there are areas of flat ground, the region is interspersed with slopes and undulations (Fahey 1893: 398). This eastern region of the county has some of the most profitable lands for pastoral farming in Galway. The 11km radius topographical map of this castle can be seen in Figure 5.14.1, but we can best see in the OSi road map in Figure 5.14.2 how the region is marked by low ridges stretching south-west to north-east. Kiltartan Castle was positioned in a river valley running between two of these hills.

The castle is located 4km north of Gort, 34km south-east of Galway City, 65km north-west of Limerick, and approximately 190km west of Dublin. It does not appear to have been an important Irish power base before the arrival of the Anglo-Normans. The geo-political history of the area was well documented in the Annals of Connacht (1224-1544), and has been recounted by Fahey, Orpen and others. But this particular study was challenged by an almost complete lack of documentary references before the 16th century relating specifically to the castle. The first mention of the castle dates from *c.* 1574 when it was included in an inventory of fortifications in Connaught (Nolan 1901: 114). It is generally accepted that Kiltartan is one of the many De Burgh castles in Connaught, built in the mid or last half of the 13th century. The probable builder of the castle has been identified as Raymond de Burgh, the youngest son of Walter de Burgh and Egidia de Lacy. Raymond, in tandem with his brother Hubert apparently took control of the district in 1264 during a dispute between the De Burghs and the Fitz Geralds (AC 1264: nos. 7-8; Fahey 1893: 161-2). The castle may have been built by Raymond to consolidate his holding (Cruden 1999: 18), and Fahey suggests that Raymond's descendants, the MacRedmonds continued to inhabit the lands and castle until the late 1600s (1893: 301).

Cruden has suggested there were two building phases at Kiltartan Castle: the construction of the gate house, curtain wall (with possibly a D-shaped tower in the south-west angle), and the great tower (to at least first-floor height) occurring sometime in the mid to late 13th century. The second phase Cruden suggests was in the 15th century when two more storeys were added to the tower, and wicker work vaulting was installed in the ground-floor. At the same time that the vaults were constructed, a dividing wall was inserted into the first-floor, with a finely pointed ashlar doorway connecting the two spaces. A similarly fine doorway can be found on the second-floor (1999: 59).

The castle has been known by several different names, Ballinamantaine or Ballinamantane (from the townland), Castletown Castle and Ballecastle. It takes the name Kiltartan from the parish and barony of Kiltartan in which was sited. The name is seen in various forms in the documentary sources: Killetaraghe, Kiltaragh, Kiltaroght and Kiltaraght (Fahey 1893: 241; Cruden 1999: 51). The church of Kiltartan which is located 1.3km to the north-west may pre-date the castle, and the Ordnance Survey Name Book suggests the name Kiltartan is derived from *Cill Tártáin* (Tartan's Church). Fahey suggested the original name was Kilattaraght, or *Cill Attaraght* and that the church was dedicated to the memory of St. Attrachta (1893: 241 & 398).

Kiltartan is currently a rural castle, however the alternate name of Castletown Castle suggests there was a community associated with the castle at some point. Directly to the west of the castle, in what is now a farmer's field, are several enigmatic rock groupings that hint that at some period a hamlet stood just outside the castle gates. Unfortunately no trace of a street pattern or any footprint of a possible market site has been found. This field can become quite marshy when the Castletown River is high, and geophysics conducted there suggests that the majority of the settlement took place along the drier south-west edge of the field (Cruden 1999; 124). The bulk of the stone belonging to this small settlement has disappeared, most likely forming the modern field wall in the castle ward, which now runs roughly east to west between the gate and the tower. Unfortunately the settlement remains are undated, and while it *may* correspond with the earliest phase of the castle and form part of the story of Kiltartan Castle during the subject time frame, it could have developed at any point in the castle's rather long period of habitation (roughly 1260 to 1690). Because of this, Kiltartan Castle was treated herein as a rural castle.

Seigneurial aspects of Kiltartan Castle are few and most of them rely on assuming that features present in the recent past were also present during the subject time frame. These include the possible town at the castle's gates, an eel weir on the Castletown River just south-east of the great tower, and a watermill (Cruden 1999: 120). The OSi map in Figure 5.14.2 reveals an area approximately 1km to the north-east of the castle called Deerpark, suggesting the presence of a park associated with Kiltartan Castle. Other than these aspects, there is no clear evidence of a managed landscape at Kiltartan.

Location and discussion of the primary components

The location of the principle components of gate and lord's chamber were not too difficult to ascertain at Kiltartan, but the position of the great hall was not straight-forward. The castle plan can be seen in Figure 5.14.3. It shows that the castle is roughly wedge-shaped with six segments; the base of the wedge facing north-west, the point towards the south-east. A double-D gate house stands along the north-west curtain wall, and the remains of what may have been another D-shaped tower can still be seen beneath the grass at the south-western angle (though it is not shown on the castle plan). The remains of a three-storey great tower stand at the south-eastern tip of the wedge, its south and east walls acting as the curtain at this angle.

Although the great tower has been suggested to be the site of a first-floor great hall, the nearly square shape of the structure and the lack of a fireplace within the extant walls on this floor would make it an unconventional layout for a 13th century hall. Resistivity surveying, magnetic susceptibility and phosphate analysis conducted in 1999 determined that two structures of some size were located along the curtain wall north of the gate house (Cruden 1999: 124). These buildings utilized the ground-level arrow-loops in the curtain as lights, and one of them may have functioned as a ground-floor hall, which were becoming popular towards the end of the 13th century in Ireland (as at Adare, Athenry, Dunamase, Limerick and Trim). However it would be speculative to assume either of the structures along the curtain wall acted as a great hall and in the face of this uncertainty, the great hall component was not analysed in this investigation.

The gate house consisted of a pair of D-shaped towers which projected beyond the curtain wall facing north-west. The inclusion of a fireplace flue in a 1793 sketch by the antiquarian Grose suggests the gate house rose to at least first-floor level as seen in Figure 5.14.4 (1793: 62). The great tower in the south-east angle, while not making a very convincing great hall, could have functioned well as a chamber and is analysed here as the location of the lord's chamber.

General views

The viewshed in Figures 5.14.5-7 was generated from the centre-point of the great tower, using an elevation offset of 10m to simulate battlement height; an azimuth of 360° was used. The topographical map in Figure 5.14.1 shows artificial lines running north-west to north-east; these are most likely associated with the edges of the digital elevation tiles. Luckily these appear to have caused no technical anomalies in the viewshed. The most far-reaching views are to the east from 1:00 to 5:00 on the clock-face. Here the rising hills in the distance are highlighted over the top of the ridge directly east of the castle. These far-reaching views stretch from approximately the 5 - 11km radius between 2:00 and 3:00. At approximately 5:00 there is a fair sized area of inter-visibility between the 8 - 10km radiuses, although there is no visibility of the intervening ground. Figures 5.14.6 and 5.14.7 illustrate how the hills funnel visibility to the north-east and south-west in the middle-distance and close-range views. The best over-all views are within the immediate 200m radius around the castle. There is no position which has an uninterrupted view farther than about 300m from the castle.

Projective views: viewsheds and personal observation

Gate

The viewshed from this component, which can be seen in Figures 5.14.8-10, was generated using an elevation offset of 10m (to simulate gate house battlement height). An azimuth of 130°-110° was factored in to compensate for the obstruction caused by the tower to the south-east which rose to 10m. In general this viewshed largely replicated the view from battlements of the tower. Figure 5.14.9 is perhaps the best illustration of the problems to the projective view caused by the ridges on either side of the castle. The gate house would not have been able to observe traffic on the Loughrea to Gort road (modern N 66) to any degree. To the south-west, the traffic between Gort and Galway (modern N 18) could have been intermittently visible (though only *very* sporadically), but this route becomes almost completely unobservable as it passes through the 1km radius of the castle. The road which runs between these two main routes, and which directly passes the castle, can be observed fairly well to the south-east (Figure 5.14.10) although to the north-west topography interrupts the projected view after only 300m.

The ruinous state of the gate house which is shown in Figures 5.14.11 and 5.14.12 made it unsafe to obtain a projective view from anything other than ground level. What was obvious from personal observation at that level was that the ridges truly do block most of the projective views from the gate house. The view to the west is restricted to the hillside itself, and while the roadway is occasionally visible to the north, heavy vegetation made projective views of that same road to the east impossible. The view to the south was slightly more extensive, though it was blocked to some extent at ground level by the tower, the modern field wall within the ward and vegetation. Photographs of these openings and their views can be seen in Figures 5.14.13-18. In general personal observation supported the viewshed.

Great hall

Because of the uncertainty of the location of the great hall; this component was not treated in this exercise.

Lord's chamber

The first-floor of the tower at the south-east angle of the castle was considered the likely location of the lord's chamber at Kiltartan Castle. At least initially (during the subject time period) this space may have been a combination of council and private chamber. The tower is roughly square, measuring 8m x 10m internally with a 45° splay at the south-east corner. Cruden has suggested a garderobe was contained in some way within the splay, though there is no internal evidence for one (1999: 69). The slightly longer axis of the building runs east to west. There were at least six openings at first-floor level. It is not clear which of these openings might have been used as the main entrance. The east opening in the north wall has been suggested as a possible entrance, the large amount of rubble in that area hinting at the remains of a fore-building (Cruden 1999: 65). The aperture on the west wall of the tower has also been suggested as the main entrance (Lynn 1985: 109). Unfortunately neither of these openings is convincing as a key entry; the opening on the north lacks the width and prestige associated with an outside door, and the opening to the west was narrow and may even have had window seats.

The viewshed from this component can be seen in Figures 5.14.19-21. It was created using an elevation offset of 6m (first-floor height) and an azimuth of 310°-290° to allow for the obstruction caused by the gate house. We cannot be sure what height the curtain walls may have reached, but in the interest of providing the maximum projective view it was assumed that these windows were able to see over the top of the curtain wall. We should be aware that as always, the actual projective view may have been less extensive.

What we see in the viewshed is that the projective view from the lord's chamber was not far-reaching, only 3.9% of the 11km radius was visible. Nor were the views in the middle-distance extensive at only 4.7% of the 4km radius. Within the 1km radius however, the situation is a bit better, with almost one-third or 29.75% of the immediate hinterland being visible. Interestingly the view did not include the possible deerpark to the north-east of the castle. These immediate views, like those from the battlements above the lord's chamber were mostly to the north, south and east. The most far-reaching views (which can be seen in Figure 5.14.19) are towards 2:00 on the clock face, stretching in a cone-shape from the 5–11km radius. Between 3:00 and 4:00 another, slightly broader cone stretches from roughly the 6–9km radius, and at 5:00 is a small patch of visibility around the 9km radius line. Figure 5.14.20 shows a similar (though reduced) projective view to that seen in the viewsheds of both the tower battlements and the gate house.

Currently it appears that the north wall had two arrow-loop width lights (the east loop illuminating the mural staircase in the north wall). The castle plan in Figure 5.14.3 does not suggest a window in the east wall, but a later (unpublished) sketch by McNeill hints at one in the north end near the top of the north mural stairway, and personal observation during field work confirmed the presence of an opening. The south wall had two openings; to the east is a square light which served to illuminate the first-floor landing of the south mural stairway, and another larger window on the west side may have been a trefoil lancet window (Cruden 1999: 59). The west wall appears to have had only one opening, towards the north end of the wall. These openings are shown in Figures 5.14.22-33.

Personal observation suggested that the projective views from the lord's chamber were sylvan, but localized. Field work took place during the summer months of 2008, and observation to the east was hampered a great deal by vegetation, which completely obscured the road only 100m away. The tower is crumbling, and while it was possible to reach the first-floor landing by way of the north mural staircase, it was not safe to access each of the openings to photograph their views. Photographs of most of the projective views from the lord's chambers were taken from the landing; photographs to the south and west were taken at ground level. These views are shown in Figures 5.14.34-38, and allowing for vegetative obstruction, support the viewsheds quite well.

Comparison of these projective views with each other and the views from other castles

Overall, the projective views from the general viewsheds at Kiltartan Castle are not extensive, especially in the farthest distances; 6.5% within the 11km radius and 8.4% within the 4km radius. The view of the immediate hinterland is much better at 43.2% within 1km. Percentage-wise Kiltartan compares well to Castleroche (4.5%, 12.9% and 45.78% respectively), Clonmacnoise (5.7%, 15.75% and 41.64% respectively) and Kilbolane (7.8%, 16.5% and 34.67% respectively).

An interesting thing about the gate house is that it faces the ridge to the north-west, and so is overlooked by higher area 325m away. The castle is surrounded to both the east and west by higher ground. It was built on a slight rise overlooking the Castletown River, and at times of the year flooding by the river of the surrounding valley forms a natural moat around the east half of the castle. However, we should not consider this to have provided any type of defensive advantage. While it is deep in places, along much of its course the Castletown River would not have been difficult to cross or even circumvent all-together; as it disappears completely not far from the castle, going underground through a sinkhole. Taking these details into consideration it would be difficult to say that the castle was sited specially for defence.

The gate house does have some projective views of the Castletown Road to the south for almost 1.3km, but examination of Figure 5.14.9 shows how the west edge of the projective view runs close to the route taken by the modern Gort to Galway road. It is tempting to imagine that if that road had originally been even 100m to the east of where it is today, it might have been observable intermittently for quite a stretch from the gate house. But of course this is conjecture.

Because Kiltartan Castle was obviously not built on the highest elevation within its immediate environment and is surrounded on two sides by higher land, it raised the question of how siting the castle at the nearest high point may have changed the available projective view. The highest point within 200m of the castle was used to generate an alternate viewshed. This point was 6m higher than the castle site, and was located only 83m east-south-east of the castle. It can be emphasised that this was not the *only* higher point within the 200m radius, indeed there were 91 pixels within the 200m radius that were higher than this, each pixel being 25m². The viewshed from the alternate site is not shown here, but the percentages of visibility were much higher than the castle. The alternate site gave almost twice again the percentage of the actual site in the far-distance views (12.54% compared to 6.5%); was over three times higher in the middle-distance (31.7% compared to 8.4%); and almost twice as high in the near-distance views (76.6% compared to 43%). Obviously the alternate site would have offered larger visibility overall. That it was not used suggests that a large amount of visibility was not the priority. It was noted that the alternate site is nearly 70m closer to the road, is equidistant to the fresh water source, but is over

50m farther away from the cross road than the actual site is. This would seem to suggest that being within close proximity to the cross road was the highest priority. Generally, we might surmise that convenience to the junction of transportation routes (economics/administration of the manor), and looking impressive were the high priorities at Kiltartan.

Reflective views: viewshed and personal observation
Approaches

The Taylor and Skinner maps did not show the approach routes to Kiltartan Castle, but the Ordnance Survey Ireland 6 Inch Map, engraved in 1840 shows both castle and road. The road seems to have changed very little since 1840. This map is seen in Figure 5.14.39.

The two modern highways leading north from Gort (the N 66 to Loughrea and the N 18 to Galway) lie just beyond the line-of-sight from the castle as seen in Figures 5.14.6 and 5.14.9, obscured by the ridges to the east and west. As mentioned the cross-road between these highways, Castletown Road, runs right past the east side of the castle. The only approach sightings available of the castle were from this road, and the reflective view does not extend far. Coming from the north-west, the castle is visible from only a 320m distance, and from the south-east it can be seen from 700m. Currently, vegetation and built environment compete visually with the castle, and unless a traveller knows exactly what they are looking for, Kiltartan is easy to miss. Photographs of the first-sightings of Kiltartan Castle from Castletown Road can be seen in Figures 5.14.40 and 5.14.41, a table showing these sightings can be seen in Figure 5.14.42.

Community

As discussed, the small community of Castletown may or may not have been viable during the subject time frame. If there had been a small village associated with the castle in the latter half of the 13th century, it would have grown up literally on the doorstep of the castle. For the inhabitants of such a community, the castle and its gates could have been a major element. The church of Kiltartan however, which may pre-date the castle (Fahey 1893: 241 & 398) has no reflective view of the castle. If the church was already established before the castle was built, Raymond de Burgh may have felt no proprietary reason to place his castle in a position of inter-visibility with it. And such is the case; the castle cannot be viewed from the parish church.

Comparison of these reflective views with each other and the views from other castles

Judging from the difficulty in sighting the castle today, it does not seem that the builders were concerned with commanding a reflective view. Kiltartan is similar to Clonmacnoise, Kilbolane, Nenagh and Ballylahan all of whose builders ignored the nearby high ground in favour of building close to either the nearest source of water or a major transportation route. Interestingly, all of these locations could be considered part of the 'frontier' of Anglo-Norman settlement; areas where the land was worked not by large numbers of immigrant Anglo-Norman settlers (as at Castleroche, Carlingford and Ferns), but by the Irish themselves, and the castle builders/landholders were in a definite minority. Of these castles (Clonmacnoise, Kilbolane, Nenagh and Ballylahan), only Clonmacnoise could be considered a military outpost. Built within the first quarter of the 13th century, the intent at Clonmacnoise was to carve out and maintain a royal Anglo-Norman presence along a lucrative trade route. Kilbolane, Nenagh and Ballylahan on the other hand, like Kiltartan were family residences; owned and operated by resident lords.

Proximity to water and/or transportation routes are not the only similarities these castles share; Kiltartan, Kilbolane and Ballylahan did not have inter-visibility with their parish churches, and all were believed to have been built within a similar time period (roughly 1240-1270). As at Adare, these castles seem to be mainly basic manorial administration centres, often with truly commodious accommodations. Reflective views, as seen in the viewshed and in field observation, suggest that visibility of the castle from the main approaches probably did not play a role in the site choice for Kiltartan Castle, and that a reflective view from the parish church was impossible. Overall, the reflective views seem to support the idea suggested by the projective views...that convenience and economics were the highest priorities at Kiltartan.

5.15 Lea Castle, Co. Laois

Note: OSi orthophotographic coordinates were used for the central point of the great tower, outer and inner gates. These coordinates were found online at: http://ims0.osiemaps.ie/website/publicviewer/main.aspx#V1,657027,712097,8.

Introduction

Lea Castle, Co. Laois was built on the south-western edge of the Plains of Lea (Adams 1904: 257), an area of flat land which extends more-or-less from the castle site approximately 22km to the north-east (see topographical map in Figure 5.15.1). The site does not appear to have been used either secularly or religiously prior to the arrival of the Anglo-Normans.

The earliest documentary source for Lea Castle appears in the Calendar of State Papers for 1203. Here the castle of Lega (Lea) is mentioned as one of the possessions of Gerald FitzMaurice FitzGerald, 1st Baron of Offaly, after his death (Adams 1904: 258; Leask 1937: 173). Sweetman has suggested that this first castle was an earth and timber fortress erected on earthworks (1999: 61-2).

Gerald's son Maurice obtained seisin of his lands in 1216, becoming the 2nd Baron of Offaly, and the great tower, which was built in two stages although to one plan (McNeill 1997: 120), was most likely completed before Maurice's death in 1257 (Leask 1937: 175; 1977: 50). The curtain and gate of the inner ward, termed by Leask as being of "doubtful date" (1937: 175) may coincide with the building of the great tower. A third phase of construction involved the building of a second gate house; large and double-D in shape. This has been seen as the fortification recorded in the State Papers for which £40 was granted to John FitzThomas FitzGerald in 1297 (Adams 1904: 259; Leask 1937: 175). At a later date this gate house was turned entirely into accommodations and a third, much smaller gate was built within the outer curtain directly to the west.

Today Lea is a rural castle, but a town was associated with the castle during the subject time frame (Grose 1793: 77; Adams 1904: 259). The position of the town of Lea is unknown, but the old graveyard approximately 300m south-east of the castle which can be seen on the OSi map in Figure 5.15.2 may hint at the location of the town's parish church. If so, this might suggest a clue to the placement of the town. The town, and parish church would have been seigneurial markers during the subject time period, but today there is no evidence of a managed landscape related to Lea Castle.

Location and discussion of the primary components

Only the primary component of the outer (second) gate is straight-forward at Lea. Unfortunately so much of the castle has been lost to time and ruin that we cannot be certain where the great hall and the lord's chamber may have been. A likely spot for the lord's chamber (and possibly an early hall as well), is the great tower which can be seen in the castle plan in Figure 5.15.3. However, as Figure 5.15.4 illustrates, only the foundations and 1½ walls of this building remain to be analysed. Whether this tower held an early great hall during its first building stage or the lord's chambers at *any* stage, we can only speculate. Lost with the fallen walls of the tower are the windows. We could assume that some type of opening had been incorporated into each wall, but this would be conjecture. Without more to go on, the views of the prime components of the great hall and lord's chambers are simply unattainable and were not attempted in this study.

What is available for analysis at Lea Castle are two gates; the inner (and most likely the earlier) gate into the enclosure surrounding the great tower, and the later mid-13th century gate house in the outer curtain. As mentioned, the inner gate has never been reliably dated, but by 1225 twin-towered gate houses were almost universal, so we might assume this simple gate was constructed before then. Another clue to the construction date of this gate lies within the curtain wall it around it, which has plunging arrow loops; a military development of the early 13th century (McNeill 1990: 333). The later gate (or rather gate house) leading into the outer enclosure may have been built in 1297 which is beyond the time frame of this study (the first 100 years of building in stone in Ireland: 1175 – 1275). However, it is a magnificent structure as seen in the plan by McNeill in Figure 5.15.5; stylistically contemporary with the gates at Ballylahan, Kiltartan and Nenagh as well as the later gate at Carrickfergus. To ignore it for the sake of a dozen years seems unprofitable. It is these two gates, the inner simple gate, and the outer double-D gate house that were examined within this study of Lea Castle.

General views

The general viewshed from Lea Castle can be seen in Figures 5.15.6-8. It was generated utilizing an elevation offset of 20m, and an azimuth of 360°. The position used was the centre point atop the great tower (battlement level). It was anticipated that the lines formed by the edges of the digital elevation tiles seen in Figure 5.15.1 from approximately 9:00–2:00 and 11:00–6:00 on the clock face would cause artifacts within the viewsheds. But although the lines themselves are still visible on the viewsheds running between these same points, they do not seem to have caused any problem in the viewsheds themselves.

There is visibility to all 360°, extending across more than half of the 11km view all the way to the 11km radius, with the farthest reaching views between 10:00 – 3:30 on the clock face (see Figure 5.15.6). The percent of visibility within the 11km view is 10.15%. The north-eastern slopes of the hills south-west of the castle show dense levels of visibility from the 2 – 3km radius between 6:00 – 8:00 on the clock face in Figure 5.15.7. In this 4km view there is 26.5% visibility. The best over-all views are seen in Figure 5.15.8, here within the closest 1km, the area seen by the castle (and which could have views of the castle) amounted to 71.4% of the hinterland. Although there does not appear to be any one direction that has an uninterrupted view farther than 500m distance from the castle, that uninterrupted view seems nearly complete for the full 360° surrounding the castle. This means that barring obstructions by vegetation and built environment, the castle could have enjoyed a 'blanket' like projective view, and observers anywhere within ½km would have had a reflective view.

Projective views: viewsheds and personal observation
Gates
Inner gate

The simple inner gate faced north-east, it offered very little in the way of protection; certainly not a gate house, and maybe not even a tower. The viewshed for this position can be seen in Figures 5.15.9-11. An elevation offset of 10m was used, this actually may have been an overly optimistic elevation, but as 10m has been utilized as the standard tower height within this study, it was employed here. An azimuth of 300° - 170° was factored in to simulate the obstruction in the projective view that the great tower would have caused.

Figure 5.15.9 suggests that the projective view from this gate would have stretched, in some patches all the way to the edge of the 11km radius between 11:00 – 4:00 on the clock face. However, these scattered areas of visibility do not often intersect with the approach roads. Perhaps the best illustration of this is in Figure 5.15.10, this view of the 4km radius around the inner gate suggests that an observer positioned at this gate could see very little of the road coming from either the west or the east. There is a bit of the route to the east (the modern R420 from Monasterevin) visible between the 2 and 3km radii, but absolutely nothing to the west (along what is the modern R420 from Portarlington) can be observed because the great tower would have blocked the projective views to that direction. The road to the south is visible for nearly 500m, but the projective view along that road becomes very patchy as the road nears the River Barrow. Interestingly, the inner gate had an uninterrupted view almost 500m long of the Barrow itself. It is doubtful that the river was navigable during the subject time frame, but hard to dismiss altogether. The opening of the Barrow Line in the 1790s brought the river into the Grand Canal system and it has certainly been navigable since that time (even though it did not appear navigable during the summer of 2007 when field work for this study was completed). The River Barrow is not considered here as a possible approach route to the castle. The percentages of visibility within this viewshed were 4.1% within the 11km radius, 7.8% within the 4km radius, and 30.1% within the 1km radius. Problems with the projective views from the inner gate stem partly from the obstruction caused by the tower, but are mostly due to the way the low elevation of the site curtails views. Not only was the castle built on flat land, the gate was simply too low to offer good views. Although access to the top of the gate would have been dangerous, access to the top of the curtain wall was possible due to a narrow original stair just north-west of the gate. Unfortunately there was so much foliage within both the inner and outer enclosure that no projective view farther than a hand's breadth was possible, and there are no photos to represent the projective views from this position. Photographs of the gate and these stairs can be seen in Figures 5.15.12 -14.

Outer gate

The substantial outer gate house faced south-east. It was approached by way of a raised causeway which ran at least 30m from the higher ground south-east of the castle (Adams 1904: 257), allowing the low-land surrounding the castle which was liable to flooding from the River Barrow to be crossed in all seasons. The extent of flooding may actually have been engineered, and Sweetman suggests a fosse was cut specifically to create an island out of the inner enclosure of the castle (1999: 61-2). This gate house measured roughly 16m wide by 12.5m deep (externally), the long axis running south-west to north-east. It consisted of two D shaped towers with a gate passage running between them. McNeill has stated that this gate shares with Kiltartan and Ballylahan the trait of having the walls towards the gate passage thinner than the other walls of the towers (McNeill 1997: 121). Much of those inner walls are gone today, perhaps owing in part to their thinness.

The viewshed from the battlements atop this gate house can be seen in Figures 5.15.15-17. This viewshed was generated using an elevation offset of 15m and an azimuth of 280° - 240° to simulate the obstruction caused by the great tower. Figure 5.15.15 does not suggest a great increase in visibility over the inner gate in the far-distance (7.4% within the 11km radius), but the amount visible in the middle-distance seen in Figure 5.15.16 more than doubles the inner gate's visibility to 18.6% within the 4km view. And Figure 5.15.17 shows dramatically denser visibility at 56% within the 1km radius. The projective view of the River Barrow increased to almost 600m in length from this gate, and the road to the south may have been visible (although somewhat patchily) nearly the length of its route within the 1km radius.

Compared with the view from the inner gate, the view to the west (towards Portarlington) from the outer gate is again hampered by the height of the great tower, and the view to the east is only slightly better; while a bit more can be seen between the 2 and 3km radii and certainly a great deal more of the road within the 1km radius is visible, it is still quite patchy. As with the general view from the top of the great tower, there is a substantial amount of visibility very close to the castle, this is best illustrated in Figure 5.15.17. Access to the top of the gate house was not possible, but photographs of the gate house itself can be seen in Figures 5.15.18-21.

Great hall and lord's chambers

The building which may have housed an early hall and the lord's chambers is the great tower. This structure is roughly 15m by 20m (external measurements), the long axis running south-west to north-east. The remaining three-quarter round tower at the north corner of this structure is approximately 8m in diameter (externally). It had five storeys; with basement, first, second and third-floors, and a fourth-floor at turret/roof level. As mentioned, the great hall and lord's chamber components were not treated in this analysis of Lea Castle. However, the extant windows (including the fine two-light trefoil window in the second-floor) of the great tower are included in the illustration chapter of Lea Castle for the simple reason that they are endangered. The castle is in private ownership and is untended. The owner kindly allowed this researcher access to the castle, and even showed an interest in the premise of this study. But as the photographs show, the ruins have been overrun and almost entirely blanketed by undergrowth. Although the vegetation (mostly thorny in nature) provides a *form* of security by deterring the casual archaeo-terrorist, the structures are extremely fragile and the masonry is eroding. Photographs of the great tower, its remaining walls and windows (as much as it was possible to photograph with vegetative restrictions) can be seen in Figures 5.15.22-31.

Comparison of these projective views with each other and the views from other castles

The overall projective views, as suggested by the viewshed from the battlements of the great tower at Lea are excellent in the near-distance, 71.4% (Figure 5.15.8). The middle and far distance are not as good, 26.5% and 10% respectively. This pattern is seen at Dunamase, Limerick and Nenagh. Of these castles, only Limerick (a royal castle) was located on similarly flat land.

The inner gate is comparable in its simplicity, shape and size to the early gates at Carrickfergus and Dundrum (built by John de Courcy). And the owners of all three of these castles either added to, or replaced their early smaller gates with larger double-D towered gate houses. The inner gate at Lea did not have extensive projective views, 4.1% at 11km, 7.8% at 4km and 30.1% at 1km. This is due to the low elevation of its battlements (10m) and the flatness of the surrounding terrain. The outer gate's percentages were better (especially in the 1km view); 7.4% at 11km, 18.6% at 4km and 56% at 1km, and the south road, which would have eventually crossed the ford, comes within 200m of the castle and is observable along most of its course.

Another castle built on flat ground which had an outer and inner gate was Adare, Co. Limerick. However, when comparing the visibility from the gates at these two castles, the extent of the projective view from Lea's gates seems high. Adare's inner gate could see less than 1% of its 11 and 4km radii, and only 2.1% of its immediate 1km radius (due mostly to obstruction caused by the castle itself). The outer gate was similarly dismal at 1.2% of the 11km radius, 0.08% of the 4km radius and barely 1.13% of the 1km radius. Again, obstruction by the castle had much to do with the poor projective view.

The viewsheds of Lea Castle suggest that neither of the gates commanded an extensive view of the east-west road that passes the castle today, or even far-reaching views of the River Barrow which passes within 100m of the outer gate. About the best that can be said about the projective views from the gates is that they had uninterrupted views of the River Barrow as it passed immediately next to the castle.

Leask contends that Lea was of considerable importance because of its frontier position, on the borders of Laois and English land, and that it guarded a ford on the River Barrow (Leask 1937: 173). This implies that guarding the ford on the Barrow was one of the reasons this particular site was chosen for the castle. However, the ford is approximately 1km from the castle as can be seen in Figure 5.15.2. This is certainly close enough to be *observed* carefully, but hardly close enough to be *guarded* from the castle. The land along the River Barrow is low lying in relation to the river, but there are several places on the south bank of the Barrow much closer to the ford which are elevated to the degree of the actual site (though none higher), which might have made suitable castle sites. And actually there are sites higher than the castle's location on the north bank of the River Barrow. One spot in particular (which is 4m higher in elevation) is less than 500m from the ford. This suggests that if *control* of the ford had been the priority of the builders, there were much better locations to place the castle. FitzGerald must have been content with mere observation of the ford. Figures 5.15.7 and 5.15.16 suggest that the ford did fall within the line-of-site from the battlements of the great tower and the outer gate at Lea. While the viewshed in Figure 5.15.10 does not suggest the ford was visible from the inner gate, there are patches of visibility to either side of the ford from this gate and it must be remembered that as this viewshed is accurate to only 25m, the ford may well have been observable from the inner gate.

The great tower appears to have been a fine building, and there was probably a great deal of accommodation added to the outer gate house long after the subject time period, but the main priority of the original builders of Lea Castle does not seem to have been to provide a comfortable family habitation. This may explain why, though the FitzGeralds held Lea for several centuries, Leask suggests it was administered mainly by absentee overlords during this time (Leask 1937: 173).

Judging from the projective views from Lea (overall and from the gates), projective visibility may have played a role in the siting of Lea Castle. The fact that the castle was built on the south bank of the river, giving it easy access to the main east-west transportation route might suggest however that FitzGerald's main priority was to establish a castle intended to act as an administrative centre for FitzGerald lands in the region.

Reflective view: viewshed and personal observations

Approaches

As has been established, Lea Castle was not built on a significant prominence; as such it is not readily visible from any great distance today. When it does come into sight, the remains of the skeleton of the great tower that alerts the modern traveller that they are passing a piece of history. The Taylor and Skinner 1778 map of the area around Lea Castle can be seen in Figure 5.15.32. It shows that in the 1700s two roads met close to the ruins of the castle. Today these have fossilized fairly well into the modern R420 running between Portarlington and Monastereven which passes by approximately 600m south of the castle; and the undifferentiated country road to the south which crosses the R420 at Lyman's Crossroads and continues north-east to the ford, passing within 200m of the castle (see Figure 5.15.2). This smaller road comes from Port Laois through Ballybrittas, headed towards Portarlington. It veers to the east at the Tohogar Crossroads about 2km south-east of Portarlington; and though the Taylor and Skinner map does not show the north-east leg, the road appears to head almost directly towards Lea Castle. The modern road continues north-east, coming within 250m of the ford and intersecting with the road which crosses the ford. The outer gate of the castle faces this road, and the causeway may have originally articulated with it.

The locations where the first-sightings were possible can be seen on each of the 1km views of Lea Castle (Figures 5.15.8, 5.15.11 and 5.15.17). Neither the eastern or western approach allowed for a reflective view beyond a distance of 750m from the castle (750m travelling west bound from Monasterevin and 550m coming east bound from Portarlington). The problem was a combination of the advanced state of ruin at Lea combined with vegetation and built environment (recognition of the castle as a man-made structure is hampered by the fact that the castle is largely covered with ivy and hawthorn vines, and merges seamlessly into the surrounding vegetation). The general viewshed seen in Figures 5.15.7 and 5.15.8 suggests that the castle might have been visible coming from the east in several small patches between the 2 and 3km radius, and then again at about the 1km radius. From the west the viewshed suggests snatches of the castle might have been possible all along the road from Portarlington, though no uninterrupted reflective views would have been available until a traveller was nearly at Lyman's Crossroads, roughly 550mm from the castle. The road from the south has an almost continuous reflective view of the castle from about the 1km radius onwards.

In general, personal observation of the reflective views of Lea Castle would suggest that the castle is neither dominant nor intimidating within its wider landscape. Indeed, currently only the very closest 1km of the castle's hinterland has *any* modern reflective view of Lea. However, the outer gate was significant in size and character, and must have been fairly intimidating to anyone within the immediate 500m or so. Coupled with this, the height of the great tower with its fashionable windows may have offered quite an impressive reflective view. Photographs showing the castle in its environment can be seen in Figures 5.15.33 and 5.15.34. A summary table of the approaches can be seen in Figure 5.15.35.

Community

Although we are aware that the medieval Anglo-Norman town of Lea existed (Grose 1793: 77; Adams 1904: 259), its location, and the position of its parish church is uncertain. There are certainly clues; the country road running towards the ford on the River Barrow runs between the castle and the graveyard shown on the OSi map (Figure 5.15.2), and the castle's outer gate and causeway point roughly in the direction of this graveyard. If this graveyard is associated with the original parish church of the Anglo-Norman town of Lea, the orientation of the outer gate would make sense.

Judging solely by personal observation of the remaining structure, a reflective view of the tower and gate within the near-distance could have been very impressive, especially (as always) if a *cordon sanitaire* had been kept about the castle. Surrounded by river water, encouraged to flow onto the flood-plain to form a moat, the castle would have appeared as an island, approachable (like Bodiam Castle) only across a causeway. However, the uncertainty of the town's location makes it impossible to estimate the impact that the castle's gates and tower may have had on the local community. We cannot know if the outer gate at Lea imposed into public life or not; or if it had any type of presence in the daily lives of the people of Lea town. The gate could easily have been as significant psychologically within its community as the gates of Trim and Carrickfergus were within their respective communities, *if* it had been inserted directly into the town, as the gates of these castles were.

The 'feel' of the castle today is a little disconcerting. It has an eerie look which is promoted by thick shrouds of hawthorn and ivy, overlying hundreds of tumbled stones and pieces of crumbling masonry. It was easy enough to gain the owner's permission to explore the ruins, but difficult to access them physically. Lea has none of the benevolent nature of Adare or Swords, but instead offered a sense of almost antagonistic malevolence.

Comparison of these reflective views with each other and the views from other castles

As mentioned with the projective view, Dunamase, Limerick and Nenagh had comparable visibility. Dunamase and Nenagh were both built on hills in otherwise relatively flatland, while Limerick and Lea were built along rivers and on low ground. Ferns Castle was built near the top of a hill, but could be seen by only 10.57% of its 11km hinterland compared to Lea's 10.15 %, by 22.6% within the 4km radius compared to Lea's 26.54 %, and only 59.4% within the 1km radius compared to 71.4% at Lea. This serves to remind us that *many* factors are involved in visibility; simply building upon a hill (or the side of a hill) does not necessarily increase the chance that your structure will be seen.

History suggests that Lea Castle (built as it was on the cusp of Anglo-Norman society), was the scene of many stirring events. By 1270 (towards the end of the subject time frame) Lea Castle was the only castle held by the Anglo-Normans in Offaly (Adams 1904: 259), and it was continuously in use or being repaired for use until at least the late 1700s (Adams 1904: 260-61). Judging from the reflective views within the immediate hinterland, visibility might have played a role in the choice of site, (however without examining visibility from the site north of the River Barrow mentioned above, we cannot be certain that this location provided optimal visibility in this area). The size and aggressive nature of the outer gate seems to suggest that psychological dominance may have been a priority of the builder; and the height of the great tower with its fine windows may hint at some desire to socially dominate by a display of wealth and knowledge of current architectural fashions. These priorities do not necessarily support or detract from the possibility that management and administration of FitzGerald lands was the main priority, as suggested by the location of the castle and the projective view.

5.16 Limerick Castle, Co. Limerick

Note: OSi orthophotographic coordinates were used for the great hall, gate house, market place and parish church, and were found online at: http://ims0.osiemaps.ie/website/publicviewer/main.aspx#V1,557698,657791,8.

Introduction

Limerick Castle, Co. Limerick is built upon flat, gently undulating terrain at the mouth of the River Shannon as it wends south-westwards toward the sea (see Figure 5.16.1). At the Shannon Estuary just north of the city of Limerick, the Shannon splits and forms an appendage (the Abbey River) which flows south-east. These two rivers flow around an islet before joining up again approximately 2km downstream. The southern end of this small land mass (known as King's Island) was the location of the original Norse settlement at Limerick in the 10th century (Killanin and Duignan 1967: 349; Givens 2008: 192). By the late 12th century Limerick had become the axis of the Irish king of Thomond: Donal Mór O'Brien.

As part of a short-lived agreement between Henry II and Donal Mór O'Brien, an Anglo-Norman garrison was placed in Limerick in 1172 (Wiggins 2000: 16-18). O'Brien re-thought his agreement with the Anglo-Normans in 1175, and put the garrison under siege. There is some thought that the site of Limerick Castle had been in use by the church prior to the erection of the present stone castle (Adams 1904: 253), but it had already become the site of an Anglo-Norman ringwork-type castle when Raymond le Gros FitzGerald marched on Limerick to support the besieged troops 1175-6 (Wiggins 2000: 16).

The ringwork utilized by Le Gros included a clay rampart, faced with a wall of rough limestone masonry and bordered by a wide ditch. It is located beneath the north-eastern corner of the castle, and stood as part of the outer defensive limits in the early Anglo-Norman stronghold until the east curtain wall was constructed (Myles 1991: 8). Although Wiggins proposes this feature represents the origins of King John's Castle and should be considered a 12th century ringwork (2000: 16), it has also been suggested that this rampart and ditch constitute a portion of the original O'Brien fortress at Limerick (Myles 1991: 4; McNeill 1997: 12). If this spot of ground *had* been the position of the O'Brien stronghold, the initial Irish siting agenda may have been quite different from the Anglo-Norman schema. We should keep in mind the possibility that the Anglo-Norman builders simply inherited the site from O'Brien, but even so, with the essential element of the natural ford (and subsequent bridge), the Anglo-Normans would have realized that this was a viable location for a castle.

Le Gros was able to maintain an Anglo-Norman presence at Limerick until the spring of 1176 (Killanin and Duignan 1967: 350). But it was not until after O'Brien's death in 1194 and the subsequent succession-confusion that the Anglo-Normans were able to establish a permanent garrison at Limerick (ibid; Wiggins 2000: 16). Construction of the castle in stone (traditionally attributed to King John) probably commenced around 1210 (Wiggins 2000: 18), and continued in spurts for the next century and a half (Myles: 1991: 2-3), though the southern half of the east curtain wall and the south curtain may not have been built in stone until after the subject time frame. This is an urban castle, and its location within Limerick town is illustrated in Figure 5.16.2. The castle was well placed to maintain a central role in the medieval town (Wiggins 2000: 15). The only seigneurial aspect of Limerick Castle is the Thomond Bridge, though there may have been a park in the Cratloe Woods, approximately 10km to the north-west. Although there is no extant evidence of a managed landscape, the length of time and extent of the city limits may have obscured any such. The OSi road map of the area surrounding Limerick Castle can be seen in Figure 5.16.3.

Location and discussion of the primary components

The primary components of gates and great hall are to be found in two principle features at Limerick Castle; the main gate and the great hall. The lord's chamber component is not treated in this examination because as a royal castle with a perpetually absent sovereign, Limerick was governed by a resident sheriff or constable who almost certainly resided within the gate tower accommodation. There simply would not have been a lord's chamber *per-se* (although Myles does suggest that the ground-floor room in the north-east tower was the finest chamber at the castle (1991: 18).

The two remaining components were easy to discern, thanks in part to excavations conducted during the 1990s. The gate (which is straight-forward enough) faces north and was built during the initial construction phase of the castle, which saw the erection of the four towers along the north curtain wall (a tower at both the east and west end, and two gate towers in the centre).

A great hall (the only stone structure known to have been built in the medieval period within Limerick Castle) was constructed along the west curtain in 1280 (Wiggins 2000: 20). However the above ground walls of this hall were demolished around 1790-1800 (Wiggins 2009: 14), and the rubble was backfilled into the building's undercroft, effectively obscuring the hall's location. Depictions of the hall in several 15th to 17th century representations of the castle suggested it had been positioned along the west curtain, but its exact location was unknown until excavations from 1993-1994 exposed the undercroft. The castle plan can be seen in Figure 5.16.4.

General views

As illustrated by the topographic map above, Limerick Castle is in an area of gently rolling land. To the north and north-west the Cratloe Hills rise to 310m above sea-level, to the east the ground undulates slightly. There are a number of much smaller hills to the south-east and south, while west of the castle the Shannon Estuary spreads across low-lying flatlands. The viewshed seen in Figures 5.16.5-7 was created using an elevation offset of 15m to approximate the height of the gate towers. Because the gate consists of two adjacent towers, it was determined that generating viewsheds from the tops of both towers was redundant, and a mid-point between the two towers was used as the centre-point for the calculation of the viewshed. A full 360° view was applied since it was unlikely that structures either within or without the curtain walls (other than the fairly slender steeple of St. Mary's cathedral to the south) were elevated enough to interfere with views from this location.

The most far-reaching views are to the north-east and south-west. Some fragmentary far-reaching views can be seen to the east and west. The densest projective views beyond 4km are to the north-west and reflect the upslope along the south face of the Cratloe Hills. Within the 4km radius (Figure 5.16.6), the most complete views are from the north-west to the east. But it is within the 1km radius that the largest projective views can be found (Figure 5.16.7). The viewshed suggests that within the 1km area, more than half of the surrounding area could be visible from the top of the castle's gate towers.

Projective views: viewsheds and personal observation
Gate

There are actually two gates at Limerick Castle; the main gate and a watergate. The watergate, an inconspicuous postern located in the base of the west curtain allowed for supplies and guests arriving by water to enter directly off the Shannon. Important as it might have been, it was mainly geared to accommodate river and service traffic, and was not considered as part of this study. The viewshed from the main gate (introduced above), stands as the main viewshed for Limerick Castle (Figures 5.16.5-7). The main gate is a double-D design with a narrow entrance passage. The gate towers are still quite tall, rising at least 20m above the River Shannon, the eastern-most gate tower sits at its original height, but the western gate tower was altered (heightened) later to support heavy ordnance (Myles 1991: 16). The castle was perfectly positioned to observe its surroundings; built at the head of the medieval town's main road (now Nicholas Street and Mary's Street), and set along the river adjacent to the rock shelf which forms the ford (which was later bridged), it could track water as well as land traffic. Indeed travellers approaching Limerick Castle during the study time frame may have been travelling by either land or water, and the viewshed suggests that the northern approaches (along the Shannon, as well as R464 and R463) could be well observed from the gate towers. Traffic coming down either of the roads could be within sight occasionally, starting at approximately the 4km radius (Figure 5.16.6), and within 2km of the castle R646 may have been visible more-or-less steadily. Both roads appear on the viewshed to be almost continually visible along the final 1km stretch. Interestingly, river travel in general cannot be quite as well monitored, for beyond the 1km radius it is only sporadically visible from the gate towers.

The most important traffic however, would have been coming from Dublin, and so from the east. We can see that the Dublin Road (now represented by the N7) does pass through the line-of-sight from the gate towers several times within the 11km radius; first between 7 and 8km away, then again at the 4km mark. There are stretches within the 2 – 3km radius where this road is not visible, but by 1km the projective view of this approach is strong.

To the south R526 (N20) passes in and out of the line-of-sight from the castle briefly many times within the 11km radius, but the viewshed suggests that it is not until within 1km of the castle that the road can be steadily seen. Likewise, water traffic approaching from the west through the Shannon Estuary is not visible beyond the 1km radius, and only sporadically visible within that closest 1km.

The main gate can be seen in Figures 5.16.8 and 5.16.9. Access to the top of the gate towers themselves was not possible; however, the north curtain wall was accessible, as were the tops of both the north-west and north-east towers. These views are shown in the photographs of Figures 5.16.10-19. Although these photographs were taken from a *lower* elevation (approximately 10m) than that used to generate the viewshed (and so cannot reflect the full extent of the projective view as suggested by the viewshed), it is obvious that much can be seen from this position, and the photos do tend to support the far-reaching nature of the projective views suggested from the gate towers. However, personal observation from the available vantage points determined that built environment (including the castle's own curtain walls) makes modern observation of the above mentioned roads difficult. While it is possible that the greater elevation of the actual gate towers may have increased the visibility of traffic on the surrounding roads, it is also possible that the few extra metres may not have made a great deal of difference.

Great hall

The great hall was located along the west curtain wall. It was a rectangular structure measuring approximately 24m in length by 15m in width, the long axis running north to south. It is depicted in the *c.*1600 representation *Pacata Hibernia* as having gabled ends (Figure 5.16.20). At the time of the hall's construction, the inner ward sloped dramatically from the north (main gate of the castle) to the south, and the entrance to the hall would have been at first-floor level, facing east. Later, the ward was filled in to create a nearly level surface from the north to the south curtains, after which the hall would have been entered at ground-floor level.

The viewshed in Figure 5.16.21-23 was generated from the location of the great hall using an elevation offset of 1.7m. The available possible views (in this case only to the west to simulate the seclusion enforced by the inner ward) were factored into the calculation. As might be expected, the projective view from this building was not extensive. There are very isolated patches of visibility to the south-west at approximately 7:00 and again between 10:00 and 11:00 on the clock-face. Figure 5.16.22 demonstrates the dramatic *lack* of a projective view suggested by the viewshed. There are of course no extant hall windows or doors (although the undercroft has four splayed arrow loops piercing the western curtain at first-floor level externally and a fine sandstone window facing east into what would have been the inner ward at the time of construction. These can be seen in Figures 5.16.24 and 5.16.25. Interestingly, the late 16th century representation seen in Figure 5.16.26 does seem to suggest that this building had four windows overlooking the west curtain. While we cannot be sure these windows actually existed, we can be fairly certain that some form of opening would have been required for light if nothing else. Windows on the east wall of the hall would only have looked into the inner ward which was surrounded by curtain walls, either of stone or timber. Projective views from west windows would have been of the Shannon and its west bank. Lacking any great elevation, these windows could not have provided very far-reaching views, as the west bank slopes up and away. Views from this window could only have extended to the top of the bank's gentle rise. Depending on what was located upon the west bank (such as a park), the view may have been interesting and even impressive, but it might not have been very informative. It was interesting to note that the 1km radius viewshed (Figure 5.16.23), suggests that the west bank of the Shannon, directly across from the hall would not be visible from this point. In fact, it *is* visible as seen in Figure 5.16.27. In this instance personal observation and photography do not seem to support the science.

Lord's chamber

As mentioned above, as there would not have been a lord's chamber as such, and this component was not treated in this exercise.

Comparison of these projective views with each other and the views from other castles

Perhaps the first comparison that can be made to Limerick Castle (at least within the context of an evaluation of the research methods of this study) is a comparison between the actual (personal) and the perceived (GIS viewshed) views from the castles. It was mentioned above that personal observation and photography did not support the results suggested by GIS; i.e. the viewshed suggested that the west bank of the Shannon (which was in fact highly visible from the great hall), would *not* be visible from this location. (In this case, the 1km viewshed from the hall at Limerick suggests that only 1.37% of available pixels are visible.) Such a discrepancy (albeit in reverse order) was seen at Adare Castle: at Adare, personal observation and photography suggested that very little of the surrounding countryside could be seen from the later great hall. The actual lack of projective views from the windows at Adare was contradicted by the perceived projective views suggested by the viewshed (which implied that 19% of the surrounding pixels should be visible). Although the cause of the discrepancy at Adare was due almost entirely to vegetation growth which completely hid any available views suggested by the viewshed, the cause of the discrepancy at Limerick is unknown.

Wiggins suggests that undoubtedly the lodgings for the constable and warders at Limerick would have been on a much more reserved scale than would have been tolerable if the castle had been intended for aristocratic habitation (2000: 20). The same probably held true for the great hall and may help to explain the poor views possibly offered by this hall. Such blunted projective views would reflect the castle's emphasis on its function as a basic military administrative centre, and underscores the fact that Limerick Castle was not intended to be an impressive *caput* or even the part-time habitation of a baron. This was not a hall built to amuse or entertain its inhabitants, or to impress and overwhelm the local community with the resident lord's wealth and power. This was simply a hall meant to fulfil the requirements of a royal administrative centre.

While the lord's chamber component is not included at Limerick Castle, and the projective views from the great hall were *very* limited, the over-all projective views from the gate towers at Limerick Castle are fairly good. Although the 11km radius allows for visibility of only 15% of the total pixels, the views to some directions are quite far-reaching. The viewshed suggests that it is within the 1km radius around the castle that the projective views are the strongest, allowing an observer at the top of the gate towers to monitor 63% of the immediate vicinity. The high percentage of projective views has much to do with the low-laying geography of the land adjacent to King's Island. Another factor influencing projective visibility is the fact that the towers are the tallest structures on the island, allowing views from this position to spread across the full 360° of the compass.

Actually it was not unusual to find that scrutiny of the closest hinterlands seems to have been a priority at these earliest castles. Other castles within this study which have shown extensive projective views of their environments within a 1km radius include Athlone (39% from the great tower), Athenry (51.7% from the top of the great tower), Dunamase (55.6% from the gate tower), and Trim which excelled in its short-range visibility (46.5% from the west gate, 57% from the Dublin gate and 70.7% from the great tower). Of these, Athlone alone shows major similarities to Limerick Castle: both were royal fortifications, built along the Shannon at natural fords. Both Athlone and Limerick were nicely placed to observe their environments; their communities, the waterways which they flanked, as well as their nearest hinterlands. At Athlone the ability to overlook these aspects was due to the height of the ridge upon which the castle was built, while at Limerick it was the loftiness of the gate towers which offered the advantage for good projective views. Although the projective views within the 1km radius at Limerick are greater than those at Athlone, the extent from both, combined with their settings adjacent to vital fords suggests two important objectives of the builders of these royal castles: the first (and perhaps most essential) objective being the monitoring of those fords; the second concern being the monitoring of the comings and goings within the towns (and the areas adjacent those towns).

In general, at Limerick Castle the projective views of the ford (and later the bridge) on the Shannon, the river itself and the town certainly suggest that visibility played a role in the siting of the castle. However Limerick's lack of emphasis on projective views from its great hall, its location within the mediaeval town, and *especially* its proximity to the ford suggest that the highest priority in the choice of siting was to allow a strategic military and administrative advantage.

Reflective views: viewshed and personal observation

The viewshed generated from between the gate towers seen above suggest that long-range reflective views of Limerick Castle may not have been a priority for the choice of location. The castle was visible to only 15% of the outer hinterland (within an 11km radius). It is true that as an observer comes closer to the castle the visibility increases until within a 4km radius of the castle 22.51% of the hinterland can offer reflective views. However, an impressive 63.41% of the land within the immediate 1km radius around the castle could have had some type of reflective view of the castle. Of course visibility of the castle as suggested by the viewshed is hindered only by topography; the reality is that modern views of the castle are nowhere near this good, due unfortunately to the built environment of the city.

Approaches

As mentioned there were possibly several approaches to Limerick Castle during the subject time frame. The Taylor and Skinner map from 1778 seen in Figure 5.16.28 illustrates just how many ways there was to approach Limerick in the 18th century. We cannot be sure which of these roads were in use during the study time frame. Because of this, this researcher tried to track as many of the likely ways into the town as possible in order to discern the first-sightings of travellers approaching the castle. Unfortunately, although the River Shannon and the Shannon Estuary would clearly have been viable routes of travel, these waterways were not assessed for first-sightings due to difficulty in accessing river transport.

The most difficult roads from which to spot the castle were those approaching from the south-east and the east. These roads include the Dublin and the Ballyshannon Roads (modern N7 and N24 respectively). Travelling north-west through the Irish Town side of the city, these routes funnel through Balls Bridge and onto Mary's Street/Nicholas Street at the south end of King's Island. At the entrance to Balls Bridge, St. Mary's Cathedral can be clearly seen, but little beyond this point is visible. As St. Mary's Cathedral was already present (in some form or other) at the beginning of the subject time frame, it might be assumed that the structure blocked reflective views of the castle at this time as well. It is not until the road comes up the hill and passes the cathedral and a line of shops that the north-east angle tower is visible...from only 90m away (Figure 5.16.29).

The High Road across Thomond Bridge (N7) coming from the north-west provides for a slightly more distant first-sighting (the castle can be seen from 250m). Interference from this direction came in the forms of topography and built environment, which effectively hide the castle until the traveller is almost at the entrance to Thomond Bridge (Figure 5.16.30). The situation coming from the south and south-west along Dock Road (N69) and Patrick Street (N20 also known as R526) is both better *and* worse. Today the castle is not visible at any point from these streets due to the intensely congested built environment. However, reflective views are available if the modern traveller actually steps a few metres off of these streets, and goes behind the buildings along the south-east bank of the Shannon (at Arthur's Quay). At this point one can get a glimpse of what may have been visible from either of these roads before the modern skyline completely clogged the view. From here the castle is quite readily visible from 500m away (Figure 5.16.31).

Killaloe Road (R463) coming from the north-east offers the best, and most long-range first-sighting (from approximately 1km), it can be seen in Figure 5.16.32. It is interesting that although the viewshed in Figure 5.16.6 suggests that the northern approaches and Dublin Road could each have had reflective views within 2km, topography, built environment and vegetation successfully compete for visibility along every approach. In general, personal observation of the approaches shows that the reflective views as suggested by the viewshed from the gate towers are not accurate *on the ground.* A summary table of the approaches can be seen in Figure 5.16.33.

Community

Limerick Castle sat within the town walls of the medieval town of Limerick. Although the town has suffered economically during the past two centuries, generally it has remained vital for a thousand years, expanding immensely. The medieval parish church was only 200m from the south curtain wall of the castle, with the likely location of the market place being near the church. A single viewshed was created to represent the reflective view from the parish church and possible market place. This viewshed, seen in Figure 5.16.34 was generated using an elevation offset of 1.7m. It suggests that (barring the presence of a great deal of built environment), reflective views of Limerick Castle were possible from the vantage point of the market place and church. The view of the castle from a market place near the church along Mary Street would have been of the south and east curtains (which at this time might have had only a timber palisade). This 'tail-end' view showed an altogether more vulnerable side of the castle, and while it would have been a reminder of Anglo-Norman strength, it was, perhaps, not a terribly *invasive* reminder.

Comparison of these reflective views with each other and the views from other castles

It is clear that the castle fixed itself into the nexus of the town, visible from more than a half of its 1km radius. The walls visible from the parish church and market place would have stood as a sturdy reminder of the constable within them. Though the gates did not jut directly into the market place (as at Trim, Carrickfergus and Swords), it would have been formidable to the north and west, to travellers and anyone crossing the bridge who would have had to traverse the length of the massive north curtain wall, passing directly beneath the shadow of its four redoubtable towers. This route could have left no uncertainty about who was in charge of the town. The castle from this perspective looms large and ominous, and is *still* in fact awe-inspiring. The over-whelming might suggested by the north curtain and the towers actually compares well to Trim and to Carrickfergus (even if Limerick didn't face the market directly and was controlled by the throne of England and not an independent baron). These three castle gates very strongly make their overtly intimidating presence felt in a psychological sense within their various communities. But, it is also possible to contrast Trim and Carrickfergus with Limerick. Limerick makes little if any attempt at social display, but instead falls back unapologetically on its function as a military facility. It had a hall because it *needed* a hall in order to be a viable administrative centre; likewise it had no great tower because it *did not need* one.

With the reflective views (as with the projective views) we can compare Limerick to Athlone. Both of these royal castles would have had an 'in-the-face' reflective presence for the stranger community. Locally, both castles were workable royal administrative centres which contained very real martial aspects that cannot be denied, and should not be downplayed. Although Limerick did have its defensive weaknesses (as the siege of 1642 proved), it stood for more than 400 years before those weaknesses were successfully exploited (Wiggins 2000: 19-20).

The reflective views (like the projective views), certainly suggests that visibility played a role in the siting of the castle. The reflective views also definitely support defence and aggression as the main priorities of the builders, as suggested by the projective view. The unrestrained 'no-nonsense' quality of the reflective view suggests that the purpose of the siting choice was not to impress the local populous or visitors with the importance or power of a particular person, but to actually command, control and intimidate as much of this part of the kingdom as possible.

5.17 *Nenagh Castle, Co. Tipperary*

Note: OSi orthophotographic coordinates were used for the round tower, the gate house and parish church; these coordinates were found online at: http://ims0.osiemaps.ie/website/publicviewer/main.aspx#V1,586648,679350,8.

Introduction

Nenagh Castle was built on a small hill west of the present site of Nenagh town. Within 10km to the west the Arra Mountains rise to 364m, to the east and south the Silver Mine Mountains rise to 694m (see topographical map in Figure 5.17.1). Killanin and Duignan suggest that the ancient name of Óenach Téite indicates that this was a meeting place of one of the principle assemblies of Munster (1962: 386).

The castle was built of limestone (Gleeson and Leask 1936: 265). It was the *caput* of Theobald Walter's chief manor in Ireland and formed the nexus of Butler activity until the second half of the 14th century (ibid). There are no written references to the foundation of the castle, and the exact dates are unknown. Several scholars agree that the castle probably dates from the final years of the 12th century, or the first decades of the 13th century (Gleeson and Leask 1936: 24; McNeill 1997: 31, Hodkinson 1999: 178). A slightly worn short cross penny of King John (minted between 1205 and 1218) was discovered during the 1996-97 excavations in a layer of clay that Hodkinson considered as trample during the construction of the gate house complex (1999: 166, 178). This coin supports the suggestion that at least some the parts of the castle were built by or for Theobald's son Theobald II, after his minority (1206 – 1221).

The sequence of construction is somewhat confusing; it appears there might have been three closely spaced phases. The first phase would have included at least the lower three storeys of the round tower, with a date before 1200 suggested by a sandstone chevron in the window embrasure on the south side of the second-floor (Gleeson and Leask 1936: 267). Hodkinson suggests that the curtain walls, angle towers and the two towers of the gate house were also erected in this first phase (Hodkinson 1999: 178). The upper part of the round tower (the third-floor), with its Gothic details suggests the second phase of construction during the early 13th century (Gleeson and Leask 1936: 267); probably after the end of Theobald II's minority (McNeill 1997: 50). The third phase (in the middle of the 13th century) saw the addition of the great hall behind the towers of the gate (Hodkinson 1999: 178).

Nenagh was an urban castle with an Anglo-Norman community; a list of documentary sources referring to the town and castle can be found in Gleeson and Leask (1936; 249-252). Although the Norman town of Nenagh was destroyed and the Anglo-Norman inhabitants 'rooted-out' sometime around 1348, the Butler lords continued to occupy Nenagh Castle until at least 1367 (ibid. 252-3). Gleeson and Leask contend that no trace of the early town of Nenagh has survived (ibid. 248), and with this in mind we cannot assume that the street pattern we see at Nenagh town today represents the community during the subject time frame.

Seigneurial aspects surrounding Nenagh Castle include the town itself and an Augustinian priory founded by Theobald I around 1200 at Tyone, approximately 1km to the south-east (Gleeson and Leask 1936: 248). The revenue from two mills is mentioned in the Pipe Roll of XVI, Edward III, and although this is after the subject time frame, it is reasonable to assume that mills were in use from the beginning of the settlement (ibid. 250). Interestingly, in 1299, shortly *after* the subject time frame, a park was licensed out of demesne land by Theobald V. A petition documented in the Justiciary Rolls (The Calendar of the Justiciary Rolls Ireland, Edward I, Part I: 234), gives an excellent account of how this lord manipulated an already established highway system to create a substantial seigneurial marker when he wanted to

"....divert a highway and prepare a new road and make into a park, a wood".

That this manoeuvre would consequently inconvenience his neighbour Robert, son of David and several others who lived

"...in the street below the Castle to the East, and they would have to make a circuit to go to their lands"...

did not appear to be much of an obstacle for Butler. The injured parties, who were forced to travel four furlongs out of their way each day to work their lands (Murphy and O'Conor 2006: 72) were simply assessed damages of 40s, and the wood became a park. Gleeson and Leask suggest that the park created included the land on which St. Mary's Catholic Church sits today (directly north-west of the castle) and the area to the north of it (1936: 250). Without a certain understanding of the exact position of the early town, it is not possible to determine if the approaches may have been part of a structured managed landscape. The OSi map of roads in and around Nenagh can be seen in Figure 5.17.2.

Location and discussion of the primary components

All three primary components of gate, great hall and lord's chambers were very straight-forward at Nenagh Castle. The shape of the castle was an irregular pentagon (see castle plan in Figure 5.17.3. A substantial double towered gate faced south-west, supposedly towards the town. The great hall sat directly behind the gate house, utilizing the gate house rear wall as one of its long walls. The lord's chambers were located in the round tower at the north angle of the pentagon. As at Dundrum and Castleroche, Nenagh had two levels of chambers for the lord: public and private, the second-floor business (public/great) chamber, with the third-floor (private) chamber.

General views

The 11km viewshed (Figures 5.17.4-6) was produced from the approximate elevation of the battlements atop the round tower, using an elevation offset of 24m, with an azimuth of 360°. When looking at the topographical map seen in Figure 5.17.1, several artifacts can be seen as straight lines which seem to criss-cross in a rather disjointed fashion. The artifacts caused a problem within all of the viewsheds seen in this study. This is seen most notably in the general view in Figures 5.17.4-5 as a sharp line of visibility running roughly east to west 1km north of the castle. An interesting aspect of this viewshed from the battlements of the round tower is the view to the north-east. Nenagh Castle sits on the south-west side of a hill, 340m short of the crest. The area of no visibility seen between roughly 1:00 – 3:00 on the clock-face in Figure 5.17.6 represents the north-east side of this hill.

The most far-reaching views from the battlements of the round tower are between 3:00 to 7:00 on the clock-face. Line-of-sight includes views of the slopes of the Silver Mine Mountains (east and south), which can be seen from the 11km radius to approximately the 8km radius in Figure 5.17.4. Towards 8:00–10:00 on the clock-face, the foothills of the Arra Mountains can be seen between 11-4km around the castle. There are patches of visibility throughout much of the 9–4km radius. The 11km radius would have allowed views of 22% of the castle's hinterland.

Within the middle distance seen in Figure 5.17.5 there is visibility to more or less all the compass points, although it is scattered. Observers from the battlements of the round tower would have been able to see 44% of the 4km radius. The best over-all views are seen in Figure 5.17.6. Here the viewshed suggests that there is an almost continuous line-of-sight between the castle and the 1km radius from 4:00 - 7:30 on the clock-face. There is some interruption from 8:00 - 10:00 but the nearly uninterrupted view picks up again from 10:00 – 12:00, where near blanket visibility stretches from the castle to the 1km radius. The view of the 1km radius would have been 79 %.

Projective views: viewsheds and personal observation

Gate

The gate house faced south-west and stretched approximately 22m in length by 11m in width. It consisted of two nearly circular gate towers connected by an arched gate passage which still retains the remains of a two-slot drawbridge pit at the front and a sandstone portcullis at the rear (Hodkinson 1999: 163-4). The great hall block was built behind the gate house, with the gate passage running directly beneath the first-floor hall. This extended the gate passage an additional 11m. While the gate passage at Castleroche ran below the lord's chambers, the arrangement of a *great hall* above a gate passage as seen at Nenagh was unique in Ireland (Hodkinson 1999: 178).

The viewshed seen in Figures 5.17.7-9 was generated using an elevation offset of 15m, (battlement level), and an azimuth of 40° - 10° was factored in to simulate the barrier to the projective view caused by the round tower. The area not seen because of the round tower is best illustrated as the slice of no visibility between approximately 12:00 and 1:00 on the clock-face in Figure 5.17.8. The percentage of hinterland observable from this location breaks down as 18%within the 11km radius, 32%in the 4km radius and 59.5%of the 1km radius.

The routes which may have been important to observe from the gate house battlements include the road from Portumna to the north, the road from Dublin to the east, the road from Cashel to the south-east, the road from Limerick to the south-west and lastly, the road from Lough Derg to the north-west. The modern road from the north (N 52) can be seen in Figures 5.17.7 and 5.17.8 at approximately 12:30 on the clock-face. This road is really only visible from the gate house at Nenagh for the initial 300m north-east of the castle. At about 300m this road crests the hill and cannot be observed for approximately the next 500m when it passes briefly through the line-of-sight from the gate house. This route was the least observable approach to Nenagh.

The road from Dublin (R 445) can be seen in Figures 5.17.7 and 5.17.8 at approximately 3:00 on the clock-face. It is more easily viewed from the gate house battlements than the N 52; although in the far-distance the projective views of this approach route are spotty, except for short stretches at about 3.5 and 9 km. From the 2 – 1km radius the road becomes more observable, but as it ascends the east side of the castle's hill it cannot be seen. It comes into sight again about 500m east of the castle and is solidly observable for the remaining distance. This is best illustrated in the 1km viewshed in Figure 5.17.9 where this road is seen at 3:00 on the clock-face.

The road from Cashel (R 488) can be seen in Figures 5.17.7 and 5.17.8 at 4:00 on the clock-face. According to the viewshed, this is probably the most observable of all the approach routes into Nenagh. Although it cannot be seen beyond the 4km radius, visibility of this road between the 1 - 4km radiuses is quite good. In the 1km viewshed in Figure 5.17.9 (where this road is seen at 5:00 on the clock-face) this approach route is almost entirely observable from the gate house battlements.

The road from Limerick (R 445) can be seen in Figures 5.17.7 and 5.17.8 at approximately 7:45 on the clock-face. It can be observed for a short distance at about the 7km radius, but then cannot be seen again until it comes within 2.5km of the castle. There are several patches where it can be observed (though not well) between the 2.5km-500m radii the castle, but the closest 500m are clearly observable, as illustrated in the 1km viewshed in Figure 5.17.9, where this road is seen at 8:45 on the clock-face.

The road from Lough Derg (R 495) can be seen in Figures 5.17.7 and 5.17.8 at approximately 11:00 on the clock-face. This route falls within line-of-sight from the gate house battlements at several points along its way towards Nenagh. It is more-or-less consistently observable between 11 and 9km away, and is visible again between 6 and 4km radii, and 2km and 500m radii. In Figure 5.17.9 where this road is seen at 10:00 on the clock-face, we can see that this route is observable in patches until it joins the road from Limerick, which is clearly observable from the castle from approximately 500m distance. Access to the top of the surviving gate tower was not possible, and there are no photographs to illustrate the projective views from the battlements of the gate house. Photographs of the openings within the gate house can be seen in Figures 5.17.10 and 5.17.11.

Great hall

As noted above the great hall block was built directly behind the gate house along the south-west side of the castle, utilizing the rear wall of the gate house structure as a common wall. The building was rectangular, the long axis running south-east to north-west. It included a first-floor hall measuring 20m x 8m internally, and two small chambers below the hall (each approximately 8m x 8m internal measurements), which were separated by the gate passage. The hall had at least four large windows in the long north-east wall. The window closest to the entrance originally had a pointed sandstone arch. The next window still has its eastern sandstone jamb, but no trace of the third window remains. The final window, which was at the high-end (north-west end) of the hall has retained its western sandstone jamb (Hodkinson 1999: 164-65). The short south-east wall of the great hall has almost completely disappeared so it is not possible to know if it contained windows or not, however there were no windows in the corresponding north-west wall.

It was determined after examination of the castle fabric that the hall could not have had a projective view reaching beyond the castle ward because it had no windows that were external to the castle ward. The curtain walls rose to a height of 12m as shown by the doorway articulating the western wall-walk with the second-floor of the round tower (12m), so the four rather fine windows in the north-east wall of the hall would have only had views of the interior of the castle. Because no exterior projective views would have been possible from this space, a viewshed was not generated from the position of the great hall. Photographs of the entrance and windows of the great hall can be seen in Figures 5.17.12-19.

Lord's chambers

The round tower at the northern angle of the curtain wall was the likely location of the lord's chambers. This structure is approximately 17m in diameter externally, and during the subject time frame the final storey would have been the battlements, reaching an elevation of roughly 24m. The tower was remodelled towards the end of the 19th century, adding 6m to the top (Gleeson and Leask 1936: 265). The walls are approximately 5m thick at the base (including batter) and thin to about 3m at current parapet level; during the subject time frame there were four interior floors. A plan of the round tower can be seen in Figure 5.17.20. Photographs of the outside of the tower can be seen in Figures 5.17.21-24.

As when dealing with the round tower at Dundrum Castle, it will be helpful to utilize the concept of the clock to describe the locations of the openings within the round tower at Nenagh. The lowest level (ground-floor) had no external openings and must have been accessed by a ladder and trap-door, as there is no extant staircase (mural or otherwise) descending from the first-floor (note that the plan shows an opening on the ground-floor, this is a modern opening). The main entry to the tower was at first-floor level through an entrance lobby on the south-east side (5:00 – 6:00 on the clock-face). This door was reached through a fore-building which is now little more than masonry rubble. The entrance lobby gave access to a winding stair on the left (leading to the upper floors), then descended down several steps before opening into a single first-floor room. This ante-room to the lord's council chamber above was 8m in diameter. It had two narrow round-arched plunging arrow loops set in deep embrasures, one facing north-west (at 11:30), and one facing east (at 3:30).

The second-floor was clearly a fine space, roughly 9m in diameter with a hooded fireplace and four windows; facing north-east (at 1:30), south-east (at 4:00), south-west (at 7:15), and north-west (at 11:00). The window on the north-east had seats; this and the south-eastern embrasure had bluntly pointed arches. The embrasure of the window on the south-west had a round sandstone arch decorated with a chevron moulding which Gleeson and Leask suggests would have been late 12th century in England. They also suggest that this window, which today is fairly wide, was probably once an arrow loop (1936: 267). This embrasure opened onto a mural stairway leading to the third-floor. The embrasure of the window on the north-west had a low round arch and led to an external opening (at 10:00), which allowed access to the west curtain wall-walk.

The third-floor, added in the 13th century, is approximately 9.6m in diameter. It had a finely columned hooded fireplace wrought in sandstone, and four windows facing roughly the same directions as the windows in the floor below. These windows were placed to the north-east (at 1:00), south-east (at 4:00), south-west (at 8:00), and north-west (at 11:00). The windows were of pointed sandstone, and wider than those of the floor below. The arches of the embrasures were irregularly segmental and quite detailed; the south-west window had a sandstone moulded segmental back-arch with angle shafts with sandstone bases and capitals (Gleeson and Leask 1936: 267). Within the embrasure of the north-west window was a small passageway leading to a *machicoulis* projecting over the space between the door from the curtain wall-walk and the north-west window on the second-floor. The viewsheds for the lord's chambers are from the second and third-floor levels.

Lord's council chamber; second-floor of great round tower

The viewshed from this location can be seen in Figures 5.17.25-27. An azimuth of 220°-190° was factored in to simulate the obstruction caused by the height of the gate house and great hall. It should be stressed here that this is the maximum projective view; the actual view from this room might have been less extensive as the curtain's angle towers may have blocked views to the east and south of this chamber. But since we do not know the height of these towers they have not been included as obstructions in the calculation of the viewshed. The slice of no visibility seen at approximately 6:00 – 7:00 on the clock-face in Figures 5.17.25-27 represents the obstruction caused by the gate house and hall.

Since the doorway leading from the curtain wall-walk directly into the second-floor of the round tower is known to be at 12m, this elevation has been used for the offset of this floor (this is 6m higher than the elevation offset used as standard for second-floor height within this study).

From the position of the round tower near the top of a hill in a relatively flat landscape, we might expect the projective view from the lord's council chamber to be good, and indeed it is. The projective views are quite far-reaching, similar to those seen in the general view from the round tower's battlements (Figures 5.17.4-6), though reduced by the lower elevation of the second-floor. However, it should be remembered that this floor was at the same level as the wall-walk on the curtain walls, it is entirely possible that the battlements of the curtain could have curtailed the majority of the projective view to within the wall itself. The curtain wall is now gone, so it is not possible to determine the degree of effect it might have had on the view from this room, but we should treat the expansive percentages with caution. Within the 11km radius 15.8%of the hinterland might have been seen, this increases within the 4km radius to 27.8%, and in the immediate distance of the 1km radius nearly half (49.8 %) of the hinterland might have been visible. Barring vegetation, obstruction by the curtain wall, and the built environment of the town, the view from this space may have been both informative and interesting.

Nenagh Castle is currently undergoing much needed restoration, and during the first field visit was inaccessible. Fortuitously, entrance to the ground-floor, first, second and third-floors of the round tower was possible during a second field visit; when Nenagh Castle was on the itinerary of the Castle Studies Group Conference in April, 2009. However, as the photos of the round tower have shown in Figures 5.17.21-24, the tower was very much a construction site at the time, and the group had to carefully pick our way through the workers and their equipment. This researcher apologises that photos of the projective views from these windows are not included in this study. It was an amazing opportunity to be able to see and photograph the windows from the inside, and photographs of the first and second-floor windows can be seen in Figures 5.17.28-37. Unfortunately time constraints on the visit did not allow photographing the views through those windows. However, the brief glimpses available from the round tower windows did suggest that if vegetation and built environment were not a consideration, the projective views could have been extensive.

Lord's private chamber; third-floor of great round tower

The viewshed from this location can be seen in Figures 5.17.38-40. The elevation offset used was 18m, with an azimuth of 360° as there would have been nothing high enough to compete with the round tower during the subject time frame. Again we might expect an extensive projective view of the hinterland and are not disappointed. A comparison between the round tower battlements just 6m above this floor (seen in Figures 5.17.4-6), and these views illustrate how similar the two positions were, varying only slightly by degrees. From the private chamber on the third-floor of the round tower, 19.8%of the hinterland would have been visible within the 11km radius, 38.5% of the 4km radius and 68% of the 1km radius. Barring obstructions from vegetation and built environment, this view could have been informative, interesting and certainly entertaining. Photographs of the third-floor windows can be seen in Figures 5.17.41-48.

Comparison of these projective views with each other and the views from other castles

Over-all the projective views from Nenagh Castle are very good (22%at 11km, 44% at 4km and 79% at 1km). Indeed among the viewsheds in this study generated using the 25m accuracy DEM, they are the highest percentages found within the 11km view (and are second highest within the 4 and 1km views). The castles which came close to Nenagh's excellent percentages are Carrickfergus, Limerick and Trim. The reason is certainly the height of the towers (great towers in the case of Carrickfergus and Trim, and the gate towers at Limerick.)

While the general views at Nenagh were extraordinary, the projective views from the lord's private chamber were excellent, 19.8% at 11km, 38.5% at 4km and 68% at 1km. And as was expected the views from the lord's council chamber were not very different from those of the floor above it (15.8%, 27.8% and 49.8% respectively). Although it seems like the obvious course, it is difficult to compare percentages of visibility between Nenagh and Dundrum, the only other round tower within this study. This is because the DEMs used on the two castles reflect different levels of accuracy (Dundrum was necessarily completed using a DEM accurate to 10m). However, this addresses only the degree of percentage, not the fact that the viewsheds were expansive in both cases. But it is possible to compare other aspects of these round towers. Both towers were built by major magnates; Dundrum by Hugh de Lacy III and Nenagh by Theobald Walter II. These were clearly lords who were aware of the greater architectural trends and fashions in England and on the Continent, and had the resources and imaginations to build adventurous and stylish personal accommodations in Ireland (Hamlin 1977: 3; McNeill 2003: 105). Both round towers had council and private chambers, one above the other. The two towers are similar in the way the windows of the lower chamber (the council chamber) appear to have looked out onto the insides of the battlements. They therefore offered much needed light, but would have offered few projective views that could attract the attention of those within the room. While

Dundrum was actually at least 10m shorter than Nenagh (13m at battlement level), it provided more accommodation space because it contained mural rooms on its private chamber floor. Even though Nenagh shared the uniqueness of a round tower with Dundrum, it was still unique in Ireland as the only castle built with a great hall straddling the gate passage. This suggests that imagination and adventurousness (at least architecturally) was a quality that Theobald II passed on to his successors.

While the great hall at Nenagh did not provide some type of projective view of the lands controlled by the Butlers, it was not unique in this. The early first-floor great hall at Athenry did not allow for much of a projective view, having windows which looked onto curtain walls. Still, this may have been remedied at Athenry by the later construction of the large structure against the south-east curtain wall, the fine windows of which pierced the curtain at ground-level. Even though this building offered only stunted projective views, these views were *exterior* to the castle and sylvan. What we see at Nenagh is the castle's final hall; a replacement for an earlier and most likely timber structure. It seems odd that having gone to the expense of building in stone, no effort was made to allow a view that heralded Butler power and/or authority in the region. However, the main entrance and windows of the great hall at Nenagh did face the round tower, framing and exhibiting that distinctive building, specifically the tower's fore-building. It is possible that the designer of this hall simply made the best of cramped circumstances, while artfully forcing the focus of those gathered and waiting within the hall onto the grand procession of the great man himself, as he left his elegant tower and advanced toward the hall (as had the builder of John de Courcy's great hall at Carrickfergus).

The projective views from the gate house battlements were some of the best within those castles examined at 25m accuracy (18% at 11km, 32% at 4km and 59.5%at 1km). The castles which were comparable were the gate tower at Dunamase (9.2%, 18.2% and 55.6% respectively) and the Dublin Gate at Trim (5.1%, 14.2% and 57% respectively). Of all the approach routes, the most observable from Nenagh's gate house was the road from Cashel. With no obstruction by vegetation or built environment, this road could have been completely visible within 1km of the castle, and retained some degree of visibility for a full 4km. All the other approaches were much less observable, none viewable beyond 500m from the gate house. This may suggest the importance of the Cashel Road to the Anglo-Norman settlement at Nenagh. The main roads approaching Nenagh are from the north, east, south-east, south-west and north-west. It is interesting to note that the windows on the second and third-floors of the round tower face north-east, south-east, south-west and north-west, *roughly* the same directions as the roads. Unfortunately time constraints during the field visit made it impossible to determine phenomenologically the degree to which these openings offered projective views of these approach roads.

Nenagh was obviously a habitation site and administrative centre, occupied for more than 200 years by a single family. After analyzing the viewsheds from the different positions at Nenagh Castle, it would be difficult to argue that projective views were not an important aspect in the choice of site. The amount and quality of the views strongly suggest that visibility was a priority in the selection of location.

Reflective views: viewshed and personal observation Approaches

The Taylor and Skinner 1778 map seen in Figure 5.17.49, shows at least five approach routes to Nenagh. When we compare the Taylor and Skinner map to the OSi road map seen in Figure 5.17.2 it is apparent that these roads have fossilized quite cleanly into modern roads. In Figure 5.17.4 the road from the north enters line-of-sight with the round tower approximately 7km away, after which it comes in and out of line-of-sight and could have had reflective views of the castle several times. The road from Dublin really only enters line-of-sight with the round tower near the 4km point, but within 2.5km of the castle it could have offered a more-or-less continuous reflective view of the tower. The road from Cashel could have had fairly solid reflective views starting at the 4km point. From the Limerick Road (as with the road from the north) the castle may have been visible as far away as 7km, but it is not until a point about 2.5km from the castle that the reflective view would have been frequent, and not until about 500m that the tower could have been continuously within sight. The road from Lough Derg enters line-of-sight about 6km away from the castle, though the reflective views would have been sporadic until about 1.5km from the castle; but by 1km visibility of the castle could have been fairly unbroken.

Even though the general viewshed seen in Figures 5.17.4-6 suggests that the castle was *highly* visible in its environment, in the course of two field visits it was very difficult to get a reflective view of the round tower (from any direction) farther away than about 50m, and the round tower is approximately 6m *higher* now than it was originally. Local industry, built environment and vegetation (in the form of tall trees) almost completely obscured both the tower and the remains of the gate, which could be seen from no closer than approximately 25m. The furthest distance from which the castle was visible was approximately 100m; this was from a small lane where today a modern park provides a picnic bench from which to view the castle. Visibly, in the modern town the castle, as a whole, is almost a non-entity. The positions of the first-sightings are marked in Figures 5.17.6, 5.17.9, 5.17.27, and 5.17.40.

Photographs of the first-sightings of the castle can be seen in Figures 5.17.50-52. It should be stressed that *none* of the photos of first-sightings were taken from an approach route. From every approach attempted, no reflective view of the castle (either gate house or round tower) was possible. In fact, it wasn't until leaving the main routes and approaching on foot that the castle (which is today sandwiched between two city streets) becomes visible. A summary table of the approaches can be seen in Figure 5.17.53.

Community

The exact location of the early town of Nenagh is uncertain. That it was close to the castle is likely, as the reference from 1299 suggests. Here mention is made of several settlers who lived "in the street *below* the castle towards the east" (Calendar of Justiciary Rolls of Ireland 1299 March 16, membrane 19 as referenced by Murphy and O'Conor 2006: 72, italics added). The term 'below' suggests the castle looked down upon the town. The present town surrounds the castle hill, but this reference suggests that there were streets to the east and that the area to the west (or north-west) of the castle was wooded.

There was a medieval church located 500m to the south-east of the castle, but whether this represents the original Anglo-Norman parish church (known to have been in existence between 1217 and 1221) is far from certain. While we might expect the market place to be located somewhere near the parish church, or between it and the castle, there are no real clues of an early market square in the present town layout. A viewshed from the location of the *possible* parish church was generated using a 1.7m elevation offset and an azimuth of 360°. This can be seen in Figure 5.17.54. Although the viewshed suggests that the ground upon which the castle sits is not visible from this vantage point; we must remember that this does not mean the round tower was not visible. Indeed, at 24m in height, it almost certainly could have been visible, barring of course vegetation and built environment. (Also, we should remember that this may not actually represent the location of the parish church during the subject time frame.)

Comparison of these reflective views with each other and the views from other castles

If we were to go solely on the modern reflective view, we might imagine that Nenagh Castle was not a potent image during the study time frame. This should not be assumed to have been the case; were a *cordon sanitaire* kept around the castle, its position on the side of the hill could have shown the castle (and specifically the round tower) off to great advantage. Just because it cannot be appreciated today from the approaches does not mean it was not so originally. Its elevated position on the hill might have made it a prominent sight for community members. An undated sketch by Fleming (Figure 5.17.55) shows both gate towers with the arch and opening of the gate passage between them; it indicates that the gate house might have been an imposing, and perhaps even an intimidating sight centuries after it fell into disuse.

The passage of time, destruction of the castle fabric and limitations caused by modern obstructions made personal observation of Nenagh Castle difficult and actually impossible at times. While personal observation did *not* support the science in this case, the viewsheds suggest that the reflective views of Nenagh Castle (like its projective views) are substantial in amount and quality. The reflective views implied by the viewsheds certainly seem to support the priorities of the builders suggested by the projective views; that visibility was a priority for choosing the site at Nenagh Castle.

5.18 Rinnduin Castle, Co. Roscommon

Note: OSi orthophotographic coordinates were used for the gate, central point of the great hall, the bay, the town gate. These coordinates were obtained online at: http://ims0.osiemaps.ie/website/publicviewer/main.aspx#V1,600588,754314,8.

Introduction

An anomaly in the digital elevation model accurate to 25m surrounding Rinnduin Castle, Co. Roscommon created 'phantom' coast-lines, as seen in the topographical map in Figure 5.18.1. This is probably the result of the castle's close proximity to Lough Ree; a fairly large body of water. The anomaly created massive digital artifacts within the viewsheds, and resulted in views that were inconsistent with personal field observation. Digital elevation data accurate to 10m was purchased from Ordnance Survey Ireland, and all analysis for Rinnduin was done utilizing this more precise data. Figure 5.18.2 represents the 11km topographical radius of Rinnduin Castle completed using the 10m DEM.

Rinnduin Castle, Co. Roscommon was a royal castle. It was built on a peninsula known as Warren Point, which reaches south from the west shoreline of Lough Ree. This is lowland, interspersed with hills; the castle was sited roughly 50m from the shore on raised, but somewhat marshy ground. It is located 13km north of Athlone and 17km south-east of Roscommon. It was constructed of coarse limestone with limestone quoins. The name Rinnduin (Rinn Dúin) means fortified headland, suggesting that the location had been fortified before the arrival of the Anglo-Normans in Ireland (Harbison 1995: 138-9). It has been known by many variations; Rendon, Randun, Randoon, Rindoon, Rinnduin and Rinndown (ibid; Leask 1977: 26; Killanin and Duignan, 1967: 346). An inlet along the east side of the promontory (to the north of the castle site), probably marks the site of a port for water-born traffic across the lough and up and down the Shannon.

The first Anglo-Norman castle at Rinnduin was erected by Geoffrey de Marisco (Justiciar) and Toirrdelbach, son of Ruaidri O'Conchobair *c.* 1227. This was in preparation for the Anglo-Norman invasion of Connacht (Killanin and Duignan, 1967: 346; McNeill 1997: 125; Sweetman 1999: 40). A later hall was added 1290 – 1302, and is probably the structure which extends outside of the curtain wall to the south-west (Sweetman: ibid). Like Athlone (another royal castle in Connacht), it was intended that Rinnduin should become an urban castle, and Anglo-Norman settlers were encouraged to immigrate into the region. The degree to which they did so is unclear. The walled town was located directly north-west of the castle, on the (upper) landward end of the peninsula, separated from the castle by a fosse dug across the narrowest part of the neck of land. Portions of the medieval town wall and its main gate are extant, and the position of the wall can be clearly seen (as a scored red line) on the Ordnance Survey road map shown in Figure 5.18.3. The area between the town wall and castle was larger (25ha), than the town area at Athlone; suggesting great optimism on the part of the Justiciar. However, the nearly constant conflict between the Irish and the Anglo-Norman settlers had an adverse effect on the growth of the town; it languished and eventually disappeared completely.

Seigneurial aspects at Rinnduin include a parish church (located 150m south-west of the great hall), a ferry, a market (Harbison 1995: 141), and a windmill. A foundation of Knights Hospitallers, or Cruciferi was located less than 1km to the north-west but cannot be included as a seigneurial adjunct for Rinnduin, as it pre-dates the castle. Other than the layout of the town (whose main gate opens directly facing the castle), there does not appear to be evidence of a managed landscape.

Location and discussion of the primary components

The primary components are seen in only two extant features at Rinnduin; a (much diminished) gate, and the remains of the early great hall; the castle plan is shown in Figure 5.18.4. There is very little fabric left of the great hall, only the north-west wall and its two windows survive. However, as any other windows within this hall would have been internal to the castle ward, only these north-west facing windows were treated in this study. The lord's chamber component was not treated; there is no evidence of the lord's chamber and since this was a royal castle, it is possible that this component never actually existed at Rinnduin.

General views

The viewshed seen in Figures 5.18.5-7 was generated from the centre-point of the great hall using an elevation offset of 10m, to simulate the height of the battlements atop the first-floor great hall. It was completed using a full 360° azimuth. It should be noted that although this viewshed suggests that all directions could be seen, this full projective view would not necessarily have been visible standing in one spot. The roof of the great hall would have interfered with a full 360° projective view. However, an observer at this location would have had the advantage of the great hall's wall-walk, and could have monitored all directions by walking around the battlements.

With the exception of some shortened views between 12:30 and 2:45, and between 7:00 and 10:00 on the clock-face, there are quite extensive far-distance views. Towards 11:00, 3:00 and 6:00 on the clock-face, the views reach all the way to the 11km radius. Views across the lough could have provided uninterrupted visibility almost to the limit of human visual acuity. Directly north (towards 12:00) there are uninterrupted views across the water to the northern shoreline of the lough, as far as 8km away. The same is true to the east, from 3:00 – 7:00. The best visibility spans from the north-west to the south.

Projective views: viewsheds and personal observation

Gate

There is not much evidence left of the gate at Rinnduin. What we see today is little more than a narrow piercing of the curtain wall with a small opening. The viewshed seen in Figure 5.18.8-10 was generated from the position of the present gate-way, using an elevation offset of 10m to simulate the height of the gate house battlements. An azimuth of 110° - 50° was utilized, to replicate the obstruction caused by the roof of the great hall immediately north-east of the gate house.

Figure 5.18.8 suggests that the far-distance view encompasses a large area of visibility. Figure 5.18.9 shows that the one land route approaching the castle was visible to some degree (with the exception of a small portion of the area directly north-west of the substantial town gate), for nearly 2km. This town gate, as suggested, faced directly towards the castle, and the road appears to have run a straight line through the town to the castle (Thomas 1992: 186). It is clear from the viewshed that the castle gate house offered a substantial projective view. Noticeable however, is the large slice of visibility removed from the view by the interference of the great hall. This covers a full quarter of the near-distance view seen in Figure 5.18.10. Because the section that was not visible includes part of the bay, this would have represented a considerable vulnerability for the castle. This suggests the need for an extension to the gate, possibly supplied by the addition of the barbican. However the complete extent of such a barbican and drawbridge are unclear and a viewshed from the position of those battlements was not attempted. Currently vegetation hinders much of the projective views from the castle, and it was not possible to determine if the viewsheds accurately reflect the possible view. No photographs are available to illustrate these projective views, however, photographs of the exterior and interior of the actual gate can be seen in Figures 5.18.11-12.

Great hall

The early great hall was located just north-east of the gate house. It was a first-floor hall with a twin vaulted basement below. This was a rectangular structure, the long access running south-west to north-east, measuring approximately 25m by 40m externally. There are two windows in the north-west wall (which is also the curtain wall at this point in the castle's perimeter.) The viewshed seen in Figures 5.18.13-15 was created from the position of the centre-point of the great hall, using an elevation offset of 6m and an azimuth of 280°-350°. In Figure 5.18.13 it appears that the far-distance view from this hall was rather limited. However Figures 5.18.14 and 5.18.15 indicate that the middle and near distance views were quite extensive; with the exception of the land surrounding the town wall, this projective view stretches 2km between approximately 9:00 and 10:30 on the clock-face. Much of the town could be seen from the hall, and it seems that the hospital might have been visible as well. Interestingly, the two seigneurial aspects of parish church and the bay were *not* visible from the hall. The whole structure is very damaged and precarious, and will not last long without attention. Photographs of these windows are shown in Figures 5.18.16-20.

Lord's chamber

As mentioned, this component was not evident at Rinnduin and was not treated in this exercise.

Comparison of these projective views with each other and the views from other castles

Rinnduin as a royal castle is in a league with Athlone, Clonmacnoise and Limerick. None of the four royal castles within this study had an evident lord's chamber component. Of these castles, Rinnduin stands out as having the largest possible projective view. One reason for this is the smaller pixel size used, which allowed for more ground elevation data to be expressed, and subsequently exposed in the line-of-sight. The location of the castle, surrounded on three sides by water, also meant there were few topographical obstructions to interfere with visibility, allowing the viewshed to stretch for great distances.

Reflective views: viewshed and personal observation

Approaches

The castle of Rinnduin was such a distant memory by 1778 that it was not included in the Taylor and Skinner 1778 itinerary maps of Ireland. While an Ordnance Survey Six Inch County Map, engraved in 1838 does show Warren Point peninsula and the castle (seen in Figure 5.18.21, no roads are shown approaching the castle. During the study time frame land-based access to the town and from thence to the castle was through the one gate mentioned above. The River Shannon was a well-used waterway, and a ferry running across Lough Ree terminated in the small harbour within 50m of the castle, so approach by water could have been common. Unfortunately such an approach was not attempted for this study.

Today a traveller approaching Rinnduin by land would do so using a country road leading to St John's Wood (purported to have been the location of the wood supply for the castle and town during the subject time frame). The road ends near the site of the medieval hospital, and the castle is approximately 1km from this point. The level of the ground begins a gradual rise at the beginning of the peninsula, and the town was on higher ground than the hospital, town wall or even the castle. This, combined with the height of the town walls (they still stand to 4.5m in some places) could have prevented a reflective view of the castle until a traveller had entered the town gate (700m from the castle). During fieldwork, Rinnduin Castle was approached on foot by following the shoreline (unfortunately modern field boundaries and a bull prevented approaching through the town's main gate). The castle was photographed from this shoreline approach, and it is this location which has been utilized as the first-siting. This photo can be seen in Figure 5.18.22. Coming from the north-west along the shoreline, Rinnduin comes into site at approximately 630km away. The problem with viewing the castle from a distance is topographical; but vegetation (both the trees surrounding it and the vines which cover and thereby camouflage it) also play a large role in concealing the castle to the modern traveller. Personal observation of the approach to

Rinnduin suggests that it is not highly visible in its modern setting; nor is it dominant or intimidating. A summary table of the approaches can be seen in Figure 5.18.23.

Community

The medieval town of Rinnduin was located directly north-west of the castle, between the castle fosse and the town walls. While the location of the parish church is known, at the time of the site visit vegetation made accessing it extremely difficult, and it was not visited. The position of the market place is as yet undiscovered; while it may have been near-by the parish church this is not known for certain. For this reason, the reflective views from these components are not available. However, the castle is easily visible at most positions within the medieval town, and because the town is deserted, obtaining a reflective view from the perspective of the local community was not hampered by built environment. Today the castle seems merely quaint and ineffectual; it is neither intimidating nor impressive (Figure 5.18.24). If we can envision it as it was in the twelfth and thirteenth centuries, with gate house, barbican and drawbridge across the fosse, it may have been, as Sweetman has suggested, considerably more impressive (1999: 59).

Comparison of these reflective views with each other and the views from other castles

Because the gate has disappeared, we might surmise that it was predominantly of timber. With the small amount of castle fabric left, it is simply too difficult to know to what degree the gate might have imposed into the everyday public life of the community. By the same token, we cannot tell if it had any kind of a psychological impact on an observer. The castle itself can be quite well seen from the medieval town, though not so well from the exterior of the town's gates. If the modern field boundary running down the centre of the peninsula *is* a fossilization of the main street, the castle would have been the visual core of the town itself. Although it does not appear today to be a dominating structure, it could certainly have dominated the medieval prospect. The castle ruins are benign in the modern reflective view, somewhat reminiscent of Swords Castle. The reflective view suggests that the castle was sited to do a job: monitor the Shannon and connect Roscommon with the Anglo-Norman lands to the east through the maintenance of the ferry. The driving priority seems simply to facilitate that goal, and the castle was extremely well placed to meet those ends. Overall, Rinnduin appears to have been serviceable, but it definitely does not seem to have been threatening nor intimidating to either the stranger or the local community.

5.19 *Swords Castle, Co. Dublin*

Introduction

Swords Castle is located in Swords, Co. Dublin, approximately 13km north of the city of Dublin. The castle was built on the periphery of a relatively low-elevation coastal strip of ground, just as it begins to rise in elevation (Figure 5.19.1). It was the centre of one of the principle archiepiscopal manors owned by the Archbishop of Dublin (Leask 1914: 259; Killanin and Duignan 1967: 433), and is believed to have been started around 1200 by John Comyn, the first Anglo-Norman Archbishop of Dublin (Adams 1904: 344-345; Leask 1914: Fanning 1975a: 205; Fanning 1975b: 47).

At the time of construction of the castle, Swords was already the location of a monastic settlement founded by St. Columbcille in the sixth century. This monastery and a round tower (which is still standing today) were sited on the side of a hill. The religious house would have had at least a small attendant village close by. The decision to make this the residence of the archbishop, by building what amounts to an ecclesiastical castle across the small river from the church enclosure, insured the growth of the town of Swords and created an urban environment around the castle. Seigneurial aspects at Swords are the town and its market place. The main approach from Dublin, which funnels the traveller directly down the main street to the market place, and behind it the gate of the castle, is the only current indication of a managed landscape at Swords.

The site was undergoing conservation work at the time this field research was conducted, and thanks are gratefully given to the Parks Department of the Fingal County Council for access to the construction site. For health and safety reasons, access to several areas was not permitted, though every effort was made to facilitate this study. Figure 5.19.2 is the OSi map, giving an illustration of the network of modern-day roads surrounding the castle.

Location and discussion of the primary components

The primary components of gates, great hall and lord's chambers are rather straight forward at Swords (Figure 5.19.3). A gate tower is built above and to the side of a fairly simple gate (Sweetman 1991: 127) which is centrally located along the south curtain wall. The great hall is no longer extant (with the exception of a window in its gabled end), but was located along the east curtain wall (Adams 1904: 345-346: Fanning 1975b: 64; Killanin and Duignan 1967: 43; McNeill 1997: 108; Sweetman 1991: 127). The lord's chambers were in the south-east angle of the curtain wall (Fanning 1975a: 209; Fanning 1975b: 64; McNeill 1997: 108). The curtain wall along the south-east and east sides of the castle was formed by the outer walls of several buildings; the chapel, the lord's chambers, and the great hall.

General views

As illustrated by the topographic map, Swords Castle is located in a low lying area next to the River Ward (also known as the Swords River (Adams 1904: 344). The general viewshed seen in Figures 5.19.4-6 was generated from the location of the gate tower battlements (the tallest element in the castle dating to the subject time frame), using an elevation offset of 10m, and a azimuth of 360°. It shows that the projective views, even from the top of the gate tower were not extensive. The most far-reaching views are to the north-west, where the view extends out (very intermittently) to the limits of the 11km radius. There are only slight views to the north-east and south-east, but directly to the east a narrow line-of-sight reaches beyond 5km. The distant south-west and west are almost devoid of projective views due to the sharp rise in elevation towards these directions.

As seen in Figures 5.19.5 and 6, the area within the 1km radius of the castle has the most complete views. But because of the low elevation of the castle site, even within this immediate vicinity the projective views are limited and scattered. The best views are directly south-west (towards the round tower and the site of the original monastery, approximately 500m away), and to the north-west. However, there are no comprehensive projective views.

Projective views: viewsheds and personal observation

Gate

The castle gate faces roughly south-south-west, towards the town. It has a fairly straight projective view of the approach route from Dublin (R132), a road which leads directly to the castle gate before diverting around the castle along the east curtain, and then heading north. Figure 5.19.5 shows that almost uninterrupted visibility of this road extends approximately 500m to the south-west, but only about 200m to the north-east.

Access to the top of the gate tower was not allowed, but the second-floor of the gate tower was accessible, and personal observation towards the town was possible from a window from this vantage point. It revealed that the projective view to the south-west, as suggested by the viewshed, is indeed representative of the phenomenological projective view (which is to say, *not* far-reaching). A photograph, taken from this point can be seen in Figure 5.19.7, and was taken through a window in what was in fact, a construction site. While this is a rather poor photograph, it nonetheless illustrates how completely the modern urban built environment and current vegetation along the street blocked the view. Personal observation from the gate at ground-level (Figure 5.19.8), determined that the modern pedestrian view from the gate was limited to the shops and buildings within the immediate area (within 10m - 500m). According to McNeill, there used to be a wider space in front of the gate, which may reflect the presence of an earlier ditch (2009: pers. com.).

Unfortunately no projective views in any direction except towards town were available from the position of the second-floor of the gate tower, so it was not possible to determine if the viewsheds to the west, north and east reflect reality as well as the view to the south does.

Great hall

The great hall appears to have been located along the east curtain, built on ground slightly higher than the buildings immediately surrounding it. This was a first-floor hall, the long axis of the building running north-west to south-east with one of its gabled ends forming part of a heavily buttressed curtain wall (Leask 1914: 262). The viewshed from this hall is illustrated in Figures 5.19.10-12. It was generated with an elevation offset of 6m (first-floor level) using an azimuth of 50°-170° to simulate that the only external window looked to the east. The viewshed is remarkable in its *lack* of a projective view; even zoomed in to a radius of only 1km (Figure 5.19.12), there was very little to see. The view from this window would have been neither informative about the hinterland, nor visually compelling. Unfortunately it was not possible to obtain a photograph from this window.

Leask (1914: 262) informs us that there are two smaller openings located directly below the above mentioned window, but construction debris and vegetation completely obscured any evidence of them. The observable window is fairly large, estimated to be approximately 1m wide and at least 1.5m in height (Figure 5.19.13). It was clearly a grand window, with rare red sandstone mullions (Adams 1904: 346), some of which remain intact. The projected views from this single window, while not far-reaching, would certainly have been framed nicely. However, it is very likely that vegetation in the form of tree cover would have prevented much of a view. There is no trace above ground of any other wall for this structure, so it is impossible to know the amount or dimensions of other windows or doors. In any case, their projective views would have been interior to the castle ward.

Lord's chamber

The lord's chambers were located in the irregular angle of the south and east curtain walls (see Figure 5.19.3). The building was rectangular, the long axis running north-east to south-west. Like the hall, this structure was built utilizing the curtain wall to the south and east. It had at least a ground and first-floor level, and may have gone even higher. Excavation of the ground-floor in 1971 uncovered an extensive portion of a medieval tile pavement in the south end of the building; believed to have paved a private oratory, presumably for the archbishop (Fanning 1975a: 209; Fanning 1975b: 47 and 62). Extant at the first-floor level are four windows; one in the south wall, and three along the east wall. The projected views from the first-floor of the lord's chambers were not expected to be expansive. The wall of a tower just to the north, and the east wall of the chapel just to the south-west of the lord's chambers would have curtailed the azimuth and truncated the width of the available view. The viewshed (Figures 5.19.14-16), which was generated with an elevation offset of 6m, using an azimuth of 80°-190° to simulate the direction of view, illustrates that the projected views were as limited as those from the great hall. The view from these windows would have been of the immediate, local area, looking directly down upon the Dublin to Drogheda road (the approach from the north).

The windows were fairly generous in size, the south facing window measures 56cm in width by 1.5m in height. The east windows vary (note that all measurements are approximate); the most southerly being 1.12m in width by 95cm in height, the central being 74cm in width by 1.85m in height and the most northerly (which had a window seat), being 32cm in width by 91cm in height. Some of these windows can be seen in Figure 5.19.17. There are no existing doors. As with the window in the great hall, these windows were not safely accessible for photographing, so unfortunately no photos are available to illustrate the reality of the modern projective views.

Comparison of these projective views with each other and the views from other castles

Several stone castles in this study were located on ground that was owned by and had been utilized by the church; i.e. Trim, Dunamase and Clonmacnoise. Whether this was because these prime locations had already been cleared and were relatively easy to obtain, or because the use of these locations was related to a desire to appropriate the authority and prestige already inherent in the sites is unknown, perhaps it was a bit of both. With this in mind, it would seem that building the archbishop's residence adjacent to the church and the monastery at Swords would have been an obvious decision, endowing the site with an extra portion of importance by 'piggy-backing' on an already significant setting. However, an alternate location was chosen. Reasons for this are unknown. It may be that the monastery was not held by the archbishop at the time, although Leask (1914: 257-8) suggests that it probably *was* in the archbishop's hands. While no documentary evidence remains to help determine whether this is true or not, in either case, archbishop Comyn chose to build his manor and castle on the low ground next to the river.

It is important to remember that Swords Castle was at the centre of a large and lucrative holding which formed 'no inconsiderable part of the archbishop's revenue' (Adams 1904: 344). The church at Swords (of which there were several at the time in question) had been named a prebend in 1211 in St. Patrick's Cathedral in Dublin, and was termed the 'Golden Prebend of Swords' (Leask 1914: 259; Killanin and Duignan 1967: 433); a substantial hint at the economic importance of the area. As the castle was actually the hub of a prosperous and busy manor, it made sense to locate it on low ground next to the ready and constant water supply of the River Ward. Leask suggests that the small projecting tower in the west curtain wall may have housed a well or even a spring (1914: 262), if so, the location of such would have added incentive to build the castle at this spot. Regardless, this lower position would have been much easier for carts, men and animals to access than a position atop the hill would have afforded.

The window of the great hall at Swords could be compared with those at Carrickfergus and at Trim, for each of these castles had great hall windows which pierced the curtain walls at first-floor height (although the great hall at Carrickfergus was a *low* first-floor). As for quality and opulence, the openings in all three castles seem to have been quite large and elaborately set; Carrickfergus was marked with fine Cultra stone frames and window seats, Trim having four-lights in each window and yellow sandstone window seats, and Swords with its fine red sandstone mullions.

However when projective views are considered, the similarities turn to contrasts. The low first-floor hall at Carrickfergus had two windows with sweeping views of Belfast Lough, and a direct line-of-sight up the sea lanes leading to the open sea. Trim had four hall windows facing the River Boyne, an abbey funded by the lord and an extensive deerpark. Both of these lordly castles seem to have great hall windows that enclosed landscapes that exemplified the power, influence and wealth of their builders. In comparison, Swords' *one* extant great hall window looked out directly onto the dusty road which ran alongside the curtain wall, and beyond that to the fronts of the houses built along that road. This window was undoubtedly attractive, would have been functional in supplying light, but it certainly does not appear to have been a 'stage-set' for a display of wealth, influence, power or strength.

One reason for this contrast may be that the power in the church did not come from overt political domination; it came subtly, through wealth, tradition and spiritual (though still political) prestige. It may not have been necessary to display the trappings of wealth and power when wealth and power were implied so clearly by the very nature of the place…the residence of the archbishop of Dublin. He did not merely have a grant of land from the king, control over the physical lives of his tenants and the right to harass the native population at will, he had the backing of God and the direct authority to act in God's name. Quite simply, the archbishop might not have felt a need for a *show* of power and influence, because everyone already *knew* he had power and influence.

That comfort and accommodation were important to the main resident seems obvious from the fine windows and large fireplace in the archbishop's personal chambers (the fireplace can be seen in the centre of the photo in Figure 5.19.17. Indeed McNeill suggests that (while confirmation through excavation would be helpful) the domestic provision within the castle as a whole 'hints' at being elaborate (1997: 166). The south-east placement of the archbishop's chamber windows did not look down on the market directly (as is the case at Trim). This might suggest that commerce, the exchange of goods, or the accumulation of wealth was not upper most in the mind of the archbishop himself; perhaps he had higher and less worldly concerns. However, like Carrickfergus and Trim the gate chambers which would have been the purview of the constable, whose responsibility it would have been to oversee and regulate the exchange of goods, *did* look directly down upon the market (McNeill 1949: 175), thus insuring that the commerce of the manor and its attendant town ran smoothly. And, like Carrickfergus and Trim, the parish church would have been visible from the gate tower, suggesting it was at least observed by (if not directly under the protection of) the castle.

All things considered, it would not appear that the provision of extensive projective views played a role in the choice of site for Swords Castle. John Comyn's priorities seem to have been first and foremost to provide for the necessary requirements of a working manorial centre, and then to make his own accommodations as comfortable as possible.

Reflective views: viewshed and personal observation

In Figures 5.19.4 and 5.19.5 the effect of the low-lying position of the castle on the projective views is apparent. These same illustrations also give us an indication of just how undistinguished the castle was in regards to reflective views. Basically, unless an individual was actually on the road to the castle and within 600m of it, the castle might not have been readily visible.

Approaches

The 1778 map of the Swords' area drawn by Taylor and Skinner (1969: 1) which is shown in Figure 5.19.18 shows four main arteries leading into Swords town, coming from each of the compass points. These roads are the approach from the north: the Drogheda road (south-bound R132, known as North Street within the town limits); the approach from the east: coming from Howth Port (west-bound R125, known as Malahide Road within the town limits); the approach from the south: the Dublin road (north-bound R132 known as Main Street); and the approach from the west: the Ratoath or Trim road (east-bound R125, known as Bridge Street within the town). Two other approach roads can be seen entering the town, one known as Church Road, which adjoins the west approach along Bridge Street approximately 200m from the castle gate, and one leading directly to the south gate of the castle from the east known as Seatown Road. These approaches can be traced more clearly in the modern map in Figure 5.19.2.

The viewshed in Figure 5.19.6 suggests that *sporadic* reflective views might be available coming from the north and east within 0.5km or so, and that from the south and west, the castle might be clearly visible within that last 0.5km. The reality is that current built environment and vegetation quite effectively block views of the castle along both approaches from the west. This is also true of the approaches from the north and the east. *None* of these roads allow for reflective views from farther than approximately 200m.

The best approach from which to get a reflective view of the castle is the north-bound road coming from Dublin (Main Street). This road is quite straight, with a slight bend approximately 600m from the gate tower on the south side of the castle. At this point the castle comes into view. The west-bound road that ends at Main Street

(Malahide Road) has its first views of the castle only at Main Street, just past this bend. Unfortunately vegetation in the form of urban landscaping hinders a clear view, and unless a viewer actually *knew* the castle was at the end of the street, it might be missed altogether. In general, reflective views approaching Swords Castle are nearly non-existent. Photographs from the various approaches were lost because of equipment failure, but a summary of reflective views from the approaches can be seen in Figure 5.19.19.

Community

The medieval town of Swords apparently lay before the castle, spreading southwards from the curtain walls towards the holy well. The market would probably have been located at the intersection of the main roads into town, directly in front of the castle's gate tower. The fact that the castle gate was backdrop to the market is a reminder that it was important to the life of the manor and the occupants of the castle. The photograph already introduced in Figure 5.19.8 of the gate house exterior, was taken from the approximate location of the market. From this position the castle would have been *more* than obvious in the community; it would have almost been palpable. It would have loomed large in every detail except menace, for the simple gate and the presence of the chapel attached to it seem almost gentle.

The parish church would have been located on the site of the monastery, near the round tower. The area is slightly higher ground than the castle, and logic suggested that, from this vantage point, at least during the winter months when the trees were bare, there should have been reflective views of the castle. However, field work was conducted in full summer, and vegetation hid the castle from the position of the round tower. Since it could not be seen (and hence photographed) from this vantage point, a viewshed (Figure 5.19.20) was generated from the location of the round tower. An elevation offset of 1.7m, and an azimuth of 360° were used. While the *ground* upon which the castle gate sits may not be within line-of-sight, the viewshed does suggest that the castle's gate tower might easily have been seen from the round tower.

Comparison of these reflective views with each other and the views from other castles

In general, the reflective views available along the approaches to Swords Castle suggest that the castle did not broadcast its location to travellers. They came upon it as the road unfolded, and could not really glimpse it a moment before-hand. It was placed conspicuously in the medieval town, and could not be missed if the viewer were passing directly *through* the town. But it was not built in a location that could be spotted from any great distance away. This is not for lack of a higher elevation, for across the river only 500m or so to the south-west, the ground begins to rise considerably. Either the castle was sited for purely logistical reasons (ease of access and proximity to a fresh water supply), or its builders simply did not care whether their presence was noted by distant travellers. Perhaps both possibilities apply, but the result of choosing this location meant that the castle would have been almost ambiguous in its larger environment.

For the local community however, a different story can be seen. The castle, as mentioned sits at the junction of four approaches, blatantly the centre of attention in the town. However, it does not 'lord' over the town, but seems to be an integral working part of it...the reason *for* the town, but also a 'good neighbour' (so to speak), *to* the town. There was certainly nothing obscure about where the market was located; it had 'pride of place' at the gate of the castle, much nearer the castle than even the markets at Trim or Carrickfergus had been. Even so the gate tower here did not have the formidable, looming presence over the market that the other castle gates had, despite its closer proximity. Actually in contrast to Trim, Carrickfergus and Dunamase, the gate tower is more like a block than a tower, with the front of the castle dominated not by the gate, but by the chapel attached to it. There is little about the gate, or indeed Swords Castle in general, that is formidable. There is a benign quality; a feeling of docility that is perhaps aided by the facts that there are no mural towers keeping watch on the town or the hinterland, and that the simple entrance is guarded by a straightforward gate that offered no barbican jutting out into the community space.

The reflective views (obtained by both GIS viewsheds and personal observation) do not imply that long-range visibility of the castle played any great role in its placement, although the prominent location at the junction of two main roads does suggest the castle was intended to be easily accessed by travellers. By and large, the priorities of the builders suggested by the reflective views support those suggested by the projective views; that this castle was both a comfortable residence and a working manorial centre; any other priorities were subordinate to those concerns.

5.20 Trim Castle, Co. Meath

Note: Many thanks are extended to OPW in general, but especially to Kevin O'Brien. OSi orthophotographic coordinates were used for all prime components, these were found online at:
http://ims0.osiemaps.ie/website/publicviewer/main.aspx#V1,680195,756691,7.

Introduction

Trim Castle, Co. Meath was built in a landscape of fairly flat terrain, bordered by mountainous and boggy areas (Hennessey, 2004: 1). The entire region is fertile, being drained by the Boyne and Liffey river systems (Givens 2008: 215), and is suitable for both pastoral and arable farming (Hennessey, 2004: 1). The topographical map of the 11km area surrounding the castle can be seen in Figure 5.20.1. The castle is located approximately 40km north-west of Dublin, 9km west of the Hill of Tara, and 35km south-west (upstream) of Drogheda on the River Boyne. Trim was already an established ecclesiastical and political centre when the Anglo-Normans arrived in Ireland. The castle was built on land belonging to the church (Potterton, 2005; 71), and was founded in 1172 by Hugh de Lacy, Lord of Weobley (Mullally 2002: 135).

Construction of the castle in stone began *c.* 1175 (O'Brien and Fenlon 2002: 12). The castle in general was built in five main phases, and three phases of construction of the great tower have been dated by dendrochronology, and are detailed by Sweetman (1998). The first stone phase (1175-1186) saw the erection of the first three storeys of the unusually shaped great tower. The exterior of the tower has twenty individual surfaces, and was built in the shape of a Greek cross; a large main square tower having smaller towers attached to each side of the main block, the whole rising as part of a single plan (McNeill 1990: 314). During this initial building phase (*c.* 1180), the core of the West Gate was erected on the site of an earlier wooden gate (O'Brien and Fenlon 2002: 10, 62). This initial spate of construction was halted around 1186 when Hugh de Lacy was murdered while working on a new castle in Durrow. Hugh's son and his heir to Trim Castle, Walter de Lacy, was still a minor at his father's death. The castle was administered by the crown, with royal officials administering the De Lacy estates during part of this period (O'Brien and Fenlon 2002: 17). While royal works on baronial castles was not common in England, the king took a different view of spending money on castles in Ireland, in hopes of retaining them in the long run (McNeill 1990: 331).

The second building phase of the great tower (starting approximately 1195/97), was the work of Walter de Lacy. This stage saw the addition of a second and third-floor with a gallery in the great tower block. An internal dividing wall was inserted in the first-floor for support of the heavy double roof. Cap houses were built atop each angle of the tower on the roof level. Another structure which may have been erected during this phase includes the north corner tower (the Magdalene Tower) at the angle of the west and east curtain walls. O'Brien suggests that this impressive structure should be considered "stand alone", and was probably built before the curtain walls were constructed in stone (O'Brien, 2008: pers. com.). The north-east end of the east curtain wall and the mural towers along the east curtain were also constructed during this period.

The third building phase, which started *c.* 1202/04, was almost certainly the work of Walter de Lacy. This included a heightening of the walls of the great tower and its side towers, the insertion of mural passages between the cap-houses, and the construction of a pair of lofty arches on the third-floor to carry the double roof of the great tower (Sweetman 1998: 224). In the castle as a whole, it is possible that the south-east and south-west angle towers, as well as the D-shaped projecting towers along the south curtain (which have basic non-plunging arrow loops) were started during this period (McNeill 1990: 333), though this may have been ongoing construction between the second and third phases. Although it is not clear when the single-towered Dublin Gate was added to the south curtain wall (1990: 325), it is possible that the gate tower and barbican were built during this phase. One reason to place this gate tower within this building period is the fact that the plunging arrow-loops on the ground-floor were a military development of the early 13th century (1990: 333).

The fourth period of construction at Trim Castle took place after the siege by William Marshal and Walter de Lacy (against Walter's own castle) in 1224. Trim was at this time being held by Walter's brothers, Hugh (the son) and William, in defiance of the crown. Walter, eager to maintain his relationship with the King, nevertheless must have been grieved about damaging his own home. This may account for the length of the siege which lasted seven weeks, although McNeill suggests it is difficult to see how the great tower could have held out for longer than a week against the well equipped and determined force besieging it (McNeill 1990: 331). When the siege was over, repairs of damaged elements as well as a general upgrading were carried out. These works included the reconstruction of the West Gate and the west curtain wall. This work can be traced through the change in the lighter grey colour of stone towards the top of the wall, as well as by the insertion of plunging arrow-loops where straight-forward loops had once been (O'Brien 2008: pers. com.). If the Dublin Gate tower and barbican had not already been built by this time, they would certainly have been done after the siege.

The fifth and final phase (~1250 and onwards) saw more or less a refinement of the great tower as a whole. This was done by the addition of a massive battered plinth to its base. Outside the great tower, a quarry was opened in the castle ward just east of the east side-tower. Stone from this quarry was used to construct several new structures within the castle. These included a mantlet wall and an extensive fore-building with administrative offices built adjacent to the great tower's main entrance (Sweetman 1998: 227). At sometime around 1270 (O'Brien 2008: pers. com.) stone from this quarry was used to build the great hall (or Red Hall). This was an impressive ground-floor structure, 25m in length (McNeill 1990: 325), built just south-east of the north corner tower. To facilitate this, the north-east end of the east curtain wall was heightened, so that it could be used as the outer wall of the hall. Its lower arrow-loops were blocked up, and the upper portion of the curtain wall was pierced initially by four massive windows looking down on the Boyne River and the monastery beyond; the great hall was extended to the south-east sometime around 1370 and a fifth grand window inserted. At the same time as the construction of the great hall, the north corner tower was extensively enlarged and refitted to become an impressive suite of lodgings; new embrasures, fireplaces, and windows with window seats were added. The result was a spacious and commodious accommodation (Sweetman, 1998: 227).

Trim was an urban castle, with a community of transplanted settlers. Meath in general was highly productive agriculturally, and the Trim market drew much commerce to the settlement (Hennessy 2004: 3). Trim quickly became a successful town; and Hennessey suggests that Meath in general was a "prize", and that whoever controlled it had great power (ibid). A plan of Trim town as it would have been *c.* AD 1200 - 1600 is illustrated in Figure 5.20.2.

A number of seigneurial markers are relatively obvious at Trim, such as the planned settlement of Trim town itself and its successful market place and parish church. There were also several religious houses established by the lords in and around Trim town. The first of these was the Augustinian abbey of St. Mary's, located on a gentle rise across the Boyne and directly north-east of the castle. This abbey may have first been established by St. Malachy *c.* 1145, if so... it was *re*-established by Hugh (the father) before his death in 1186 (Hennessey, 2004: 10). During the time of Geoffrey de Geneville and Matilda de Lacy (granddaughter of Walter), two houses of friars were founded. These included a Franciscan friary founded in 1260, located just north-east of the castle's West Gate, opposite the market place. Shortly thereafter (in 1263), a Dominican friary was established on the northern edge of the town, outside the town walls.

Seigneurial signs that are *not* plainly visible at Trim are hinted at by both later documentary sources and the current landscape. These include a mention of the "Mills of Trim" in the *Calendar of patent and close rolls of chancery in Ireland* 1364-7, and reference to two mills (locations unknown) *c.* 1425 at Trim (Hennessey 2004: 12). A large, round dome-shaped granite millstone was recently found in what would have originally been the castle moat (Potterton, 2005: 162). According to O'Brien and Fenlon (2002: 66) the castle also had an orchard and a dovecote, possibly located outside the castle's walls. Parks are often difficult to trace in the modern landscape, as the value of the land precludes them from remaining open and obvious. The earliest documentary reference to a "parker" (manager of a park) at Trim is from 1388 (Potterton 2005: 127), but the references then become plentiful. Hennessey gives no fewer than five documentary references referring to a park, specifically "a King's park" before 1550 (2004:12). The current backdrop across the Boyne to the north of Trim Castle includes a relatively large plot of land (approximately 80 acres) known as the 'Porchfield', between the Sheep-gate in the town wall and Newtown Trim. Although not labelled, this area can be seen in the Down Survey plan of 1655 (Figure 5.20.3) as the open region directly below (actually north of) the Boyne, between Newtown Trim on the left and Trim town on the right. As far as documentary evidence can show, this area has remained unoccupied and open since the initial town charter. The east side of the great tower at Trim, as well as the grand windows of the great hall look directly onto this area. Although it is not referred to as a deerpark, nor even as a park, it has been suggested that what is known today as the Porchfield may have intentionally been left open to allow an unhindered view of St. Mary's abbey (and eventually of St. Mary's cathedral), or may have been the site of a deerpark (O'Brien, 2008: pers. com.). This idea was furthered by excavations at the castle which uncovered some of the earliest evidence for fallow deer (a species brought in by the Anglo-Normans themselves) in Ireland (McCormick 1999: 360-1).

There is strong documentary and circumstantial evidence for a structured approach at Trim. The Down Survey plan of 1655 in Figure 5.20.3 illustrates how the road from Newtown Trim skirted the Porchfield area, entering the town from the north. It would then have passed the parish church and St. Mary's Abbey before approaching the bridge over the Boyne and the castle just beyond. All along this distance there would have been strong reflective views of the castle. The impressive and obviously structured approach from the south would have brought travellers around the perimeter of the castle, from east to west, to enter the town through the *town's* Dublin Gate. Throughout this whole distance, the *castle's* elegant Dublin Gate (with the massive great tower of Trim looming behind it), would have been visible. The OSi map of the modern roads surrounding Trim Castle can be seen in Figure 5.20.4.

Location and discussion of the primary components

The primary components of gates, great hall and lord's chambers are to be found in five principle features at Trim; two gates (the West Gate and the Dublin Gate), a great hall, and both a council chamber and a suite of private chambers (found within the great tower). These are illustrated in the castle plan in Figure 5.20.5. There are actually three gates at Trim: the West Gate which faces the town and sits on the north-west side of the castle; the Dublin Gate which faces south-south-west and is located outside the town walls; and the River Gate which is south of the great hall, leading onto the Boyne (however, this was essentially a service entrance and was not examined here). While the gates are uncomplicated at Trim, the locations of great hall and lord's chambers are not so straightforward. The extant great hall in the northern end of the castle ward was, as mentioned, built *c.* 1270. Exactly where the great hall was sited *before*

1270 is uncertain, and is complicated by the continued erection of the tower during the first fifty years of the castle's existence.

There is a good possibility that had Hugh de Lacy (the father) not been murdered in 1186, the great tower would have been constructed in one effort. As it was, at his death work on the great tower halted and the finished portions (ground and first-floor in the central block, and three complete floors in each of the side towers) were utilized in a 'stop-gap' manner until construction could continue.) At least four locations for the hall during this time period have been suggested, these include the following:

1. The first-floor of the great tower (considered the likeliest spot for a temporary hall before the second building phase) (O'Brien and Fenlon 2002: 30, 36). While this location *is* possible it could only have been a short-term measure, for the scale upon which Hugh de Lacy operated was grand and his son Walter exhibited an equal taste for opulence. It might have served as the hall for a brief period, but could never have been intended as the final great hall.

2. O'Brien and Fenlon have further suggested that a great hall in the east side of the second-floor of the great tower might have replaced the temporary hall in the floor below it *after* the second building phase (2002: 30, 40). This is also an unlikely location for the great hall of Trim. The room itself is not particularly large, but this does not preclude its use as a great hall (it is in fact, fairly similar in size to the great hall that De Courcy built at Carrickfergus.) However, access to it was not straightforward (there would not have been direct communication between this location, the chapel or the north-east stairs, and from thence to the entry of the great tower). Access would have required crossing the first-floor to reach the south-west spiral stair, ascending that stair and passing through another room before reaching this room. While this could be seen as a deliberate funnelling of guests through parts of the castle meant to impress, this is questioned by the fact that the south-west spiral stair was the *smaller* of the two tower stairways and not the best choice if the motive was to make an impression upon the visitor. There also does not appear to have been provision for kitchen facilities within the great tower, and food coming from an outside kitchen through this meandering route would have arrived cold.

3. Possible locations *outside* the great tower include the building directly north-east of the tower (O'Brien 2008: pers. com.). This is another unlikely location; for the structure is very small, and although it had ample windows, these are at sub-ground-floor level (it simply does not fit the convention for a great hall of the time period).

4. Lastly, the site of the extant hall could be considered as a possible location of an earlier timbered hall structure, and while this *is* likely, it is unfortunately, not provable. While all of these suggestions are possible, they are either not provable or unconvincing. Because of this, only the extant hall was treated in this study.

The exact setting of the lord's chambers has been equally controversial, again complicated by the continued construction of the tower. And, like the great hall, the lord's chambers may have conceivably been in several *different* locations during the course of construction. However, McNeill, O'Brien and Fenlon all seem to agree that at the conclusion of the erection of the great tower, the rooms of the second-floor of the great tower were used as private chambers (McNeill 1990: 323; O'Brien and Fenlon 2002: 40-42), while the expansive and grand third-floor was the location of the lord's ceremonial chamber (McNeill 1997: 23; O'Brien and Fenlon 2002: 43). With this in mind, the second- floor of the great tower was treated as the lord's private chambers, and the third-floor and gallery as the lord's council chamber.

General views

Because the great tower at Trim sits on a very slight rise in a relatively flat area, it can command excellent projective views for long distances. The viewshed used to illustrate the general views from Trim Castle was created using the location of the battlements of the great tower, and can be seen in Figures 5.20.6 - 8. This viewshed was generated using an elevation offset of 24m and an azimuth of 360°. In the topographical map in Figure 5.20.1 the edges of the tiles can clearly be seen as discrete lines running from roughly 12:30 to 6:30, from 11:30 to 1:30 and from 4:00 to 9:00 on the clock-face. We would expect these lines to cause artifacts in the viewsheds, and indeed they do; these artifacts are most noticeable within Figures 5.20.6 and 5.20.7.

While there are some far-reaching views towards 12:00, between 2:00 and 6:00 and again at approximately 11:00 on the clock-face at the 11 - 9km radii, these are patchy and very limited. Within the 11km radius, the percentage of visibility was 12.24%. There is continuous (although scattered) visibility throughout the entire 360° within 4km of the castle. Within this view the percentage of visibility is 29.9%, suggesting that the great tower had line-of-sight with almost one-third of its 4km radius. However, the most impressive views are within the 1km radius where 73.5% of the immediate area could have seen, or have been seen from the battlements of the great tower. There is an area to the north-east and east (between 12:00 and 3:00), which has almost uninterrupted visibility for more than 500m. Access to the battlements was provided as part of the regular tour offered by the Office of Public Works at Trim, but Kevin O'Brien himself kindly provided valuable information and a lengthy opportunity for taking photographs. His efforts and knowledge are much appreciated. The photographs taken from the battlements of the tower support the viewshed and can be seen in Figures 5.20.9-13.

Projective views: viewsheds and personal observation
Gates
West Gate

The West Gate at Trim Castle actually faces north-west. It consists of a gate passage running through a rectangular building (approximately 12m in length by 8m in width), whose shape changes to a polygon once it rises above the gate passage. This is a style that had been used twice at the De Lacy castle at Ludlow, and was reused at Trim (McNeill 1997: 24). The base of the West Gate was elevated (approximately 3m), above the street level of the town, and the battlements of the West Gate (which had two floors above the gate passage) rose well above that. The stumps of two walls projecting outwards from the front of the gate passage suggest that there may have been a barbican in front of the gate house (McNeill 1990: 327-8; 1997: 24; O'Brien and Fenlon 2002: 62).

This was the public entrance to the *caput* of the liberty of Meath, and it was expected that this gate would have good views of the south-west part of the town as well as the market to the north-west. The viewshed from this gate illustrates that this was indeed the case. The viewshed in Figures 5.20.14-16 was generated using an elevation offset of 10m (battlement height), and an azimuth of 154°–124° to simulate the obstruction the great tower would have caused in the projective view. Figure 5.20.16 shows that this gate was well placed to observe the entire medieval town, especially that part of the town west of the Boyne River. What was not expected were the views of a portion of the approach along Dublin Road (R154). The percentage of hinterland observable from this location breaks down as 5.38% within the 11km radius, 13.74% within the 4km radius, and 46.55 within the 1km radius. To the north the low hills create pockets of no visibility (this is even more true to the north-east). Directly east there are only patches of visibility, caused by the hill north of the Boyne. To the south-east the great tower entirely removes a significant portion of the hinterland between 4:00 and 5:00 on the clock-face from view, and to the south and south-west (between 5:00 and 8:00 on the clock-face), the rising hills do the same. What the viewshed shows is that though this gate was well placed to observe the town and a portion of the Dublin Road, it was not in a position to actively observe much more. Access to the top of the West Gate was not available, but photographs showing the exterior and interior of the gate can be seen in Figures 5.20.17 and 18.

Dublin Gate

The Dublin Gate faces south-south-west. It is a graceful structure consisting of a circular tower of three floors, pierced through by a narrow gate passage. This passage continues exterior of the castle, beyond the tower, and extends another 10m over the moat before culminating in a three storey rectangular barbican. The whole structure measures approximately 23m in length in its present state. The viewshed seen in Figures 5.20.19-21 was generated from the position of the battlements of the Dublin Gate tower; an elevation offset of 13m was used, with an azimuth of 335°–310° to simulate the obstruction caused to the north-west by the great tower. The great tower had a large impact upon visibility to the north-west from this gate (a significant slice of the hinterland is simply taken out of observation). The percentage of hinterland observable from this gate breaks down as 5.16% within the 11km radius, 14.22% within the 4km radius, and 57% within the 1km radius.

Since the Dublin Gate was poised specifically to greet high profile visitors to the castle arriving from Dublin, it might be expected that the gate would have good views towards this direction. Indeed the viewshed suggests that this gate was very well placed to observe traffic coming from Dublin. Sporadic views of approaching travellers could have been possible for a distance of 5km, though the density of the line-of-sight increases drastically within 2km of the castle, and travellers could have been completely within sight of the Dublin Gate during the last 1km of their journey. What was *not* expected was the excellent view this gate tower had of the Navan Road (R161) as well. Travellers approaching along this route from the north-east could have been visible within 1km of the gate tower.

The Dublin Gate opens to the south, and although it sits on relatively flat ground, it faces the up-slope of the hill immediately south of the castle. This slope results in a significant area of no visibility directly south of the gate. This would have kept travellers coming along the southern Kilcock Road (R158) out of sight until they crested that hill, less than 500m from the gate. Those approaching along the Kinnegad Road (R161) from the south-west could have been seen sporadically within 1.5km of this gate. Lastly, the immediate approach to the *town's* Dublin Gate would have been visible from this castle gate tower no matter which southerly road the visitor had taken. Access was not available to the top of the gate house so there are no photographs to illustrate these views, but photographs of the exterior and interior of the gate itself can be seen in Figures 5.20.22 and 5.20.23.

Great hall

The viewshed seen in Figures 5.20.24–26 was generated from the ground-floor of the great hall. An elevation offset of 1.7m was used, with an azimuth of 300°-120° to simulate the obstruction caused in the projective view by the great tower. We would expect the projective view from this hall to focus on the wealth and power of the lord, and the view does not disappoint. There are no appreciable far-reaching views, and the percentages of the views are not high in any of the distance indexes: 0.5% within the 11km radius, 1.07% within the 4km radius, and 9.57% in the 1km radius. This low percentage of visibility is due to the facts that the hall had windows exterior to the castle on only one side, and the up-slope of the hill north-east of the Boyne greatly curtailed all further views to that direction. However, it is *what* was visible that mattered; the view to the north-west took in a portion of the town (including the seigneurial bridge), the windows directly faced St. Mary's Abbey (a De Lacy foundation), and the prospect to the east allowed for sweeping views of much of the Porchfield (the possible park) area. The added bonus was that while these views were certainly impressive and powerful statements about the De Lacy family, they must also have been very aesthetic.

The great hall sits almost at the northern apex of the castle, just south-east of the Magdalene Tower. The exterior windows face the north-east. Originally the hall was almost square in shape (approximately 25m in length by 22m in width), with the slightly longer axis running north-west to south-east. After the subject time frame the south end of the building was later extended by 10m, making the hall rectangular. It was a ground-floor hall (the fashion of the time), but because the ground sloped off rapidly to the north-east towards the Boyne, the windows would have been seen at first-floor level from the exterior of the castle. Photographs of the exterior and interior of the great hall can be seen in Figures 5.20.27 and 5.20.28.

Built in the 1270s, it originally had four large windows piercing the east curtain wall; approximately 1.25m to 2m in width and between 2.4 to 2.5m in height. Each window would have consisted of four narrow lights with simple heads (either round or semi-elliptical), with central mullions (O'Brien 2008: pers. com.). There would most likely have been matching windows on the south-west wall of the hall (although during the later extension of the hall, these inward looking windows were updated to reflect the change in window styles). These interior windows would have framed the great tower which would have been an excellent reminder of the power of the lord. However, as mentioned above, the projective view from such windows would have been entirely of the castle ward. For this reason the view to the south-west was not factored into the viewshed seen in Figures 5.20.24–5.20.26 as.

Photos of the windows can be found in Figures 5.20.29-36. Such large windows would have been inviting to stand or sit at, and the evidence that they were encouraged to do so can be seen in the remains of the sandstone window seat, still in place at the base of the second window from the right (see Figure 5.20.37). As seen in the photos above, the four north-east windows have been substantially blocked, making photographs of the projective view from them impossible. A partial view taken from the southern end of the hall (looking east) can be seen in Figure 5.20.38 and certainly supports the viewsheds.

Lord's chambers

Lord's council chamber; third-floor of the great tower

The viewshed from this vantage point can be seen in Figures 5.20.39-41, which was generated using an elevation offset of 20m, the estimated height of the gallery around the lord's council chamber. An azimuth of 360° was used to represent the fact that at least one window faced each compass point. The percentages are quite impressive for a lord's council chamber, 10% within the 11km radius, 26% within the 4km radius and 68% in the 1km radius. But it must be stressed that these percentages are a *maximum* possible projective view, and would not have represented the projective view as seen by council members while 'on the job'. Also, not all directions would have been available for casual observation as the windows in the south-east corner were not served by either gallery, nor was the north-west facing window.

The third-floor of the great tower is a splendid room, and makes a convincing ceremonial chamber for the lord's of Trim. It consists of one large square-shaped room with a grand fireplace at the high-end in the south wall. It is accessed by both of the great tower's spiral stairways; the smaller private south-west stairway opening at the high-end and the more public north-east stair opening at the lower-end of the chamber. There are no doors coming off this room to the third-floor rooms of the side-towers. Instead, the side-towers are accessed from two unconnected mural galleries, one of which is reached by way of the public stairway. It runs the length of the north wall, past the stairway and around the north-east angle to the east side-tower (see Figure 5.20.42). The other gallery is accessed by the private spiral stairs on the south-west. It connects the private chambers on the second-floor to the south and west side-towers on the third-floor (and thus might be seen as a private gallery overlooking the council chamber, perhaps meant specifically for family members or intimate guests). This intriguing arrangement separates the two galleries except by way of the lord's ceremonial chamber below them.

A central feature of this chamber and the galleries are its windows, which (with the exception of a simple light within a circular mural room in the north-west angle of the central block, and another within the gallery on the north-east side), are treated here. These seven windows were originally substantial medium sized windows (90cm to 1m in width and 1 to 1.5m in height), which may have had two-lights. Because of the large number of windows within the lord's council chamber, they have been numbered here from 1-7 in Figure 5.20.42. Each compass direction is represented in these windows, suggesting fantastic projective views; but in actuality the windows are somewhat complicated. Although they are substantial in size and would have allowed a considerable amount of light to enter the room, they are set quite high in the wall and would not have been convenient for casual observation from the council chamber itself. This would not have been a room where a guest could informally peer out a window and see the embodiment of the lord's authority in the landscape (or even see much of the *landscape* for that matter). Indeed, the position of the windows would have forced the council's attention to remain within the room (perhaps exactly as it should have been). However, four of these windows were accessible for viewing from the galleries. To take a glance out of one of these windows, an observer would only have to climb half a flight on the spiral staircases. The views would have been stunning; the elevation of the third-floor would have ensured the quality and extent of the views. Basically, with the exception of the view from the roof, the third-floor gallery was the most valuable location in the great tower for casual visibility.

Unfortunately the current bridge system for moving visitors about within the tower does not allow for a close-up inspection of every window, or an examination of their projective views. When it was possible, interior photos of the windows (or at least the opening of the embrasures) were taken, these are seen in Figures 5.20.43-49. This researcher apologizes that not every window within the lord's chambers are represented here by internal photographs. However, the 20 external surfaces of the great tower (and all its windows) are represented in Figures 5.20.50-57.

Lord's private chamber; second-floor of the great tower

The viewshed from the lord's private suite on the second-floor of the great tower at Trim Castle can be seen in Figures 5.20.58-60. It was generated using an elevation offset of 12m, and an azimuth of 130°-50° to illustrate the fact that there was not an east facing window within the lord's private chambers. The percentages of projective views from this location are not high: 3.94% within the 11km radius, 11.85% in the 4km radius and 37.76% within the 1km radius. The lord's private chambers consisted of two large rooms within the central block, and two private apartments (one within each of the west and south side-towers on the second-floor level). There were two private apartments on the third-floor as well (also within the west and south side-towers). The floor plan of the second-floor can be seen in Figure 5.20.61. Most of the windows within the central block on this floor were included photographically in this study, and are illustrated in Figures 5.20.62-65.

From the lord's private chambers we might have expected recreational views of the town, the abbey and the deerpark. What we have are views of the Dublin Gate, the hinterland to the south, the Boyne River, and some views of Newtown Trim to the east. Although the town and abbey are well framed by the windows, only limited views of the Porchfield area can be seen from these chambers, which was somewhat of a surprise. (The best room on the second-floor from which to observe this area was the chapel, which was *not* part of the lord's private chamber.) Perhaps the most intriguing projective view from the lord's private chambers is from the north-west window which overlooks the market place.

Comparison of these projective views with each other and the views from other castles

Overall the projective views seen from Trim Castle are very good (the percentages within the general view from the battlements of the great tower are the fourth highest within the study in the farthest distance and third highest in the middle and closest distances). In reality, there were no poor views from Trim Castle, and although the best views were from the roof of the great tower, *every* window from the great tower and the great hall would have offered a scene that could appeal to a specific interest. Without a doubt the most potent 'lordly' display (geared to exhibit power, wealth and piety), would have been the view from the great hall which framed a significant part of the town, the abbey and possibly the most prestigious piece of hinterland, a deerpark. This was a projective view of great significance, enclosing a landscape that exemplified the influence and assets of the Trim lords.

As mentioned, the West Gate was subjected to a destructive assault on at least one occasion during the study time frame, and it would be safe to consider it a defensive gate as well as a civic gate. If there was an urban *cordon sanitaire,* every millimetre along the north-western curtain could have been observable from the battlements of the West Gate. We cannot be certain of the height of the buildings in the town during the first 100 years of the castle, but the roof of the gate house should have been elevated enough to give some view over even two and three storey structures in the town.

The arrangement of ceremonial chamber and gallery within the great tower is worth discussion. This same arrangement is seen in the great tower at the baronial castle, Castle Hedingham. Built approximately fifty years before Trim, it also contains a majestic upper chamber and gallery combination. As at Trim, this gallery gives views down into the ceremonial hall, as well as projective views of the surrounding countryside (Dixon and Marshall, 1993: 304-306). Other English castles with this same arrangement are Rochester where a mural gallery surrounds the archbishop's grand suite (Brown 2004: 46), and the royal castle of Dover where a most elaborate floor (presumably intended for the king himself) is overlooked by a mural gallery (ibid. 48-50).

This raises some interesting issues about the galleries at Trim (which were a later modification and most likely had been the wall-walk on the roof of the great tower before being enclosed and opened to the third- floor). The galleries were excellent viewing platforms from which to observe events taking place within the room below, but also allowed observation of the spectacular views from the windows as well. The view from the roof of a great tower was usually available to only a privileged few; guards on duty on the wall-walk, the lordly family that lived in the tower, and perhaps their privileged and invited guests. Taking the views from a roof, as impressive as they might have been, would often have been a windy (even dangerously so), chilly and/or wet event. The De Lacys' utilization of the former wall-walk of the great tower as an enclosed observation gallery may have been designed to make such privileged encounters safer and more comfortable for themselves and their guests. The necessity of actually leaving the chamber to look out the windows may have provided a welcomed moment of privacy or intimacy, and the windows of the gallery might be seen as a deliberated measure to ensure such moments. Another interesting aspect of the galleries is that they focused on different aspects of the hinterland; the north-east looked out over the town and abbey while the south-west gallery viewed the southern end of the town and its wall, as well as the town's Dublin Gate. While the private castle community could probably view at leisure from all available windows, the public portion of the castle community (council members or guests invited to view from the north-east gallery), would have had a view only of the town, its successful market and the abbey.

Using the great tower as the main illustration, we can trace some of the changing priorities of the builders of Trim Castle through time. We can see how windows, as an interface between the interior world of the lord and his family and the outside world reflected periods of peace or conflict, and even signalled economic changes. Initially the window openings of the great tower were small windows or simple lights, set high in the walls (in the first-floor of the great tower), suggesting a need to secure the great tower against the world and to protect the inner safety of the tower from what has been perceived the violent era of the Anglo-Norman conquest of Ireland. However, this proved to be a temporary measure, for within only a few decades of the castle's foundation these windows and lights began to be changed and enlarged to varying degrees. This would seem to suggest that peace had been established, and that the succeeding lords of the Liberty of Meath were not as concerned with restricting the size of the openings as Hugh de Lacy (the father) had once been. Sometimes these window changes seem confusing, and even suggest conflicting priorities. For instance, during the first 50 years the second-floor east window on the north face of the central block was made into a door, and used to reach a newly installed hoarding (O'Brien and Fenlon 2002: 35). The hoarding, which was accessed from the lord's private chamber on the second-floor, ran around the exterior of the tower both north and east, suggests the intent to increase the over-all defensibility of the great tower. This might have been due to a perceived threat, but during the same period some of the existing smaller windows were enlarged, and even larger windows were being newly installed. This would suggest confidence in an established climate of peace, and great faith that this peace would continue. Perhaps then we should not assume that the construction of hoarding was necessarily a military response to a threat. It might have been intended as a display of the might, vigilance and strength of the castle and its lord...meant more for show than for defence. It might even have been used as a viewing platform for the lord from which to view the Porchfield area, which otherwise was not easily viewed from the lord's private chambers.

We might examine the size of the great tower openings from the very start of the construction of the castle in the light of 'display versus response to an actual threat'. McNeill gives a good list of reasons why several of the castle's defences, though impressive *looking,* may not really have been militarily sound (1990: 322-324). Particularly the defensive arrow loops at the base of the tower. At the same time as the high and narrow windows and lights were installed in the first-floor...arrow loops were built into the ground-floor rooms of the side-towers. These arrow loops measure approximately 15cm in width to 1 to 1.5m in height, and have wide embrasures internally. The walls in the side-towers are 2m thick (as compared to the walls of the central block which are nearly 4m thick), and as McNeill states, "Removing a dozen or so stones from the outside could allow a man to get through any of the loops" (1990: 324), and thereby gain access to the tower itself. It seems likely that these impressive openings at ground-level were for display and were simply lights for these rooms 'dressed up' to appear defensive. In fact, 'show' loops were installed near the ground on the west side of the north-east corner tower which was built about the same time, as well as in the curtain wall at ground-floor level below the great hall as seen in Figure 5.20.66 (O'Brien 2008: pers. com.).

The windows in the great tower suggest that all the lords of Trim Castle during the study time frame were highly concerned with lordly display and personal comfort. For Hugh de Lacy (the father), comfortable accommodation for himself, his family and his guests while projecting an aura of strength and style seems to have been important. Later, although he had an obvious interest in protecting his castle from the crown and his peers in Dublin, Walter de Lacy's priorities appear to have been similar to his father's. We see this in the lordly control of access to the views from the great tower, his private chamber, as well as the lord's great chamber, its gallery and the roof; all of which are 'deep' into the tower in access terms. We also see it in the way Walter enlarged the existing openings on the lower floors and installed generously proportioned windows on the upper floors. Matilda and Geoffrey de Greneville mounted windows the same size and scope as those in the Red Hall into the first and second-floors of the great tower near the end of the subject time frame. This too seems to speak overtly of display and personal comfort. In this way Trim has parallels with De Courcy's Carrickfergus, as both castles seem to have comfortable accommodation as one of their lord's first priorities. This is evident by the use of the entire upper floor of the great tower at Carrickfergus as the lord's chamber, as well as the impressively large windows (especially for the time period) that De Courcy had installed there. At Trim we see it in the size and commodiousness of the great tower, the number of rooms (even before the tower was completed), as well as the inclusion of lights in ground-floor rooms. These elements point to comfort and upscale lodgings as a very high priority. Indeed, like the great tower at Carrickfergus, Trim's great tower might be best thought of specifically as the *lord's* tower, and the possibility of the great hall being located therein dismissed altogether. Display also seems to have been as high a priority for De Courcy at Carrickfergus as it obviously was for De Greneville at Trim, for (although the two men were separated in time by nearly a century) both lords compromised their curtain walls by piercing them with their grand hall windows.

At both Carrickfergus and Trim, the lord seems to have exhibited his authority and 'muscle' by imposing his power right inside the town in the guise of a castle gate. Trim's West Gate and Carrickfergus' main gate display military might and are positioned almost identically in relation to (and distance from), their respective marketplaces. This can have left no doubt to all beneath their shadows as to who was in authority, and where exactly the powers lie. It is probably not a coincidence, and might even be significant that the impressive West Gate at Trim was built by Hugh de Lacy (the father) and the even more massive main gate at Carrickfergus was built by his son, Walter's half-brother Hugh de Lacy (the son), who had been granted Carrickfergus after John de Courcy's fall from royal grace.

Finally, it was exciting to discover the dynamic role that economics seemed to play in the priorities of the lords of both Trim and Carrickfergus. De Courcy had a harbour which pumped the economic life blood of his lordship; that this was a priority is suggested by the largest windows in his chamber being poised to observe the hustle and activity of that enterprise and the sea lanes that approached it. Although Trim did not have a bustling sea harbour, it did have an economic life-line. Trim was at the centre of a thriving and productive liberty. The castle was located in an important town with a very successful market, which was, interestingly enough, overlooked from the lord's chamber. Perhaps *both* lords watched eagerly as silver and goods changed hands, quite literally beneath their noses.

Reflective views: viewshed and personal observation
Approaches

Because Trim was not built on a significant prominence, the curtain walls and mural towers are not readily visible from any great distance. Driving along the modern roads to Trim, visibility of the castle was determined by the sighting of the great tower.

Today the castle is approached by the following route-ways: The Navan Road (R161); the Dublin Road (R154); the Kilcock Road also known as the Maynooth Road (R158); the Kinnegad Road (R161); and the Athboy Road (R154). During the subject time frame, the Boyne would have also been a viable approach route. Hennessey suggests that some type of protective circuit of walls and gates were in place from the inception of the town (Hennessey 2004: 3-11). If we assume that these early town gates were placed to manage traffic that was advancing along established routes, it follows that the Navan Road would have entered Trim from the east through the Navan Gate, the Dublin, Kilcock and Kinnegad Roads would have entered from the south via the Dublin Gate, river traffic along the Boyne would have entered through the Water Gate, and the Athboy Road would have entered from the north by way of the Athboy gate. A comparison of the Taylor and Skinner map of 1793 seen in Figure 5.20.67 with the modern roads seen in the OSI map in Figure 5.20.4 illustrates that these roads have changed little in the past two hundred years. However the actual modern entries into Trim have altered dramatically and most of the gates survive only in document form today, so it was impossible to follow exactly those final mediaeval approaches into the town.

Note that, although the viewshed in Figures 5.20.6 - 8 suggests the great tower could be visible from several points as far away as 11km, those points are not necessarily on an approach road. Along each of the roads listed above, the viewshed in Figure 5.20.7 suggests several patches where a reflective view of the castle might be possible. However, it was generally not until those patches were very frequent that the great tower was actually seen. The Kilcock Road (travelling north-west) proved to be the most difficult for catching a reflective view; there the great tower was not visible until almost literally at the castle's Dublin Gate (200m away). The main reason was topography (which, though relatively flat, is undulating enough to cause problems with the reflective views from each of the approach routes). Vegetation and built environment also played havoc with the reflective view from this road. Coming from Athboy (travelling south-east), the tower was visible within approximately 1km. In this case, built environment was the main disruption of the reflective view, though topography and vegetation also caused problems. Reflective views from both the Dublin (travelling west) and Kinnegad (travelling north-east) roads were possible at 1.3km, with topography and vegetation being the main hindrances to visibility. The route that allowed the farthest view of the great tower was the Navan Road (1.9km), here only topography interfered with the view. Photographs of the first-sightings from these approaches can be seen in Figures 5.20.68–72. A table showing the factors relating to the initial sightings can be found in Figure 5.20.73.

Community

Positioned approximately 100m from the imposing West Gate, the market place was located securely under the watchful eye of the castle. Even in a ruined state, the slight elevation of the ground and the height of the tower make the West Gate an imposing sight from the street (see Figure 5.20.74). However, the gate would have appeared even taller medievally than it does currently, as the medieval cobbled surface of Market Street was approximately 1m *below* the present street level (Potterton 2005: 153). The proposed barbican tower (McNeill 1990: 327-8; 1997: 24; O'Brien and Fenlon 2002: 62) would have positioned the gate still further into the town, and closer to the market place. The presence of the castle would have been impossible for anyone standing within the market to ignore.

Today it is difficult to know *exactly* what an individual standing in the market place in the 12th century would have seen when looking at the castle. Currently a courthouse and a line of shops block any possible line-of-sight between castle and market (Figure 5.20.75), however, during the time frame of this study, this spot of land was apparently relatively open, as there was still enough space to found a Franciscan friary there in 1260 (Hennessey 2004: 3). What we *can* visualize today is the reflective view of the great tower from the parish church and from other sites within the immediate hinterland. Figures 5.20.76–79 illustrates how even today, the great tower continues to maintain a presence within the local community.

Comparison of these reflective views on each other and the views from other castles

The GIS viewshed generated from the battlements of the great tower in Figure 5.20.7 suggests that the tower can be seen fairly well from all directions within the 4km radius. Overall, Trim has extensive reflective visibility according to the viewshed. However, personal observation suggested that even on the Navan Road, from which the castle was visible the farthest, it was not *strikingly* visible. In fact, recognizing it in the landscape required a fore-knowledge of the shape and size of the tower, and identifying it in the photograph taken at the initial point of sight is frustratingly difficult (see Figure 5.20.69). This is because the Navan Road (like the North Road from Larne to Carrickfergus) is dropping in elevation, and the castle sits lower than the road at the point of first-sighting. This allows the great tower to blend into the surrounding urban landscape. Unlike Carrickfergus, which sits in a semi-bowl shaped area on the shore-front, the land around Trim is fairly flat and the great tower has a distinctive shape. This enables it to be somewhat silhouetted against the sky from the east, south and west approaches (Figures 5.20.68, 5.20.70-72) although *identifying* it in the photographs taken at the first-sighting is still difficult. It was interesting that, what was quite clear visually at the time of sighting is not nearly as clear in a two-dimensional photograph.

We can continue the comparison of Trim with Carrickfergus. GIS suggests that Carrickfergus could be visible from both its land approaches from as far as 4km away, and certainly should be visible from 1km. But observation has shown that built environment, vegetation and the choice of stone used in the castle's construction reduce its reflective views on land to *less* than 1km. The viewsheds of Trim suggest that there could be reflective views of the castle within the first 1km radius, and spotty (though significant) reflective views available within a 4km radius. However, unlike Carrickfergus, Trim, despite being surrounded by a good sized town, actually *can* be seen from its hinterland (although admittedly not as far out as the viewshed suggests). Three of the five approach roads offered views of the great tower from farther away or just at 1km. The reflective view most interrupted was from the south, and this corresponds to the disruption in topography within the 1km radius shown in the viewshed in Figure 5.20.8 First-sighting from this direction was almost literally at the curtain wall, but it must be noted that vegetation and built environment also had a part to play in that poor reflective view.

The Dublin Gate is worthy of discussion as far as reflective views are concerned. This gate was poised specifically to greet visitors to the castle arriving from Dublin; honoured and privileged guests, representatives or messengers from other barons or from the government in Dublin. Its location facing the main route to the town's south gate meant that both traffic heading to the town and leaving town were presented with the graceful side-view similar to that seen in Figure 5.20.80. This gate was a potent symbol of power, illustrating Walter de Lacy's intent to impress his peers. It was an imitation of the gate at the castle Coudray-Salbart in Poitou, France, and demonstrated to the nobles living in Ireland that De Lacy was familiar with the latest styles and building ideas from the Continent (McNeill 1990: 335; 1997: 24). This placed him firmly on the international stage. Although the plan did not utilize the twin-towered gate house that was more defensively effective, it was replete with defensive measures; a pair of murder holes, a portcullis, two gates and arrow loops which plunged two metres below the sill of the embrasure (ibid. 333, 335). It was strong, and it looked good. This illustrates that De Lacy knew his audience (his peers) and what would impress them; and built to that affect. However, we can be certain that De Lacy's peers were not the only individuals impressed by the Dublin Gate. This effective statement of power and sophistication would not have been lost on either the English colonists in Ireland, or the Irish themselves. The Dublin Gate was as sleek and stylish as the West Gate was bold and oppressive.

Part III

Chapter 6 Implications of the analysis

The following chapter includes an examination of the strong points and limitations of the two divergent approaches. Concerns that developed during rigorous field testing and subsequent analysis are examined, and the manners in which they were addressed are explained. Comparative analysis of the sample castles show commonalities which suggest an emerging pattern in castle siting.

6.1 Strengths of GIS viewshed analysis

Some of the major strengths of GIS were actually brought to light by experiential observation. At Adare, personal observation from the windows of the great halls suggested that the views were essentially the same; both halls looked out on to the River Maigue to the south and were separated by approximately only 15m to 20m. Currently, the projective views from these windows are completely obscured by the prolific vegetation on the other bank, and it was not physically possible to determine which hall could have offered a larger projective view. If anything, it was thought that the views from the early hall would be superior, as this was a first-floor hall and the later hall was at ground-floor level. The viewsheds generated from these two locations illustrated how much difference only 15m can make in the quantity of the view. What was discovered is that, even with the seeming elevation disadvantage of a ground-floor location, the projective views from the later hall were considerably *more* extensive than from the earlier hall. This was helped by the fact that the ground on which the later hall was built is roughly 2m higher than the early hall (25m versus 23m above sea level respectively).

At Ballylahan Castle, it was noted by personal observation that the castle was overlooked by several points within the immediate vicinity (within 200m). It became apparent at several succeeding castles that this was actually quite common. This raised the question "how many of the subject castles were built on the highest points in their area?" It was a simple procedure within GIS to determine accurately that only three castles within this study were located on the highest positions possible (Carrickfergus, Dundrum and Greencastle). This in turn trigged other questions, "why were so many lower sites utilized?", and "what made those sites attractive to the builders?" By using the 'measuring tool' within GIS, distances from the castle sites to amenities such as sources of fresh water, transportation or communication routes, fords and cross roads were readily available. This is something that would have required hours to accomplish in the field, but took only moments in the lab. While motives behind site choices have been assumed in the past, they have never been explored in any systematic manner; with GIS it was possible to do so using just a few key-strokes.

When compared with field observation, the viewsheds offered reasonable descriptions of both projective and reflective visibility. Although in some cases the reflective views offered by GIS tended to be somewhat optimistic, suggesting visibility of the castle before it was actually experienced, this probably had more to do with human visual limitations than GIS error. There was only one instance where the viewshed did not offer a maximum view, and indeed offered a view that was diminished from reality. This was at Ballylahan, where the viewshed suggested the castle would not be visible from the R321 north-bound approach road until within 200m of the castle. In fact, the castle could be seen from 360m away along this road. This difference (160m) is well beyond the allowable 25m accuracy limit of the DEM, suggesting that the LoS (and hence the DEM itself) was inaccurate. This was the only instance where there was an obvious inaccuracy within the LoS dealing with topography. In general, the viewsheds were useful and realistic.

6.2 Issues and responses dealing with GIS viewshed analysis

Scrambling

The first issue encountered with the creation of the viewsheds within this study was an occasional 'scrambling' of the geographic coordinates for the locations of prime components which had been recorded on-site. This scrambling was the result of the satellite signals 'bouncing' from one mass of stone to another in castles that still retained a large amount of extant castle fabric. The shuffling of coordinates was detectable as soon as they were plotted on the DEM because the points were not close to what in reality had been observed in the field. Such was the case at Adare, Athenry, Greencastle, and to a smaller extent at Ferns. This was dealt with by the use of high-resolution orthophotographic information obtainable on the web using the Ordnance Survey Ireland website: http://www.osi.ie/en/alist/aerial-photography.aspx. Coordinates from this site are given in ITM format (Irish Transverse Mercator). However, the rest of the data sets were in the Irish National Grid Reference System. To keep all data sets compatible required a conversion of the OSi data from ITM to the Irish National Grid. This was done using *Grid InQuest*, a transformation software programme for the UK (using Belfast Lough as vertical datum) and Ireland (using Malin Head as vertical datum). Converted coordinates were then utilized in place of the unusable data obtained on site. Orthophotographic coordinates were not initially available for the castles in Northern Ireland; where for security reasons data has not been freely accessible, only recently becoming openly available under the Northern Ireland Mapping Agreement (NIMA). Starting 25 November, 2009 the Land and Property Services supplied digital mapping data and a 'free to use' web-based mapping interface to educational establishments, including Queen's University, this interface is located at NImapWeb; http://www.nimap.net/Mapviewer/Default.aspx.

In order to ascertain how well the converted orthophotograhic coordinates related to usable coordinates taken in the field, a comparison test was conducted between the primary (on-sight) data from a castle whose coordinate points did not scramble (Carlingford Castle) and satellite orthophotography. The Carlingford components tested included the main gate (a centre point along the west exterior of the gate), and the great hall (centre point along the west exterior of the curtain wall). It was found that the two coordinate sources produced locations which were within 3m for both components (2.9m and 2.6m respectively). With the extent of distances involved with a study of this scope, this is almost seamless, and suggested that OSi orthophotographic coordinates could offer a reliable substitute if and when the primary sourced coordinates were unusable.

These on-line coordinate sources came in very helpful, and were eventually used almost routinely: to check the accuracy of data obtained on-site; when GPS coordinates had been unintentionally missed during field work; to obtain accurate centre-points of castles and/or great towers; whenever there was a question of precision in the field data; and especially whenever analysis of the DEM suggested that an 'alternative' castle site was a possibility (as mentioned above this was any location within 200m of a castle that had a higher elevation than the castle). It should be noted that while personal assessment of the projective/reflective views and the wider environment surrounding each of the castles was essential to the phenomenological portion of this research, it would have been possible to obtain the GIS coordinates of the prime components *entirely* within the lab. The castles which eventually benefitted from these on-line resources are: Athlone, Carrickfergus, Castleroche, Clonmacnoise, Dunamase, Dundrum, Kilbolane, Kiltartan, Lea, Limerick, Nenagh and Rinnduin.

Data availability, quality and resolution

This limitation was not apparent at the outset of the study, but was discovered during the course of the project. Quality output depends upon quality input, and beginning with a foundation which is flawed builds flaws into the resulting data obtained (Conolly and Lake 2006: 100). Initially this study intended to utilize only the Landmap DEM; the standard DEM available without cost for student use at Queen's. Sixteen castles were examined using the Landmap data without any problems, but because of two significant errors in the basic satellite images it was not possible to process each castle using this one DEM source. These flaws were not akin to the minor artifacts which were mentioned in Chapter 3: section 3.1, but were actually fairly significant faults. The first flaw stemmed from the large resolution (pixel size) of the DEM. MIMAS Landmap data is accurate to within 25m on the ground. Although the coordinate point entered into the program will be located in a fixed position on the DEM, in reality its location is only accurate to 25m; it could actually be *anywhere* within a 25m distance. This can allow for a large margin of error. In one instance a hilltop location (Dundrum) was represented as a *hillside* position; vastly changing the viewshed interpretation of what could be seen from that point. On a national level, a 25m DEM is adequate, but on a local level is can be quite coarse.

As with the much smaller artifacts, the second flaw was related to large bodies of water. The castles involved were the coastal castles of Carlingford and Greencastle, and the inland castle of Rinnduin which is located on the western shoreline of Lough Ree. In the Landmap DEM the outline of each of the shorelines was shown correctly, but the shorelines themselves had been 'ghosted' or projected out into the water several kilometres. This had nothing to do with the size of the pixels as seen at Dundrum, but was an actual defect in the satellite image. The land had been superimposed, but with a slight offset; this had created an odd double-vision effect on the illustration. The viewsheds were jumbled and unusable.

In each of these cases an alternate DEM source was needed. This was relatively easily addressed using replacement DEMs obtained from Ordnance Survey Northern Ireland (OSNI: http://www.nimap.net/Mapviewer/Default.aspx), and Ordnance Survey Ireland (OSi: http://maps.osi.ie/publicviewer/#V1,591271,743300,0). Both sources were accurate to a higher resolution (10m on the ground) and are considered much 'tighter' DEMs (i.e. the pixel sizes are smaller, so there are more pixels in any given area compared to the 25m DEM). This results in a more precise representation of land elevation than is possible with the Landmap DEM due to the *considerable* increase in the number of pixels within each of the 11km, 4km and 1km radius views. For example, the OSNI and OSi DEMs accurate to 10m contain almost four million pixels in the 11km view; while the Landmap 25m DEM contains approximately only six hundred thousand. The smaller pixel areas covered by the 10m DEMs are able to pick up nuances of terrain elevation that the larger pixels (due to their sheer size and relative crudity) are unable to illustrate. It was anticipated that the 10m DEMs would give a more in-depth accounting of visibility than the Landmap DEM, and this indeed proved to be the case. In general the 10m DEMs gave sharper illustrations of the terrain, sharper viewsheds, and higher percentages of visible pixels.

This prompted the question of exactly how *much* of an enhancement the 10m DEMs might offer over the Landmap 25m DEM, and spurred a comparison between the 25m and 10m DEMs to see what the differences might be under a controlled situation. Two castles in the Republic were chosen: Dunamase, which is in somewhat hilly terrain, and Trim, which is located in gently rolling but fairly flat land. The 10m DEMs for these castles were purchased and the castles were analysed using identical constraints (distance ranges, elevation offsets and azimuths) within the two different resolutions. The viewsheds themselves can be seen in Figures 6.2.1 and 6.2.2, while the percentage results are illustrated in Figure 6.2.3. In short, at the 11km radius the 10m DEM was able to 'see' an average of 55% more of the hinterland than the 25m DEM; an average of 33% more was visible at the 4km radius; and 24% more was visible at the 1km radius. It was interesting to note that generally the benefit of the tighter data is greater in the far-distance and lessens considerably the closer the radius is to the castles.

This was not necessarily the case at Dunamase; the 25m DEM suggested that the village of Dunamase, located approximately 600m west of the castle did not fall within the LoS of the castle. However, the 10m DEM portrayed the village as being just within LoS. This small difference can have a larger effect on the story offered by the viewshed; suggesting on one hand that visibility of the town from the castle might have been important, and on the other that it might not. We can do little about this, except to be aware that both scenarios are a possibility. However, it does point out the fact that data precision can certainly alter results. As the use of the tighter 10m DEM on Dunamase was meant only to be a comparison of the two resolutions and not to replace the original data, the information obtained from the Landmap 25m DEM is used in this study. It seems obvious that it would have been advantageous to analyse all twenty of the subject castles using the higher resolution DEMs, but the cost was prohibitive, making that level of precision unfeasible considering the large number of subject castles.

Datums

The inclusion of these two new data-sets raised another issue. We now not only had two different resolutions, but we were working in three different datums. Each of these DEM sources is based on a different 'datum' (a vertical and horizontal control point). OSNI and OSi are both derived from local frames of reference which define the surface of the earth by measurements that are linked horizontally and related vertically to mean sea levels; as mentioned above, OSNI is based on Belfast Lough, Co. Antrim, and OSi is based on sea level at Malin Head, Co. Donegal. These two frames of reference are completely compatible and inter-relatable. Landmap on the other hand is based on the World Geodetic System for 1984 (WGS 84) which is *not* drawn from a fixed earth-bound point, but from a point located in space, using measurements and triangulations taken from an extraterrestrial perspective with the mass centre of the earth as the absolute basis (ESRI www.esri.com/news/arcuser/0401/datum.html). In a sense, the project was now working in two different mediums; mediums which were not compatible and could not be easily compared.

There is a process available within GIS with which to determine the differences between the different datum sets in order to make them more comparable. However, this problem arose after the study was started, and had certainly not been anticipated. One of the aims of the research was precision, and it was attractive to pursue the issue of compatibility to its resolution. However, it was determined by the project committee that the individual case studies comprised the research units, not the island as a whole, and that the separate viewsheds are essentially guided by heights relative to each local situation. Furthermore, the application of an absolute standard could prove a distraction rather than a refinement to the research. So, it is acknowledged that the base-line ground-level elevations are different on four of the twenty castles. The absolute numbers of pixels visible are not used herein as a measurement; it is instead the final percentages which are of interest, not the different constituents which led to those percentages. The standard became accuracy rather than complete precision, something which is in line with archaeology as a discipline.

Percentages of visibility

This study used the percentages produced by the two datums to establish a scale of visibility. It was done rather coarsely, using the lowest and highest percentages of each view as the limits of the scale, grouping the castles in between into the high, medium and low percentages. The pixel percentages for every component at each castle (sorted by distance indices) and the resulting visibility scales can be seen in Figures 6.2.4 - 6.2.51. Accompanying these tables are bar charts of each distance index, sorted by pixel sizes.

An important trend can be seen in the scales of visibility for the general viewsheds (Figures 6.2.4 – 6.2.11) which were generated from the highest point at each castle. From these illustrations we can get an idea of which builders might have placed a high priority on large-scale visibility. It is obvious that the percentage of visibility increases with proximity to the castle, but what is perhaps significant is the *amount* of increase that is seen. In the far-distance view (11km radius), only five castles in twenty can see more than 20% of their hinterland. In the middle-distance (4km radius), twelve castles have percentages higher than 20%, and in the near-distance, (1km) *every* castle has percentages over 20%. Actually, more than half the castles are able to observe more than 50% of the nearest 1km of land surrounding them. Nearly one-third of the castle sites (six castles) allowed for visibility of higher than 75% of that radius, suggesting that certainly projective visibility within the *near-distance* was a priority to the builders. This underscores Fisher's concept of a "zone of perfect clarity" for the 1km radius surrounding a monument in the landscape (Gillings and Wheatley 2001: 12).

Initially it was assumed that the larger the percentage, the greater would be the indication that visibility was important to the builder; and conversely that the lower percentages would indicate that visibility was not a priority. However, it became apparent that we should not assume that a lower percentage of visible pixels in the viewshed meant visibility was not an important issue to the builder. The question of "what could be seen within the view?" must be considered in each instance. The case in point is Castleroche, which had relatively low percentages of visibility within the 11km and 4km views (4.5% and 12.9% respectively). But, what was visible within the rather meagre views was quite telling: the lord's settlement of Dundalk, the ancestral seat of power and the richest lands under the control of the castle's builder. It seems quite possible that the builder knew *exactly* what was important to see, and to show to her guests and peers. The issue then becomes "not how much of a view, but how *focused* was the view, and on *what* was it focused?"

One final matter in relation to GIS is more of an interesting oddity than an issue, and deals with the way each of the data-sets dealt with boundaries between land and water. Ireland is not only entirely surrounded by water; it also has numerous rivers, loughs of various sizes and some very large bays. Generally, the DEMs do not include views over the ocean, instead they extend only to the coastline, tracing the shape of the island and separating land from water in a sharply defined manner. To be visible within GIS, a point must have an elevation value (any value, even a negative value). Areas which have no value at all are identified as '*nodata*' and are not factored into the viewshed equation. Pixels over land are given an elevation value, but ocean surfaces and some bays are not allotted a value. These *nodata* areas are simply ignored during the generation of a viewshed, and the ocean as a mass simply does not exist as far as the DEMs are concerned. Three castles in Northern Ireland are located on the coast (Carrickfergus, Dundrum and Greencastle) and the 'cookie cutter' nature of the edges of the DEMs are illustrated in Figure 6.2.52. This was encountered (with a twist) at Carlingford. Here the OSi 'cookie cutter' DEM assigned a data value to Carlingford Lough (which is actually, like Belfast Lough, a large bay), as seen in Figure 6.2.53, giving the water in the lough a value, while almost haphazardly shearing off the area around the shoreline to the east and south of the lough, designating it as *nodata*. This gave part of the water a data value and part of it an assessment of *nodata*. This had the effect of showing the LoS over the lough, but not towards open sea.

Of course this does not mean the castle cannot see the water, only that the program does not recognize the water. For instance, in the average 11km radius viewshed generated over land (using the Landmap DEM) there are 608058 pixels, while the average 11km radius of coastal castles (using the same DEM) includes only 473483 pixels; this is a difference of 134575 or 22% fewer pixels. For this reason, when looking at an illustration of views possible from, and of, these coastal castles we must realize that the actual visibility might be considerably greater than the viewshed illustrates. It is important to remember that white areas which indicate *nodata* may very possibly be within the castle's LoS. In fact, at Dundrum and Greencastle (see Figure 6.2.52 above) most of the area that is represented in white (*nodata*) does fall within the castle's LoS. While these difficulties cause problems with the visual representation of the LoS, they do not alter the final percentages of visibility, as these are based only on the total pixel count.

6.3 Strengths of phenomenological analysis

Personal visual assessment has the advantages of being readily understood by anyone, and requires no computer training or special equipment. Several times in the course of field work, observations were made that might not (or could not) have been obvious simply by analysing a viewshed. For instance the almost ubiquitous presence of an alternative site (site with a higher elevation) which has been mentioned; the manner in which even low ground-level buildings can obstruct a reflective view; the nature and 'feel' of an imposing gate house from the viewpoint of the market place or church; and the optimism of the viewshed in that it assumes perfect conditions and visibility acuity for all viewers. Wheatley has stated that within a GIS context "...the only genuinely reliable source of error estimation lies with field observation" (1995: 182). This research has emphasized the fact that not only GIS errors but also *human limitations* can be estimated and better understood using experiential analysis.

6.4 Issues with phenomenological analysis and responses

Concerns that were not fully anticipated with regards to the phenomenological analysis of the subject castles centred first on the physical act of 'seeing' the castles within their landscape. Second, there was the issue in the resulting 'reading' of what was seen; and third, the issue surrounding the perception of prominence. To a degree, the issue of seeing and recognizing the castle within its environment had been expected. These were the reasons that during the subject time frame surface treatments such as lime-wash were used, and why banners, flags and pennants had flown above the towers and walls (see section 3:1 in Chapter 3). But such devices are not currently commonplace, and the difficulties resulting from vegetation (highlighted by conditions at Dundrum Castle in Chapter 5: section 5.10, and illustrated in Figures 5.10.21-29), built environment (discussed in relation to Nenagh Castle in Chapter 5: section 5.17), weather conditions (as at Greencastle in Chapter 5: section 5.12, and shown in Figures 5.12.27 and 5.12.28), and human visual acuity were reiterated during both the initial observation and the analysis of those observations. Added to this, was the discovery that the settings of the castles often distracted from the reflective visibility of the castle. This was especially true for castles sited in bowl-like situations, as at Trim, Adare and Carrickfergus. It was in the search for a first glimpse of the castle, and the subsequent comparison of the field information about those first-sightings with the probable first-sightings suggested by the viewsheds, that the 'optimism' of the GIS became most apparent.

First-sightings

Between the twenty castles, there were a total of eighty-five possible approach routes (some castles like Greencastle, had only two approaches while others such as Trim and Dundrum had five or six). Fourteen of these eighty-five approaches were problematic, involving water approaches which were inaccessible, as at Carlingford, Rinnduin and Greencastle (the water approach to Carrickfergus Castle, *was* completed by means of commercial ferry crossing between Belfast and Stranraer). This left a total of seventy-one viable first-sightings to analyse. The overwhelming majority of first-sightings took place within less than 1km of the castles (forty-five sightings or 63% of the viable approaches). Fourteen first-sightings (20%) took place between 1-1.9km from the castles, six sightings (9%) within 2-2.9km, three (4%) within 3-3.9km, and one each between 4-4.9km, 5-5.9km and 7-7.9km from the castles (1.4%

each). Figure 6.4.1 expresses these first-sighting distances in graph form and clearly illustrates that reflective views (at least on the modern roads approaching the castles) are significantly better the *closer* the traveller is to the castle. This again underscores the concept of a 'zone of perfect clarity' for the 1km radius surrounding a monument in the landscape, and lends credence to the suggestion that clarity decreases substantially beyond 4km (Gillings and Wheatley 2001: 12). More than this however, it mirrors the importance placed on the projective views of the nearest 1km distance discussed in section 6.2 above. Visibility of and visibility from the *immediate surrounding area* appears to have actually been a fairly high priority for the castles builders of this study.

Creative reconstruction

The second issue was encountered in the resulting 'reading' of the extant castle fabric. While most of the castles in this study remain in various states of disrepair, those that saw continuous (or near continuous) usage through the past nine centuries have been fairly well cared for. Athlone, Carrickfergus and Limerick have in common the fact that they were either built by the crown or appropriated by the crown not long after their construction. As royal castles, they saw almost constant use and experienced varying degrees of change, with the result that some of the basic elements remain intact. Each of these castles is now in some form of State care, and after various restorations has been opened to the public. Other study castles have come under the attention of heritage services and have recently undergone considerable restoration works; these are Adare, Athenry, Ferns, Nenagh, Swords and Trim. These castles have benefited from the 'commercialisation of heritage' by having vegetation removed; crumbling walls shored and endangered portions stabilized. Such attention is a boon not only for the castles, but also for members of the public who are now (or soon will be) able to experience these monuments in relative safety.

Some restoration works have been sensitive. One example is Trim Castle where no attempt was made to reconstruct the interior of the great tower. Instead a marquee roof was placed over the top of the structure to halt interior destruction by the elements, and a series of suspension walk-ways were inserted to allow visitors to experience the shell of the tower as it is today. Dangerous areas have been stabilized but not renovated, leaving the history of the structure exposed and 'readable'. At other castles efforts have not been so tasteful, and attempts were made to try to completely recreate the past for the visitor; such is the case at Carrickfergus Castle. Unfortunately reconstruction by sometimes overzealous heritage service can be baffling for serious students of castle studies who want to understand the unique history of a castle by tracing the changes seen in the extant fabric.

Two examples of this within the subject castles are Ferns and Adare. The considerable degree of change seen at Ferns in the past thirty years was illustrated in Figures 5.11.3 and 5.11.4. This is significant to this research in that it offers a good example of how much change (or unfortunately even damage) can be done under the guise of *creative* reconstruction. At Adare, repair work was undertaken by the Dunravens in the 19th century and more recently by the Office of Public Works. Work done by the Dunravens was tasteful and sensitive to the original style and the OPW has tried to rebuild as closely as possible to the original standard set by Lord Dunraven. This actually created a problem; the effect of the (*too careful*) modern repair work made it difficult for this researcher to determine some of the modern rebuilds from the 19th-century work. Fortunately a Master's thesis undertaken in 1997 (McCleary 1997: 77-127) captured much of the castle in film and made it possible to determine what was original or Dunraven repair work, and what was OPW's conception of what the castle might have been like.

Speaking specifically about medieval garden features, Creighton has pointed out how reconstruction efforts for the purpose of the heritage industry often lack authenticity (2010). But Creighton can recognize what might be authentic and what is a reconstruction; if a visitor does not already know the actual sequence of construction events, the additions made by creative restoration can be very misleading, clouding the real history of the monuments as told by their extant fabric. Achieving a sensitive restoration requires a careful balance between an attempt at a re-creation of a past reality (as at Ferns and Adare), and the creation of an entirely false reality. An example of such work can be seen at Carrickfergus Castle where enthusiastic efforts at creative restoration sparked the sandblasting of the dark grey stone of the great tower to lighten its appearance. Whether this was an attempt to make it more noticeable to the public or simply to produce the quintessential medieval castle is not known; but the result is that the castle now offers to the public someone's 'ideal' castle, not the reality of the Carrickfergus Castle that has been experienced in reflective views for nine hundred years.

Prominence

Thirdly is the topic of the perception of prominence at the subject castles. Liddiard has discussed several castles which were sited either on hilltops or on "false crests" (a location removed from the highest point but which is still highly visible from the valley floor) (2000: 50), suggesting that the placement of a castle atop an elevated site was a metaphor for physical strength (ibid. 126). Llobera, paraphrasing Higuchi has suggested that hierarchy can be addressed using the feelings of prominence a location engenders within a viewer "In a sense (hierarchy, rank and significance) is connected to the symbolism associated with the vertical scale and the fact that prominent locations are related to visual and physical control" (Higuchi 1983; Llobera 2001: 1007). Within the context of this study this suggests that the more prominent a castle is, the higher up the hierarchal scale the owner could be considered.

Although Llobera has suggested that prominence is due primarily to the geology of the terrain (2001: 1010), sites do not *have* to be high on the landscape to have commanding views (Barnatt and Pierpont 1983: 104), or, as has been realized from personal observation in this study, to inspire a reflective view. Along this vein, while there has been little formal analysis of the possible impact on visibility of the size of an object (Bishop 2002: 707), Ogburn has stated that "size does matter" when it comes to visibility, and is in fact probably the most significant variable influencing the visibility of something (2006: 405). It might then be safe to say that size could impact prominence. This view is supported by several authors who have suggested that at times geography seems to have shaped the ideal place to construct a castle (or other monument), but at other times, the sheer scale, monumentality or architectural distinction of the structure itself seems to create a "sense of place" (Tilley 1993: 79; Barnatt and Pierpont 1983: 105). Clearly both topographical prominence and *architectural prominence* are important.

The subject castles can be separated into four groups of prominence:

1. Those topographically prominent: Carlingford and Dunamase
2. Those architecturally prominent: Athenry, Lea and Trim
3. Those both topographically and architecturally prominent: Athlone, Carrickfergus, Castleroche, Dundrum, Ferns, Greencastle, Limerick and Nenagh
4. Those which were not prominent: Adare, Ballylahan, Clonmacnoise, Kilbolane, Kiltartan, Rinnduin and Swords

Topographic prominence

Several of the subject castles were built on outcrops, hilltops, false crests, or at the very least on higher bits of ground enabling the castle to rise somewhat above the surrounding landscape. Only two castles in the study (or 10%) relied on a natural promontory alone for prominence. At Carlingford Castle, the choice of site seems to have been a compromise between the every-day needs or roles the castle was built to perform and an elevated position, with practical needs coming out ahead. Built at the base of a hill that rises steadily to over 500m elevation within 2km, the castle was constructed only 15m above sea level on a bed of greywacke sandstone which juts into the lough. As such, it is slightly elevated from its attendant town to the south-east which sits just above sea level. In this case there was an entire hillside on which to build, where the castle could have been highly visible from both land and sea. The builder chose instead to build close to his economic amenities: the town, the deep harbour and the ships which connected Dublin to the Earldom of Ulster. Something that was noted by personal observation is the fact that even though the outcrop of sandstone was only 15m above sea level, from the vantage point of the harbour, Carlingford Castle was a *considerably* prominent structure.

Another example of practicality trumping prominence is at Dunamase Castle. However, as regards a prominent position, Dunamase and Carlingford could hardly be more different. Dunamase was built atop a massive rocky protrusion rising above a flat plane. Even though it was not built on the highest point in the area, the castle is visible for kilometres. The great hall constructed at the summit of the hill was a squat ground-floor rectangular hall, with accommodation over the service rooms towards the north end. While the roof may have risen 15-20m, the hall only reached approximately 10m at battlement level. The building could easily have gone higher; the stone source was the 'Rock of Dunamase' (the hill itself) which was both immediate and plentiful. The likely builders (Geoffrey de Costentin, Meiler fitz Henry or William Marshal the son) had enough resources to build whatever form of hall they desired. Seen from the floor of the surrounding plain, the whole thing must have been substantially prominent. It simply did not *need* to be taller to impress people; if the viewer was not already impressed by the appropriation of the ancient Irish *dún* or the castle's four unassailable gates, then they probably were not important enough to worry about impressing. Interestingly, the viewshed draws attention to the fact that the direction which the great hall faces (north-west), is the direction most extensively visible from the castle as a whole. It is also (perhaps strategically) the direction which faced the Irish enemies of the castle's builder.

Architectural prominence

It is reasonable to assume that the desire to be more prominent played a role in castle construction, even if topography could not be (or for some reason was not) utilized; and that a wish or a need to extend the projective and reflective views also had importance to the builder. When this was not accomplished by the choice of castle *site,* it could be attempted in the *style* of castle built on the site. At four subject castles the builder chose castle elements that went 'up', either with great towers that rose 20m and more into the air or massive gate houses with towers that were elevated to 15m. But at three of these castles (or 15%) there was no other choice; there simply was not a prominence to utilize (Carrickfergus, though it did build upwards, was located on the highest point within 200m and is not included here). First-floor great halls seemingly became towers themselves after the construction of soaring pitched roofs and battlements, allowing a portion of the castle to see (and be seen from) farther away. McCarthy, in a viewshed study on Irish tower houses suggests that an increase of elevation of only 5m could influence the degree of visibility of a tower house by as much as 13% (2007: 45). Even if this were not true in every case, it is obvious that the higher a structure is, the greater the chance that it could not only be more prominent, but provide larger projective and reflective views as well.

Topographic and architectural prominence

The use of both an elevated position and a distinctive style is the largest of the prominence groupings and is seen at eight castles (or 40%). Examples of this dual use of prominence can be seen in both medieval art and literature. The images from the *Book of Hours* presented in the illustrations for Chapter 3 (Figures 3.2.3 - 3.2.8) show several castles which are depicted in naturally prominent positions using tall turrets and towers. And Sir Gawain was quite impressed with the "huge" height of Bercilak's castle, and the lofty roof of the great hall. The whole of which was built, we are told, upon a mound (Anderson 2005: 199-200). Castleroche is one of the most dramatic castles in this study. Like Dunamase it is perched on a hilltop overlooking a fertile plain. Unlike Dunamase, Rohesia de Verdun, builder of Castleroche combined the striking prominent natural position with a sizeable three-storey (12m) gate house and a large two storey great hall. Interestingly, while the projective views from the castle (as mentioned) seem to have been focused meaningfully, the reflective views from the fields of rich farmland below the cliff were stunning. Carrickfergus Castle was built on the highest bit of land in the *immediate* vicinity, but this was a scant 5m or so above sea-level. The interesting thing about Carrickfergus is that it was built in a bowl-like setting, surrounded in the far-distance by several tall ranges. It was one of the castles which nearly 'melted' into its surroundings: both the dark grey of the lough and the dark green of the hills behind it. If the castle had not been built *upwards*, it could have been completely swallowed by visual competition within its environment.

The final castle to be noted in this group is Greencastle, which like Carrickfergus, had *very* little to work with by way of prominent positions. This was a substantial grey-stone castle built on the eastern shoreline of Carlingford Lough, attributed to Hugh de Lacy (the son). The castle was constructed on a shelf of land which rises only 10m above sea level across the length of the point (the highest elevation within approximately 5km). De Lacy utilized what little elevation there was to good effect, creating a massive first-floor hall with a sizeable forebuilding. It was not an overly tall building, but without doubt the largest built structure within many kilometres. The great hall could have gone higher, or De Lacy could have chosen to build a great tower, but instead he created high-class accommodation within the four D-shaped angle towers of the curtain wall. The whole must have been (and still is) impressive, even though no one part rose to any considerable height.

Not prominent

The second largest group in regards to prominence is the group that apparently did not seek any prominence at all (seven castles or 35%). At each of these castles, the visitor needs to be almost at the castle gate (or alongside the curtain walls) before the castle becomes apparent. Two of these castles were royal (Clonmacnoise and Rinnduin), both of which were quite short-lived. However, four of them (Adare, Ballylahan, Kilbolane and Kiltartan) were among the longest inhabited castles in the study, retaining their status as family domiciles into at least the last half of the 1600s. These four castles appear to have all been first and foremost estate centres.

The above examination has suggested that thirteen of the study castles (65%) utilized prominence, either topographical or architectural to project their presence or to create a 'sense of place'. This implies that prominence may have been a priority in either the choice of site or the style of construction. It has also been demonstrated that the builders appear to have focused specifically on the viewing from and presentation of their castles within the *immediate* hinterland. We can, with some degree of confidence suggest that visibility (especially near-distance projective and reflective views) did indeed play a role in the siting of the subject castles. We now proceed to an examination of the sites themselves to determine if there is a pattern to the siting of the subject castles.

Chapter 7 General siting of castles

This chapter addresses the general settings of the castle sites to determine the type of locations most often selected by builders. It examines the almost universal disregard for building on the high ground, and establishes the physical and psychological amenities that might have attracted lords to a site.

7.1 Water and routes

Malcolm has suggested that in Ireland the three ideal locations for an Anglo-Norman castle were, in order of importance: firstly, an easily defensible site; secondly, a site that was in close proximity to good pasture or farmland; and thirdly, a position at a cross-road of land routes, land and riverine routes, or land and sea routes (2007: 198). The analysis of the 200m radius around each of the subject castle sites (stemming from the observation that Ballylahan Castle was overlooked by higher ground) suggest that castle builders may, in fact, have had different notions of what the 'ideal' site for a castle might be. Figure 7.1.1 illustrates the significance of ease of access to these castle builders. Creighton has said that a castle site's accessibility and viability as an estate centre were possibly the most important criteria for the physical setting of a castle (2002: 64). This certainly seems to be the case here, for above all, castles were sited to be in close proximity to a road. This was almost always a main road; in fact, the only castle that was *not* within 200m of the main transport route was Dundrum (although Dundrum was only 60m from a road, the use of a promontory setting placed it 360m off the main Greencastle to Downpatrick route). While the distance to good farmland was not measured using GIS, it certainly did seem to have been an important consideration. While 'good' is a subjective concept (what is good farmland in Roscommon might not be considered good farmland in Meath), none of the subject castles can be considered to be in *poor* agricultural areas. Of the twenty, perhaps Clonmacnoise, surrounded by esker, river and bog had the worst bit of land agriculturally. But we could conservatively say that nineteen of the twenty castles were located in close proximity to productive (arable *and/or* pasture) land.

Liddiard has demonstrated that the majority of castles in England and Wales were raised in river valleys (2005: 24), and Creighton suggested that access to a dependable source of clean water had to have been a crucial dynamic in the choice of where to site a castle (2002: 54). Fresh water would not only be necessary for the human needs of drinking, cooking and washing; but a healthy horse requires 10 to 12 gallons of water daily, and a lordly household could have had *many* horses. While this aspect was not mentioned in connection with Malcolm's ideal locations for a castle, it was clearly evident within this research. Sixteen of the subject castles had an obvious water supply; twelve were less than 150m from a fresh water river, and two of these also had interior water sources (Adare had at least two wells within the castle walls and Trim had a cistern engineered within the construction of the great tower). Three other castles had wells within the castle compound: Carrickfergus, Dundrum and Kilbolane. Dunamase had a cistern built into the walls of the great hall, however, this, like the seepage well at Dundrum would have been entirely dependent upon rainfall, so may not have been sufficient to provide all the water needs. We do not yet know the arrangements for water provision at Carlingford, Ferns, Greencastle and Nenagh, or indeed how Dunamase met all its requirements; but it is not impossible that each of the castles had internal wells which have yet to be located archaeologically. Even though comparatively little research has been done on the systems of water supply to castles (Creighton 2002: 54), we are aware that wells have been quarried within improbable landscapes, and water sources have been tapped in some incredible circumstances. Knight has described the well at the royal castle of Montgomery, Montgomeryshire, Central Wales which had been dug through solid rock to a depth of at least 63m (after removing rock and civil war relics from 62.4m of the well, excavators finally gave up; they never reached the bottom! (1992: 146). It is not inconceivable that the builders at Dunamase or elsewhere could have been as intrepid and unrelenting in their efforts to provide a reliable source of water.

Fourteen of the castles were at a cross road of some type (within this study the term cross road is used for any type of junction between land, river or sea routes), which Malcolm had suggested was important (2007: 198). As for Malcolm's *first* ideal location (a defensible site), it is perhaps significant that the seventeen subject castles mentioned in section 6.1 which were overlooked by higher ground could be considered 'militarily compromised' by the standards of medieval warfare. It is interesting to note that in *every* case (including the royal sites of Athlone, Clonmacnoise, Limerick and Rinnduin) the builders of those seventeen compromised castles appear to have selected sites that were easily accessible and close to fresh water *over* the more elevated positions. The evidence of the environments surrounding the study castles strongly suggests the following revised 'ideal' locations for castle siting. Firstly, a site located on a significant route-way; secondly a site in close proximity to good pasture or farmland; thirdly, a site providing a reliable source of fresh water; and fourthly, a site providing access to a cross road (combination of road, ford, rivers, ferries or sea lanes).

Alternative sites

Figure 7.1.2 illustrates how frequently castles were built on lower sites. This table gives the ground elevation of the castle feature chosen to represent the castle's elevation as a whole, and then shows the highest elevation within 200m of that feature. Sometimes the difference in elevation between the two sites could be seen to be negligible (3m as at Rinnduin), but sometimes it is very high (37.5m as at

Carlingford). However, it is the shaded column giving the *number* of pixels higher than the castle within that radius that is most telling, especially when it is considered that the pixels represent parcels of land from 10m² to 25m² in size. It is obvious that there are many points higher within a 200m radius of the majority of the castles. The final column gives the distance from the castle to the highest pixel. At times the distance is minimal (32m at Swords), and sometimes it is at the farthest edge of the radius (as at Ferns); but here it should be stressed that these locations represent the highest points, not the *only* higher points. What we have is an indication that to the builders of the castles in this study, the concept of an 'elevated' site may have meant a site that was little more than out of flood range.

Elevated sites are generally more prominent, offer more expansive views and are easier to see. Certainly the three castles which were built on the highest points in their immediate areas all had large visual catchments in their general views, as suggested by the percentages of visibility seen in Figures 6.2.4, 6.2.6 and 6.2.8. And each of the three consistently fell on the high end of the visibility scales seen in Figures 6.2.5, 6.2.7 and 6.2.9. This sparked the question of "why were the more elevated spots *not* used?" Once again, the measuring tool within GIS was utilized; checking distances between the alternative sites, the roads, water sources and cross roads. The results seem to add weight to the revised ideal locations listed above; and helped refine the image that was developing of the priorities and the motives of the builders. When distances from the alternative sites to the amenities were considered, it was found that the castle sites were usually closer to the amenities; and often *much* closer. The exception to this is at Dunamase, where the alternative site was closer to both the road and the cross road than the castle site was. Figure 7.1.3 gives a comparison of the distance measurements from each castle and its highest possible alternative site (if applicable) to each of the amenities.

The picture that emerged from this exercise is this: for the thirteen castles which had both an alternative site *and* a known source of water, 84% of the time the actual castle site was closer to the water than the alternative site would have been. Castle and alternative sites were equidistant 8% of the time, with the alternative site being closer to a water source 8%. Nearly the same degree of percentage is seen within the sixteen castles with alternative sites that were near a cross road: 81% of the time the site utilized was closer to the junction than the alternative site would have been. Castle and alternative sites were equidistant to a cross road 6.25% of the time, with alternative sites nearer the cross road 12.5%. This suggests a great deal of weight was given to ease of access for both a water source and a major transportation link. The story is not so clear-cut at all seventeen castles with alternative sites in regards to roads in general. In several cases, the *alternative* sites were significantly closer to a transportation route, i.e. Kiltartan, which was 123m from the road while the alternative site was only 54m from the road. Interestingly, Athenry Castle and its alternative site were equidistant from a road, but they were two *different* roads; the key factor there seems to have been the water supply which was four times closer to the castle site. Castles were built on sites close to a road 47% of the time, while alternative sites were nearer a road 35%, and 18% of castle and alternative sites were equidistant to a road. This suggests that the mere *presence* of a road was not enough of a draw; obviously water and junctions were considered very high priorities.

The consequent question to this of course is "would the alternative sites have offered wider projective or reflective views than the sites that were used?" The logic being that if seeing 'more' was a priority for the builders; they would have used the location that would offer the more expansive view. If the alternative sites would have given significantly larger percentages of pixels visible but were *not* used, then seeing more was probably *not* a high priority for the builder. But, if the sites used offered the larger viewshed, then it was possible that the sites were chosen with that attribute in mind. Viewsheds were generated for the alternative sites, with the same constraints of elevation offset, azimuth and distances as were used for the General Views. These pixel percentages were then compared with the pixel percentages from the General Views. The results are shown in Figure 7.1.4, and suggest that of the seventeen castles with alternative sites, at one castle (6%), use of the alternative site would have had little to no effect on the size of the percentage visible (Dunamase); at two sites (12%) of the castles, using the alternative site would have actually been *detrimental* to the overall size of the viewshed (Carlingford and Swords).

However, at five castles (29%) the alternative sites would have *significantly* increased the total visibility, usually doubling it (though at Ballylahan and Kiltartan, it would have actually tripled the views). For the largest number of castles (nine, or 53%), using the alternative sites would have provided a slightly higher visual advantage (increasing the percentages between 1 – 11 percentage points) over the site used. The core of this is that for *most* of the castles (72%), using the higher spots would have given wider visibility (and sometimes significantly wider visibility) than the actual site did. The fact that these sites were not utilized suggests that such visibility was not a priority for the builders. On the other hand, at three castles (Carlingford, Dunamase and Swords) the lower sites offered the same or even better visibility than the higher sights, and may have been the reasons these lower positions were used. It seems evident that the use of an elevated site was *not* a priority to the majority of the builders of the subject castles; neither was having a projective or reflective view that was wide in scope. Instead, proximity to water was the most attractive quality of a possible site, access to a major junction or cross roads was also significant, as was having a major road nearby. Figure 7.1.5 illustrates graphically the relationship of both the actual site and the alternative sites to these amenities, as well as which location offered the largest scope of visibility.

7.2 Pre-existing centres

One factor influencing the placement of a castle may have had a more emotional and psychological basis than convenience to physical amenities. The land itself; its use, conquest and defence have been a catalyst throughout history. Humans often imbue landscapes with meaning: sacred, historical, and political, even financial. But in a medieval feudal context productive land meant wealth, and control of it both required and generated power. The ability to govern and control the movements of individuals in 'space-time', as well as to shape their understanding of the social and natural world around them was an aspect of this power (Tilley 1993: 82). One interesting by-product of this research was discovering the importance of the landscape, both virtually and mythologically to the contemporary medieval indigenous Irish and the Anglo-Normans. Liddiard has suggested that Anglo-Norman lords often appropriated Anglo-Saxon lordly residences in East Anglia (2000: 36-7, 49-51). This practice was continued in Ireland with the reuse of Irish royal and ecclesiastical sites by Anglo-Normans. Ireland was, for all intents a land full of 'green-field' sites on which to build. But slightly over half of the subject castles opted to recycle the power-bases of the Irish: eleven sites were appropriated/reused, one more site was possibly appropriated/reused, and nine were green-field sites, this is shown in Figure 7.2.1.

Motives for appropriation

Perhaps it would be helpful to establish the types of reuse that occurred, and the reasons why the appropriation of a previous power-base, either secular or ecclesiastical may have been attractive. Types of reuse can be categorised as:

1. Appropriation of the actual physical site for occupation. This can be seen among the study castles at Dunamase, Dundrum and Rinnduin which were all built on previous duns; Limerick, which was built on O'Brian land that had already been ceded to the Church; Clonmacnoise (where the castle appears to have been built in the location of the Bishop's house); and possibly at Adare (this remains as yet unproven archaeologically).
2. Appropriation of the indigenous name of the location with construction of a castle in the general area of an Irish power-base. This is more of an alignment with the power than a direct appropriation of site (however, that may also have occurred). This type of reuse was seen at Athlone (which had been the *caput* of Turlough O'Conor *c.* 1129); Nenagh (which used the general location and name of the meeting place of the one of the principle assemblies of Munster); Ferns (this was an instance of inheritance from Dermait MacMurrough to his daughter Aoife, and hence to her husband Strongbow rather than appropriation, however, the resulting power exchange was the same); and Trim (Hugh de Lacy re-founded a flagging monastery and built his castle at the ford just below it).
3. Appropriation of a mythological concept and/or the site associated with the myth. This was a commandeering of the mythology or psychological power of the landscape, and can be seen at Carrickfergus (supposedly built atop a holy well); Swords (which was associated with the cult of St. Colmcille); Castleroche (i.e. the original castle of Dún Dealgan. Rohesia de Verdon later rejected Dún Dealgan as too small and constructed Castleroche several kilometres to the north-west. However, the original castle and settlement of Dundalk were framed by windows of both the council chamber and private chambers).

Four sites within this study that were appropriated for castle construction have been documented in the annals, and are often considered examples of 'contentious' Anglo-Norman acquisition of land. These include Athlone, Clonmacnoise, Limerick and Trim. While these appropriations are of both a secular and ecclesiastical nature, at each, the land actually built upon was owned by the Church. Despite their litigious aspect within the annals, the land might not have been taken under contention. They may in reality have been reasonably negotiated and noted down for formality sake. Although the documentation suggests conflict, this might have been nothing more than the acknowledged settlement of a land exchange.

The reuse of sites had many shades of motivation and varied according to the instance. These motivations included the very practical use of pre-existing earthworks, specifically ring forts for the base of mottes. However, although this motivation has been cited as being very common (Ó Drisceoil 2002: 189), it cannot be assumed, and caution needs to be applied when it is suggested, as reuse can be confused with the 'ring and fill' method of ring work construction (McNeill 1997: 64). Motivations also included stepping into an established society, and absorbing a known economic network (Liddiard 2005: 30). Much of the indigenous population was already accustomed to a centralized authority, and the Anglo-Normans were able to simply displace that authority and claim it as their own. Malcolm has suggested that the seizure and reuse of Gaelic sites was especially important to lords who did not import settlers and colonists, but instead imposed themselves as the apex of an existing pyramid (2007: 215). Another reason for reuse was as a symbolic action, representing, as Barry has phrased it, a "triumph of the new invaders over the indigenous population" (2008: 117). It was also an obvious way to remind everyone of who was now in charge. Along this line, there is also the possibility of an attempt on the part of the Anglo-Norman's to degrade the vanquished ruler and clan, a symbolic 'thumbing of the nose' at the conquered. One last motivation is that some places were just too good *not* to use as a castle site, no matter their previous importance. Carrickfergus for instance seems 'tailor-made' for a masonry castle; a solid stone base surrounded on three sides by water, next to a deep water port with a fresh water spring.

It is obvious that appropriation of some aspect of pre-Anglo-Norman authority was a significant factor when choosing a castle site. We can see hints at many of these motives within eleven of the subject castles (55%), and when trying to determine the possible priorities of the builders, the nuances of appropriation should be considered.

7.3 The seigneurial landscape

Liddiard has suggested that a castle rarely "stopped at its moat", meaning that the impact of the lord on the countryside surrounding the castle could have been considerable, and that the lord's symbolic role was often enhanced by utilitarian aspects of the lordship (2005: 97). He described a "package" of external lordly components which could create a "Landscape of Lordship", able to proclaim the power of their creator (2000: 64). This package included fishponds, warrens, dovecotes, bridges, mills, parks, religious establishments (including parish churches), and planned settlements (ibid. 2, 60-62, 67, 78 and 84, 97-98: 2005: 97); gardens, orchards and vineyards were later added to this list (Creighton 2002: 3). As this idea of designing the landscape around castles was apparently already well developed in Anglo-Norman England (Liddiard 2000: 64), there is every reason to expect that the medieval Anglo-Norman lords who established manorial estates in Ireland would have brought these ideas with them. While the implications of designed landscapes have not been examined in any systematic manner in Ireland, Murphy and O'Conor have recently listed both Dunamase and Nenagh among 14 manorial centres with documentary evidence for managed landscapes in Ireland (2006: 67). They have called for a systematic program of research, fieldwork and excavation to this end. While none of the castles within this study had *all* the elements of an elite landscape mentioned above, many had several of these markers. In fact, eighteen of the twenty subject castles exhibited either documentary or landscape evidence (townland, parish, gate and street names, extant remains, etc.) of elite elements (see Figure 7.3.1).

Seigneurial markers within this study

One of the drawbacks in determining the physical locations of past seigneurial markers such as fishponds, rabbit warrens or dovecotes is that at the time of this study, no comprehensive research excavation had as yet taken place at an Anglo-Norman manorial centre in Ireland (Murphy and O'Conor 2006: 57). But evidence in England suggests that often these elements were present within a park. Perhaps if parks can be identified on the ground and investigated in Ireland, the locations of these smaller markers might be determined (ibid. 59). Along this vein, the author was present during the 2009 Castle Studies Group Expedition to Shanid Castle, Co. Limerick, during which the existence of a rabbit warren was revealed on a near-by hill (see Figure 7.3.2). The presence of the warren in such close proximity to Shanid Castle suggests that these seigneurial markers were also placed to be seen and enjoyed as part of the projective views from a castle. If this is the case, viewsheds, which can highlight those areas which were visible from mediaeval castles could help landscape archaeologists focus on areas of special interest; and may prove to be a valuable method of teasing the locations of these enigmatic features from the landscape.

That there was a socially constrained "right way" in which to move within the landscape (Tilley 1994: 28), passing these markers in the approach to a castle, was well understood by castle builders in England before the subject time frame of this study. Liddiard has suggested that to prepare the traveller to experience a reflective view of the castle properly, the planning and structuring of access routes into the castle began far out in the wider landscape (2000: 62). He gives as an example the planning and structuring of the view seen along Peddar's Way, Norfolk, as the road approaches Castle Acre from the south-east. Here the route had been diverted *c.* the 1140s to present a priory, the town wall and the castle fishpond. Next it entered the town through a gate house before proceeding directly past the castle itself. Coming from the north-west, a traveller was brought past the lord's park, the rabbit warren and dove cote. All contrived to "deliberately manipulate the way in which people experienced" Castle Acre (2005: 134-139).

We might anticipate that such structured approaches were seen in an Irish context as well. While the goal of this project did not include establishing these approaches as such among the subject castles, it was noted that there was at least a suggestion of a managed approach to each of the castles. Sometimes this was nothing more than the fact than that the castle funnelled all approaching traffic to one managed point, often quite near the castle gate house. Sometimes there seemed to be evidence of more contrived approaches and these are listed in Figure 7.3.3. While seigneurial markers have not been treated extensively here, it should be noted that only Anglo-Norman lords had the affluence to afford, and the right of privilege to enjoy these seigneurial symbols. The documentary and landscape evidence from the subject castles seems to suggest that there was certainly a consciousness of the seigneurial environment. That there was also the ability, knowledge, wealth and most likely the desire to alter the environment to fit perceived needs and priorities is evident in Ireland.

This chapter has revealed commonalities in site locations amongst the subject castles which may suggest the set of standards used by lords when determining a suitable castle site. It has also established elements in the landscape which might have been considered significant to observe and enjoy visually from a castle. We shall now explore the projective views that were possible from the subject castles, as suggested by this research.

Chapter 8 Projective views of the three communities within the castle

The phenomenological aspect of this study began by isolating five specific communities of individuals, both within and without the walls of the castle, who would have been exposed to the castles in one dimension or another during the subject time frame; as a reminder, these are the garrison, the public and the private communities within, and the stranger and the local communities outside the castle. It was hoped that by focussing on groups of viewers (identifying the positions from which they would be actively viewing, and then by illustrating the scope and focus of such views) we might gain a greater understanding of the viewers as people within their own personal landscapes. This chapter examines the various projective views possible from the prime components within the castle, and determines the patterns the castle builders deemed appropriate for their various communities to observe. The use of the concept of communities proved to be helpful by reminding this researcher at every turn that these were fellow humans. These were, in fact, the very individuals the castle was designed explicitly for. In a powerful way, utilizing the idea of communities of individuals humanized the viewers, in that it acted as a reminder that the geographic coordinates which make up a visibility study are more than mere "points or locations because they have distinctive meanings and values for *persons*" (Tilley 1994: 15 emphasis added).

8.1 Garrison community

The viewing point for the garrison community was the battlements of the gate or the gate house. Not all of the castles within this study had an extant gate, and at times those gates that were extant were ruinous. But whenever it was possible to identify the location of the gate house, the possible views from the battlements were examined. All-told, there are a total of twenty-four extant (or known locations of) gates within the study; of the fifteen castles with gates, six had multiple gates. (Several castles did have service posterns, and/or water gates, however, with the exception of the Postern Gate at Dunamase, the service entries are not considered here as a main entrance, and were not treated in this study.) Some multiple gates were successive along one line, with one or two external entrances, as at Dunamase. Some castles had a main outer gate leading into the ward, but once within the curtain walls, an inner gate led deeper into the castle complex. This was seen at Carrickfergus and at Lea which had impressive outer gates with inner gates guarding the great towers. It was also seen at Adare, however, at Adare the more impressive gate was the inner one; this gate and its drawbridge were placed so as to be admired from the interior windows of the great hall. Only Trim had two external gates of equal status allowing views of two separate approaches.

It was established in section 2.4 that the significant questions to be asked of the gates were "how many of the castle approaches could be observed from the battlements?", and "to what degree were those approaches visible?" This study has mainly concerned itself with the observation of traffic approaching from the hinterland, and not from an attendant town. However, regardless of whether a castle was urban or rural, or whether the gate was formal or informal, visibility would have depended to a large degree upon the extent of the *cordon sanitaire* (the area immediately surrounding a castle cleared of vegetation or buildings and intentionally left open in order to maintain an unhindered field of view from the castle, as well as to separate the urban community and garrison (Creighton 2002: 139).

Fourteen of the twenty castles in this study (70%) were sited within 200m of the convergence of several routes, and three others (15%) are within approximately 500m of a convergence. Of the fifteen castles with extant gates, ten (67%) were sited within 200m of the convergence of several routes, and three others (20%) are within approximately 500m of a convergence. Every gate house in this study seems to have been carefully placed to observe approaching traffic. Indeed, with the exception of Adare, all of the subject gate houses were able to view at least one of their approaches to a distance of 1km, and many had projective views that reached much farther than that. The chart in Figure 8.1.1 shows the number of approaches each castle had, and whether these were over land, water or both. This chart also gives an indication of how many of the approaches could be observed and to what extent. Of special interest is the 'dominant route'. This is the route which would have been most critical for the garrison to observe, either politically or defensively. Politically: because it came from a regional centre of government (i.e. the road leading from Limerick to Adare), or from the principal political centre in Dublin (i.e. Swords and Trim). Defensively: because of a perceived threat (i.e. an enemy approach across Lough Ree towards Rinnduin). The distance to which those routes are visible is shown, as is the quality of those projective views (whether patchy, dense or uninterrupted).

Five castles (33%) have patchy visibility of their dominant routes over a 4km distance from the castle. Five more castles (33%) have patchy visibility that reaches less than 4km, and one castle (7%) has dense visibility to approximately 1km. This suggests that only slightly over a quarter of the castles had excellent uninterrupted visibility of their approaches. The quality of the views was significant. While some views are rather short and patchy (Castleroche could only observe its dominant route for the final 216m, and Adare and Ballylahan were only able to view 500m or less of their dominant approaches), the majority or seventeen of the twenty-four gates were able to view their dominant approaches to a distance farther than 1km. An interesting discovery was that only four castles (27% of the castles) or nine of the twenty-four gates (37.5% of the gates) had views of the dominant routes that were uninterrupted. At Carlingford, Carrickfergus, Dunamase, and Dundrum the gates allowed uninterrupted visibility of the dominant routes stretching from 1.7km distance from the castle (at Dunamase) to views that surely would have been past the limits of human visual acuity (at Carrickfergus). While extensive uninterrupted views are perhaps not surprising over ocean vistas, at Dundrum, markedly far-reaching views were seen over land approaches as well. Here the road from Dublin (via. ferry between Carlingford and Greencastle) runs almost completely uninterrupted within the entire 11km projective view.

At some castles it is possible to see how a later gate was built, or a barbican added specifically to supply an observation point for a route that had not been sufficiently visible with the original gate. This was seen at Carrickfergus where the harbour was hidden from the first gate, but was well within the viewsheds of the subsequent gates. This was also seen at Rinnduin, where the position of the gate house next to the great hall meant that the battlements and roof of the hall interfered with the gate's projective view of the harbour. Although the *entire* harbour was not obscured, a significant portion of it could not have been visible; any water traffic approaching directly from the east on Lough Rea could not have been observable. This could have been rectified with the construction in 1277-8 of the barbican and drawbridge which extended outwards from the gate (Harbison 1995: 143), (however since the exact length of the barbican is uncertain, viewsheds from it were not generated). No information suggesting the minimum required distance for *defensive* observation of an approach from a castle was found, but it seems quite clear that the majority of the gate houses in this study should have been adequate for their function of monitoring the traffic approaching their castles. While it was not within the remit of this study, the issue of the general defensibility of the subject castles was raised by the ability to easily measure accurate distances within GIS, and the matter was explored with some interesting results.

Security or defence

Modern homes are equipped to offer varying degrees of protection from the elements and from intruders. However, it is doubtful that homes today could be considered completely secure; our dwellings offer us an element of *security* but are not intended to be absolutely impenetrable. The study castles seem to have been similarly secure; perhaps not entirely impenetrable, but probably sufficient for the requirements of the time. Just as there are modern ways to increase either our actual security, or to project the *image* that we are more secure (i.e. house alarms, security guards, CCTV cameras); some of the castles appear to have been equipped with extra defensive measures (portcullis, machicolations, arrow loops), which may, or may not have been viable. (An excellent example of a pseudo-secure castle is Castleroche, discussed in some detail in Chapter 5: section 5.7). However, in all probability *none* of the castles in the study would have been defensive enough to face all the dangers that we traditionally equate with medieval life. Even the castles with gates that were undeniably secure were flawed in some manner; for instance Dunamase would have been difficult to take in a frontal attack, but would have been a terrible place to defend during a siege, as apparently it had to rely on rainfall to supply its water needs.

Because of the trend to move away from the traditional military purpose of the castle, it was not assumed at the outset of this study that all the castles would be defensible in nature. However, there was the expectation that a fair number of them would have been. It was thought that those castles particularly prone to taking defence seriously would be royal castles, and castles located in contested areas (along borders and frontiers, or in what would have been considered the 'Marches' in England). It was also somewhat assumed that while the lords who arrived in Ireland early on might not have comprehended the perils of living amongst the 'wild Irish', castle builders during the second and third generation of Anglo-Norman colonization might have learned the dangers and been keen to make adequate defensibility an important priority. It came as quite a surprise when one-by-one those assumptions were not supported. It has already been shown that the majority of the castles in this study did not hold the 'high ground'. This is significant defensively because, as Liddiard has pointed out, discourses on military theory available to the contemporary medieval lords stressed the importance of terrain when choosing a field of battle or when erecting defensive structures (2000: 50). The reason usually given to suggest that building on an elevated site was important was highlighted by the Roman military treatise *The Military Institutions of the Romans:*

> "Good generals are acutely aware that victory depends much on the nature of the field of battle. When you intend therefore to engage, endeavour to draw the chief advantage from your situation. The highest ground is reckoned the best. Weapons thrown from a height strike with greater force; and the party above their antagonists can repulse and bear them down with greater impetuosity, while they who struggle with the ascent have both the ground and the enemy to contend with." -Flavius Vegetius Renatus, quote found online at: (http://www.pvv.ntnu.no/~madsb/home/war/vegetius/)

The firing capability of medieval archers and artillery ranged between 150 to 200m (McNeill 1992: 94), and Malcolm has suggested that any castle site overlooked by higher ground within that radius would have constituted a poor defensive proposition (2007: 195). We have seen in Figure 7.1.2 that of the twenty subject castles, only Carrickfergus, Dundrum and Greencastle are *not* overlooked by higher points. It may be significant that the majority of these high points (fourteen of seventeen lcoations) are between 150 and 200m from the castle site, and this might suggest that these alternate sites were at the outer limits of the range of fire by archers. But it should not be forgotten that each of the remaining seventeen castles had *multiple* higher points within the 200m radius, many of which were much closer to the castle site. Admittedly, some castles are more thoroughly overlooked than others, i.e. Rinnduin is only 3m lower than its higher site, but some alternative points rose markedly higher; and with dire consequences. At Limerick Castle, the entire south end of the castle was so low that people within the buildings around it were able to gaze down into the castle ward. An archer on the roof of the cathedral could easily shoot down into the ward and, according to Wiggins (2000: 85-6) that is *exactly* how the Irish harried the English during the siege of the castle in the 1700s. (The south portion of Limerick Castle was not only much lower than the north portion of the castle; it also remained *unfinished* throughout the entire study time frame! This makes a statement about the perceived need for defensiveness in and of itself). We simply cannot dismiss the fact that having these higher positions within their vulnerable perimeters compromised the general defensibility of the vast majority of the subject castles.

The reasons these seventeen castles chose compromised positions are varied, but certainly revolve around the priorities mentioned above; access to a convenient water supply, transportation routes and a desire to appropriate a pre-Anglo-Norman power-base (no matter how defensively weak the site might have been). Malcolm found this was also true of the Anglo-Norman castles he studied in Co. Mayo, and gives the example of Inishcoe, a pre-Anglo-Norman power base on the western shore of Lough Conn reused by the Anglo-Normans (2007: 199-200). Like Rinnduin, it was overlooked by a higher spot of ground, but was utilized anyway because it offered one of the best landing places on the western shore of the lough. At both Inishcoe and Rinnduin the Anglo-Normans chose proximity to marine transport (and also fresh water) and the opportunity to reuse an Irish site over a higher elevation. Another example cited by Malcolm is Dookeeghan Castle (a pre-Anglo-Norman promontory fort reused by the Anglo-Normans). It can be considered problematic from a defensive point because it is located towards the bottom of a downward sloping hill. Even with substantial (7m high) curtain walls the whole of the inner ward could be easily observed by a person standing just 30m outside the defences (ibid. 201). The same scenario is seen at Carlingford Castle (though not as extreme). Here the castle was placed on the western edge of the Carlingford Lough at the base of Carlingford Mountain which rises over 100m within 500m of the castle. Actually, standing 130m from the castle on this hillside (well within the range of a medieval arrow) places an observer 20m higher than the castle. The draw at Carlingford was economic (ferry and harbour), not appropriation of a site, but the result was the same, a compromised castle.

The final (but certainly not the last) example of a subject castle which made an obvious compromise in its choice of site between defence and access/control of physical amenities is Clonmacnoise. This low-lying position alongside the Shannon River was overlooked by monastic lands 3m higher than the site of the great hall within just 25m of the castle; and by a position that was 8m higher within 130m of the great hall. At Clonmacnoise, the castle had appropriated monastic land, thereby attaching itself to the acknowledged power of the site. This location also enabled the castle to claim easy access to the river to provide water for the horses and men, and to fill the moat. But, best of all, the castle gained control of an extremely valuable harbour and jetty along Ireland's major north-south transportation route. This seems to suggest that the eleven subject castles which re-used Irish power bases did so 'warts and all', occasionally accepting militarily inferior sites simply because they came with either instant status or politically strategic locations. This may certainly be true. But what about the green-field sites utilized within subject castles? Can we see militarily superior sites being used within a green-field context? Malcolm has suggested that when Anglo-Norman castles built within the Uí Dhubha kingdom after *c.* 1235 were constructed on green-field sites, they were "defensively uncompromised", while still maintaining their convenience to route-ways, river crossings and church sites (2007: 205). However, this uncompromised status did *not* hold true among the subject castles in this study; of the nine castles that were clearly on green-field sites, *only* Greencastle was not overlooked by a higher position.

Before leaving the subject of defence and defensibility, we should also consider that any castle that did not contain a reliable water supply *within* the castle walls would constitute a poor defensive proposition. As noted above, only three of the subject castles (Adare, Carrickfergus and Kilbolane) had an obvious safe and *reliable* access to water which was not reliant on rainfall. Since Adare and Kilbolane were compromised by higher ground, only a stark one in twenty of the subject castles (Carrickfergus) can be considered *not* defensively compromised in some manner. But even Carrickfergus had issues with defence in its earliest phase, the gate was little more than a minor gated entryway, which was not visible from the great tower (not to mention the two great hall windows which pierced the curtain wall at a low first-floor level just north of the gate). It seems evident that the subject castles were *not* built with any great concern for defensibility; though they were for the most part secure, these castles (almost without exception) were flawed defensively.

8.2 *Public community*

Knowing that the great hall was intended to provide display, we can reasonably expect that the windows within these buildings would have offered the *exact* view that the lord wanted his guests to see. Hall windows were sometimes large with carved ornamentation, often with window seats which can be seen as private places within this quite public space. A window seat would have provided a spot where people could have taken themselves out of the throng to talk quietly, all the while enjoying the display of power on the other side of the glass. In an effort to describe the projective scenes available to the two remaining inside castle communities (public and private), a system is suggested to illustrate the landscape within its cultural framework.

Landscape framework

Most of the great halls had multiple windows, which often faced in different directions. As many different scenes could be presented within only one window, the possibility of multiple windows greatly increases the total possible projected landscapes. The surroundings of each castle are separated into two possible contexts: urban and rural. These are further divided into sub-contexts as follows:

1. Rural Sylvan/Dramatic: A view that would have been what we would now call aesthetically pleasing or picturesque; mountains; forests; expansive ocean vistas etc.
2. Rural Economic: This is land under cultivation; pastoral scenes; the wealth of the land which could power an estate; observable transportation routes (land or water) which could facilitate the operation of (or be under the control of) a manor or estate (in the case of maritime castles this could include shipping lanes).
3. Rural Seigneurial: Rural lordly seigneurial markers associated with the land such as parks; rural mills; fishponds; eel weirs; rabbit warrens and dovecotes.
4. Rural Religious: A rural religious house; holy well or other religiously important site
5. Urban Settlement: Towns planned (in the English sense) or unplanned; boroughs; villages.
6. Urban Economic (mercantile): Market place; harbours or ports; burgages; places of trade and/or industry.
7. Other Urban Seigneurial: Urban mills; bridges; ferries; fishponds; fisheries; dovecotes (there was little physical evidence for these markers because of the lack of field work done on these features thus far in Ireland).
8. Urban Religious: Parish church or religious house founded (or funded) by the lord; holy wells.

A table showing the presence or absence of projective views of these different landscapes from the great halls of the subject castles can be seen in Figure 8.2.1. Twelve of the castles within this study had extant great halls, some (as at Trim and possibly Carrickfergus) could have had two great halls during the subject time frame, though only the one whose location is known for certain is considered within this section. Adare also had two successive great halls, and both locations are known, however, as they both looked out onto the same setting they are considered jointly here. Athenry had two structures which could have both been *either* a great hall or a lord's chamber, and both are considered in this study (in the context of both great hall and lord's chamber).

The total number of views that took in rural aspects were twenty, the number of views that consisted of urban aspects were seven, giving a total of twenty-seven projective landscapes. The most common projective view from a great hall was Rural Economic; ten of the twenty-seven projected landscapes (37%) looked out upon a rural scene that would have related to the lord's wealth or source of wealth. One castle that exemplifies this is Castleroche, where the hall was oriented to exhibit dramatically to the guests and inhabitants of the castle the richness of the fertile river valley to the south-east; Rohesia de Verdon's greatest resource. This is also the case at Clonmacnoise where the hall windows overlooked the Shannon River and the jetty which was one of the few seigneurial aspects of the castle. The next most common view was Rural Sylvan/Dramatic; seven of the twenty-seven projected landscapes (26%) had projective views that were pleasant to look at, some of them amazingly so. An example of this is De Courcy's great hall at Carrickfergus, where two grand windows with seats opened onto a view of little more than the Belfast Lough stretching away to the open sea. This also qualifies as an example of Rural Economic, as the severe but striking projective views were actually of the sea lanes which enabled the trade that fuelled De Courcy's mini-kingdom.

Creighton has mentioned that at Castle Rising and New Buckenham (Norfolk) the great halls faced towards the castle's associated boroughs (2010: 7) this was seen in three of the twenty-seven projected landscapes (11%) which framed views of Urban Settlement. Such projective views could easily be seen as reminders of the lord's political and social power (and might possibly be considered as simple entertainment). Views of Rural Seigneurial and Urban Economic landscapes were also seen, but each at only two of the twenty-seven projected landscapes (7% each). The least common prospects were those relating to Rural Religious, Urban Seigneurial and Urban Religious which are each seen in only one in twenty-seven projected landscapes (4% each).

The great hall at Trim Castle is a classic case of the seigneurial projective view from a castle in Ireland; it encompasses the broadest spectrum of landscape diversity within the subject castles. The entrance of the hall was at ground-floor, but because the land sloped down towards the River Boyne on the north-east, the windows were actually at the first-floor level along the castle's north-east exterior. Four large windows flanked by window seats pierced the curtain wall; in this way, Trim compares well to Restormel in Cornwall; a fortified 'shell-keep' castle whose great hall windows also pierced the curtain wall (something unusual in a shell-keep). Restormel's windows were large and provided views of the lord's parkland spreading across the valley as well as the probable site of the castle garden (Creighton 2010: 12). At Trim the scope was even wider; the grand windows were aimed directly at St. Mary's Abbey, a religious house re-founded by Hugh de Lacy (the father). The projective view of St. Mary's from this great hall not only highlighted the piousness of its builder, but also his generosity, religiosity and wealth. Just beyond the curtain walls of the castle, a seigneurial bridge crossed the Boyne linking the north and south sides of the town. Past the bridge to the north and north-east of the abbey the town of Trim climbed the hill in full view of the great hall. East of the abbey sat the Porchfield, an area believed to have been a medieval park during the subject time frame (O'Brien, 2008: pers. com). The prospect from the great hall would have been a powerful statement of authority and status by De Lacy. Even if the Porchfield had *not* been a park, its open expanse from the hall windows would have been aesthetic (as it still is today). What this analysis has shown is that whilst some baronial halls in this study did look at the castle's attendant town, they most often looked out on to bucolic scenes of the lord's economic base (the wealth of the land which could power an estate, and the water routes under the control of the lord), or a landscape whose value was purely attractive.

8.3 Private community

The term lord's chamber has a broad definition encompassing at least three different parts of lordly life: family, business and ceremony. These separate facets and their functions could be accommodated in one generic venue, but within the study time frame, fashion was beginning to dictate that separate chambers be used when space and finances allowed for it. Within this study only eleven castles contained a lord's chamber; but five castles contained more than one room that could be termed a lord's chamber. This meant that there were actually a total of eighteen chambers to analyse within those eleven castles. Within the five castles having more than one chamber, access analysis was used to distinguish which chambers would have required the most effort to reach. These chambers were then considered the lord's 'private' chambers. The rooms which were not as difficult to access were determined to be the council chambers. These five council/private chamber combinations were all designed as an upper and lower-floor pairing. Four had the private chamber one level above the council chamber: Castleroche and Dundrum both had a first and second-floor paring while Carrickfergus and Nenagh both had a second and third-floor pairing. The fifth castle with a council/private chamber combination was Trim. Trim was unusual in several ways: of all the castles it had the most provision for privacy, as the private space could not be accessed by the public stair and was located on the *lower* level, rather than above the council chamber. Also, only Trim had a council chamber that could obviously be considered ceremonial; two galleries overlooked the large council chamber, one accessible via the public entrance to the great tower, and another that could be accessed only by a private staircase from the lord's private chambers.

There was a case of multiple chambers at Ferns, where there is evidence of four fine rooms; an upper and lower pair along the east side of the castle, and another upper and lower pair on the south side. While it is entirely possible that either pair of rooms might represent a council and private chamber combination (as seen at the five castles discussed above), it was not possible to determine which level or which chamber had been used as the council space and which as the private space. However, two rooms seemed to be slightly superior in quality; the first-floor chamber on the east side which had grand windows (which appear to have had seats), and a massive fireplace; the first-floor chamber on the south side lacked a fireplace but had more and larger windows than the chamber above it, and all windows in that room appear to have had seats. These first-floor rooms were selected to represent the possible lord's chamber components at Ferns. Similarly, at Athenry (as has been mentioned) there were two spaces, each of which could have been used as either great hall or lord's chambers. At both Ferns and Athenry the likely rooms were analysed as simply the 'lord's chambers' (undifferentiated). This same designation was used for those castles which had only one obvious chamber (Adare, Dunamase, Kiltartan and Swords), as it was likely that the room did 'double-duty', acting as the core for both business and private life. All told there were a total of eight undifferentiated lord's chambers.

Figure 8.3.1 is a table illustrating the presence or absence of projective views of each the different landscapes introduced above, from the vantage points of the lord's chamber. The most common projective views from a lord's chamber (whether council, private or undifferentiated) were of Rural Sylvan/Dramatic scenes which were seen at nine castles (27%). The next most common prospect was Rural Economic, seen at seven castles (21%). Rural Seigneurial and Urban Settlement were jointly the third most common landscapes, seen from four castles each (13%). Urban Religious views, seen from three castles, came next (11%); an Urban Economic view was seen from two castles (6.5%), and Rural Religious or Urban Seigneurial views were seen from only one castle (4% each).

The Lord's Council Chamber

However, the real story that comes from this analysis only becomes apparent when the chambers are viewed in relation to *function*. The views from the five council chambers (located as part of council/private chamber combinations) are shown in the table in Figure 8.3.2. This table suggests that the landscapes viewed from these rooms were varied, but this is somewhat deceptive. While there were indeed windows in those chambers looking out on those landscapes, it was noted that in several cases the number of windows was sparse, and often those windows were either very narrow or set deliberately high in the walls (allowing light, but thwarting projective viewing). Nonetheless, the council chamber windows that *did* offer projective views often had views that seemed to be carefully choreographed. A case in point is Castleroche. There, the first-floor council chamber was fairly well lit, having five (quite narrow) windows. These openings faced north-east, east and south-east. The views they framed were of the castle's bailey and the lush land between the castle and Dundalk. The actual amount of hinterland visible from this room was *very* small, though the patchy views were far-reaching, spreading beyond the 11km radius. However, it was *what* was visible within the hinterland that was important: the inner workings of the bailey (real and immediate proof of the Lady Rohesia de Verdon's substantial resources), and the original motte castle of Dún Dealgan (ancestral seat of the De Verdon lordship). In a sense, Castleroche 'double-dipped' into the power source that was Dún Dealgan; drawing on the full weight of the Anglo-Norman council's belief in its lordly significance, as well as highlighting it as a symbol of the Anglo-Norman triumph over the indigenous population. The mixed use of old and new emblems of power was nothing new to the builders of the castles in this study, as

Barnwell has recently suggested. In the 12th century donjon of Peak Castle in Derbyshire, the two main windows within the lord's chamber seem to deliberately draw attention to old and new symbols of power within the surrounding landscape. There the old symbol was the prehistoric hillfort of Mam Tor, and the newer symbol was the tower of the seigneurial church (2007: 32-33).

The council chamber on the second-floor at Carrickfergus had quite good window arrangements; with six windows, five of which were wide and well placed to offer projective views. These views looked out onto the sea lanes, the lough and across to the eastern shoreline; the whole space was light and airy. There are no windows in the north or west walls however, and it is in these directions that there would have been the most activity; Carrickfergus town is to the north of the castle, and the busy Carrickfergus harbour to the west. While the inhabitants of the room would not have felt contained within the space, there would have been little movement in the projective views that could pull their minds away from the business at hand.

At Trim, the lord's council chamber had seven substantial-sized windows. Every compass direction is represented in these windows, suggesting fantastic projective views; but that suggestion is slightly misleading. Although the windows are sizeable and would have allowed for a considerable amount of light, they were set high in the wall and would not have been convenient for casual observation from the floor of the council chamber (at sitting or standing height, only sky could have been visible out of these windows). The placement of the windows would have encouraged the council's attention to remain within the room itself. (However, four of these windows *were* accessible for viewing from the galleries. Taking advantage of the tower's elevation to enjoy its considerable projective view would only require climbing half a flight of stairs; and the views would have been stunning.) This was also true at the council chamber in the round tower in Nenagh Castle. There, although there were ample windows in the council chamber (and even window seats), these windows were at or *below* the level of the battlements of the curtain wall; this would have effectively obscured or even curtailed the projective view to that point. Dundrum Castle had only two small openings (basically lights within embrasures) in its first-floor council chamber. Not only were they narrow, they were below the level of the curtain wall, and could have offered no projective views external to the castle. All these chambers would have been able to force the attention of the lord's council to remain in the room.

The Lord's Private Chamber

For lords at rest within their private spaces, looking out upon rural features still seems to have been of high importance. Figure 8.3.3 suggests that Rural Sylvan/Dramatic (landscapes that were simply lovely to look at) were as valued as Rural Economic views. Also important were views suggesting the settlement, economic and religious aspects of the lordship, i.e. the view from Carrickfergus of the bustling harbour (which was emphasized by the private stair to the roof level), and the view of the bridge, church and market place from Trim. However, the interests and perhaps the opportunities for particular views varied. For instance Castleroche and Dundrum were rural castles, so it is not surprising that their projective landscapes were rural; Ferns, Nenagh and Trim were urban castles and their projective landscapes reflect this with varying urban elements. But what seems significant is that urban castles such as Carrickfergus, Nenagh and Trim seem to have deliberately looked towards rural prospects. Creighton has given the example of Okehampton, a late 11th century castle in Devon where the commodious lord's chambers viewed a "designer-wilderness" which included the chase of the deerpark. Yet, as interesting as what could be seen is what could *not* be seen...because the castle's settlement was located on the opposite side of the castle, most of the private views from these rooms did not include the town. (2010: 10).

This 'turning of the back' to the public aspect of the lordship, while embracing views which suggest privacy as well as power is exemplified within the subject castles by Carrickfergus. While there are windows in every wall of the private chamber, and the occupants of the room *could* view the town if they desired, the windows facing the town were narrow, and unless one is standing within the embrasure (actually with head to the glass) little more than light can be seen. In comparison, the windows in the other walls were very large; these looked out onto the rural coastline of Belfast Lough, the sea lanes leading to open water, the lough itself with County Down as the backdrop, and the busy port. That De Courcy chose so obviously *not* to make viewing the town easy might suggest that observation of it was not a high priority (although the town was certainly visible from the roof of the great tower, which could be reached via a private stair from the lord's private chamber). The reverse of this of course is that observation of the sea lanes, the lough and the harbour (all of which are elegantly framed by the expansively large windows in the south-half of the tower) must have been very high priorities indeed.

There is the possibility, however, that the lack of windows on the north side of the great tower at Carrickfergus had less to do with visual priorities than with common comfort. There may have been very sensible, *practical* reasons why the largest windows of the keep would face the south. With the De Courcy's obvious interest in commodious accommodations, these large southern windows may have been an attempt to take advantage of the winter sun (windows facing south would increase the amount of passive solar heat, as well as the amount of light available within a room during the daylight hours in winter). Whether or not these practical considerations were a conscious effort, the great tower is set up to be quite energy efficient.

It seems worthy of note that out of the five private chambers within this study, builders of four of them chose to see something beautiful, and perhaps soothing when they were relaxing. But we cannot dismiss the fact that they seemed equally interested in their rural economic base (i.e. Co. Down, highly visible from the lord's chamber at Carrickfergus, was an important part of the economic base of the De Courcy lordship.)

The Lord's Chamber undifferentiated

There were eight undifferentiated lord's chambers, found at six castles: Athenry and Ferns (which each had two possible undifferentiated chambers), Adare, Dunamase, Kiltartan and Swords. While the windows within these rooms varied from very narrow to grand, on the whole they were generous sized openings, not dissimilar to windows within modern homes today. The windows at Dunamase are no longer extant; however, Hodkinson has suggested they were merely lights within embrasures (2003: 37). If that is so, these would the narrowest within this group, and possibly the least adorned. And while windows within the lord's chamber at Kiltartan were not large by today's standards, they were numerous (at least five), with the possibility of one of them having a window seat. The other four castles had fine windows, all on the large side, most of them with window seats and many with extant ornamentation. Figure 8.3.4 illustrates that two of these undifferentiated lord's chambers looked out upon their Urban Settlements, and one had a view towards an Urban Religious landscape. However, as with the differentiated council and private chambers, these rooms *most often* overlooked Rural Sylvan/Dramatic and Economic vistas. Perhaps gazing out on pastoral or riverine scenes was especially relaxing because these viewers *controlled* the resources in view.

While it was not necessarily a remit of this study to assess the type and quality of private accommodation within the subject castles, when a castle contained a lord's chamber component or obvious lodging space it was not difficult to do so (i.e. Lea whose later gate house, added at the end of the subject time frame had provision for fine accommodation, and Greencastle which appears to have had a substantial amount of lodgings built within its angle towers. It is assumed that if such was available for a portion of the household, the lord's accommodation would be commensurate.) Figure 5.3.5 ranks the quality and/or the amount of accommodation among the thirteen castles having either a lord's chamber or other accommodation. The scale was determined using several indicators: the provision for privacy allowed by the castle design (limitations on access to the more privy chambers, as well as separate venues for business and personal use), as well as the size and quality of extant features (windows, window sills and seats, fireplaces, garderobes etc). This simple scale suggests that of the thirteen examples, by far the greater number of castles had high-end lodgings (nine castles or 69%). This certainly suggests that personal comfort was a priority among the builders. Having established the extents, depths and types of projective views available to the communities within the subject castles (as well as the types of accommodation on offer within the castle), we will now examine the reflective views of the two communities outside of the castle.

Chapter 9 Reflective views of the two communities outside the castle

It has been suggested that all that is required for a castle to possess meaning is for there to be a spectator to bring meaning to it (O'Keeffe 2001: 73); this is the essence of the reflective view. Speight has humorously suggested that castles are innately conservative in their positions: "...they tend not to move and it takes a concerted effort to remove them" (2008: 389). Her point is that the castles which we study, explore and write about today have remained in their landscapes long after their periods of original usage. They may have gone to ruin, but they have not gone *away*; each succeeding generation that has lived in a castle's vicinity or passed by a castle on the road has incorporated the tumbling medieval fabric of the castle into their world view. The projective views may no longer be valid because there is no lord within the castle looking out upon his hinterland; but *reflective* views of castles have been valid since the time the castle was built, and will remain valid as long as any part of the castle fabric remains. The medieval writer John Taylor (1578-1653) wrote about the topography and the inhabitants of the English countryside, and as Speight notes, divided the contemporary society into "the viewers" and "the viewed" (2008: 388). Although Speight mentions this division in the context of the elite castle dwellers looking out upon the peasant workforce "scurrying" below, the idea of the viewers and the viewed is a lens that works equally well from either side. Strangers passing through and members of the local community were also viewers, and they would have seen the castle, its lord and his family (as well as those whose lives were spent in service of one form or another at the castle) as a distinct group: the viewed.

The perceptions of the individuals in these communities would have been influenced by many factors: political or social status, even age and gender (Liddiard 2005: 118). Shared histories would also have influenced perceptions of castles; and medieval groups with diverse self-identities could have had completely different thoughts when looking reflectively at a castle. O'Keeffe suggests Trim as an example of the numerous ideas just one castle might provoke. The great tower might have conveyed power, influence, hostility, apprehension or wonder simultaneously, depending on the status or life-experiences of the spectator. For instance, the twenty-sided Greek Cross design of the tower which has no parallels in England (2001:82), may have reinforced Christian identities for some viewers. The concept of a shared history (and even to an extent a shared destiny) is typified by the way in which even modern spectators perceive Trim as an 'Anglo-Norman' or an 'English' structure (ibid. 79, 82); we can be sure that contemporary medieval observers would have come to the same conclusion. In much the same vein, Liddiard has suggested that because castles may have represented power, resources and lifestyles denied to them by their social superiors, for the communities surrounding them, castles and their designed landscapes may also have been *contested* landscapes, even possibly being seen as "landscapes of exclusion" (Liddiard 2005: 118). However, one aspect that should be considered is that the individuals within the reflective communities might not have had the strong egalitarian views that modern individuals in western societies have; class structure was not questioned by either the contemporary medieval Irish or the Anglo-Normans.

Heuristic responses

One of the advantages of using a phenomenological approach is that human emotions can be given some weight. So when analysing the reflective views, an attempt at objectivity was made by noting the presence or absence of specific features such as a massive double-D tower, a barbican or a drawbridge that might seem threatening or over-awing; but the *feelings* engendered by these features were also noted. The feelings experienced by the stranger community, which would have been shaped by the factors mentioned above, are loosely grouped into three categories based mainly on the presence or absence of specific castle features:

1. Benign: engendered no threat (as a consequence of either the absence of a gate, a gate that was originally very unobtrusive or which has been much reduced through time)
2. Impressive: due to the social/political power and/or wealth displayed by the castle (as a consequence of either being the largest built structure in an otherwise predominantly rural area; having castle features which would have exhibited the builder's knowledge of architectural fashions in the wider European forum; or aspects which were overtly or obviously expensive)
3. Intimidating: due to aggressive or threatening power aspects of the castle (as a consequence of having a massive set of double-D gate towers, having multiple gates, gates which intruded physically into the urban space, or, as in the case of Athlone Castle, no gate at all, but simply having menacing characteristics)

We may be emphasising medieval vertical social relations when we use the word "power" in reference to castles (O'Keeffe 2001: 75), but in attempting to describe the experience of viewing the subject castles reflectively, the word power seemed valid in almost every context. Of course what was intimidating to one individual may have been seen as reassuring to another; but the basic definitions seem useful and have been applied here in relation to the subject castles. Any castle could have provoked more than one emotion (i.e. Trim Castle, whose West Gate could have been experienced as intimidating from the market place, but whose Dublin Gate could have been seen as simply an impressive display of social and political power and wealth; or Carrickfergus whose gates may have inspired intimidation or seemed threatening from the market while the entire castle as a whole (especially when seen from the distance) was also an impressive display of social and political power and wealth. All impressions experienced were noted and given equal value.

9.1 Stranger community: analysis of the approaches

Liddiard has said that the experience of the castle would have been different for those who lived "within the shadow of the walls compared with others who lived some distance away" (2005: 147). And we can include that it would have certainly been different for those who saw it only once or twice in a lifetime. Bishop suggested that the visual and emotional impact a structure has on a viewer may be modified by the activity of those viewers, and the frequency with which the structure is encountered visually; for instance the resident looking at a wind farm for the thousandth time would be affected differently than the tourist cycling past it just once (2002: 715). Likewise, the experience of a traveller who suddenly came upon a castle in the landscape, either seeing it for the first time or using it as a way-point in their journey would have been different than that of a villager who lived with the castle always within sight.

Using the three categories mentioned above (benign, impressive and intimidating) the castles were considered from the perspective of the stranger community. As each of the castles was visible at some point within their hinterland (either far away or up-close) all twenty castles are available for analysis from the stranger's view. The results can be seen in Figure 9.1.1. There were a total of twenty-seven impression responses for the twenty castles. Four castles (15%) were simply benign; one example is Kiltartan. Although it had a respectable double-D shaped gate house, the gate faced the drumlin to the west of the castle, looking *away* from the main road which flanked the castle just to the east. From that road the castle seems to be a little more than a country estate in a bucolic setting. Eight castles (30%) were intimidating or threatening. This is due in part to the nature of the gates themselves, and also the orientation of the gate houses which often directly faced the road which the long-distance traveller would be using. Of these eight castles, seven *also* engendered a feeling that the viewer was seeing a structure that was socially and/or politically impressive (the eighth castle, Athlone, had a threatening aspect that was not tempered by a sensation of admiration for worldly success; it was simply menacing). Two castles which elicited multiple reactions are Carlingford and Limerick. At Carlingford the gate faced the road that came from the north (and the neighbouring Irish); as the visitor came around a bend in the lough it would have been the first and powerful indication that the road was entering an Anglo-Norman town. At Limerick, the castle stretches along the bank of the Shannon, the road coming up from the ford (and later the bridge) leads directly in front of the massive double-D shaped gates, which loom over it. The strength is inescapable; all travellers going either east or west on this road are obliged to pass the towers in order to cross the river.

Overwhelmingly however, was the sense (felt at fifteen of the castles (55%) that a castle was socially and/or politically impressive. The impressive facets of many of the castles had to do with their placement on the vertical scale; what is tallest is most easily noticed. At Tintagel in Cornwall, Richard, Earl of Cornwall erected a stone tower on top of a Norman motte. This insured that the castle was visible for many miles, and while it certainly would have allowed excellent projective views from within, it also created an elevated visual focus of the castle as a whole from the hinterland, unmistakably "proclaiming the builder's status" to all who saw it (Liddiard 2005: 126). The construction of a high focal point is seen at four of the subject castles: Dundrum and Nenagh (both of which were circular towers), and at Carrickfergus and Trim. Elevation of the entire site is equally important, as seen at Castleroche and Dunamase, but as noted in section 6.4, explicit displays of social and political power or wealth can have an impact regardless of the height or elevation of the castle. An example of this is Ballylahan Castle, even sited in an unassuming low lying position; it was a large anthropogenic feature with a substantial double-D gate house, set within a picturesque idyllic setting. Even in ruins it appears quite striking in its river valley setting. It seems fairly evident from the reflective views of the subject castles, that while the main intent was to impress the casual stranger with the social or political power and wealth of the builder, several builders appear to have been interested in looking menacing as well.

9.2 Local community

Reflective views from the local community were taken from either the parish church/religious house or the market place of the castle's attendant town (for urban castles). There are only twelve examples of reflective views from parish churches. This is partly due to the facts that not all the castles were urban, not all had extant parish churches, and sometimes reflective views of the castle from the church were simply not possible due to distance or topography. It is also because the exact location of the parish church was unknown in several instances (Kilbolane and Castleroche for example), so there were no localized reflective views of the castle to analyse. These twelve castles elicited a total of thirteen impression responses from the position of the parish church. The same issues surround the locations of the market places, and we only have seven examples of reflective views from these (giving a total of eleven impression responses).

Figure 9.2.1 suggests that the most frequent sensation from viewers within the local community may have been that of being impressed by the displays of social and political power (eight castles, or 62% for the parish church component and six castles, or 55% for the market place component). Interestingly, the castles that were both impressive and threatening for the stranger community (Limerick and Carlingford) appear only impressive in the local view. At Limerick, the bulk of the castle's area slopes downward from the gate house, and as mentioned in section 8.1, the *back* of the castle (the curtain wall of which remained a wooden palisade until after the subject time period) would have been the reflective view seen from the parish church and market place. So, although from the front the castle was overtly intimidating, from the rear, peering into the inner workings of the castle ward, it would have been much less threatening. However, the resources and muscle of the garrison visible from this (almost intimate) reflective view would certainly have been notable. At Carlingford the castle appeared impressive, but hardly threatening from the vantage points of parish church and market place, for precisely the same reason that made it intimidating to approaching strangers; the position of the gate. From the town and harbour the gate house is side-on, and comes across as quite regal, but surely not menacing.

One interesting outcome was that from the local reflective positions of parish church and market place, slightly more castles appeared benign than threatening. Three castles (23%) seemed benign from their parish churches; and three castles (27%) seemed benign from the position of their market places. This is certainly the case from the parish church at Adare, where the castle seems harmless to the point of being a 'good neighbour'. Even its attempt at displays of power were internal (i.e. the drawbridge at the inner gate), and could only have been appreciated by those invited within the ward.

The least frequent response from the local reflective components was intimidating or threatening. This was quite low in both cases; two castles (15%) appeared menacing from the parish church, and two castles (18%) were likewise threatening from the market place. The one castle that was *only* intimidating or threatening (as opposed to being intimidating but also impressive) was Athlone, whose unquestioningly militarily stance seems a deliberate attempt at psychological dominance of its viewers. There is little doubt that it was meant as a strong reminder to the Irish inhabitants of the regions to the west about exactly who was in charge of the ford.

From the reflective views then, we see that for both the stranger and the local communities, the majority of the subject castles were impressive and perhaps overawing. While the stranger community may have found more castles threatening than benign, from the local aspect, many castles appear benign or even gentle. Few of the subject castles seemed to show a threatening or intimidating façade to their local community.

Chapter 10 Conclusions

This final chapter re-examines the research objectives as stated in Chapter 1 to determine whether these aims were met. To do so it concentrates on what the study has highlighted regarding a broad scale application of GIS within archaeology; it assesses the effectiveness of the method used to combine the approaches of GIS and phenomenology; it examines the manner in which viewshed analysis and phenomenology are able to coalesce to form a powerful analytical tool; it then reviews the results that have been gained from the use of the two approaches within castle studies. The chapter concludes by offering suggestions for further research. This investigation is the first of its kind within castle studies, and it is hoped that this work will aid castle scholars, but also archaeologists in general who desire to utilize GIS within the context of visibility studies.

Research objectives revisited

As stated in the introduction (Chapter 1), within a GIS framework the aims of the research include the following:

1. Investigate the archaeological application of GIS and viewsheds as wider decision-making tools, not anecdotally through a "high-tech fishing trip" (Lock and Harris 1996: 239); but in a systematic and rigorous manner, across a broad spectrum of physical environments
2. Give an assessment of the technical issues related to the use of GIS viewshed analysis on a series of sites requiring comparability, in contrast to single-site studies
3. Explore the use of GIS viewsheds in a valid *historical* context, using a time period about which much is known and documented, so that we may consider the validity of the conclusions reached by matching it against other sources
4. Assess the results of viewshed analysis when used as a tool to investigate the role of visibility of sites in the landscape of the period (the first century of English lordship in Ireland)

Working within a castle studies framework added the following separate goals:

1. Seek to explore the priorities of the first three generations of Anglo-Norman castle builders in Ireland, through the statements they made by their choice of building sites
2. Determine if visibility may have played a role in the siting of these castles by answering several research-driven questions:
    - Can we detect a prevailing norm, or set of norms, by which a lord fixed on the most suitable position in the landscape for a castle?
    - When faced with the possibility of being able to choose a castle site at-will, unconstrained by the density of peers, did the choice of site address the issue of visibility by:
        a. Taking in a broad spectrum of visibility (projective views which were large in scope)?
        b. Offering a visually pleasing projective view?
        c. Commanding a militarily advantageous view?
        d. Producing a psychological impact on the outside viewer?
3. While these questions define the general siting aims of the castle, this research also aimed towards refining this further, by considering how the views from differing spaces within the castle may have varied according to the specific spectators who were intended to view them
4. Conversely views of the castle from the outside, as seen by residents of communities associated with the castle as settlement, and more distant visitors to the castle, were examined to consider how they may have varied according to the specific spectators that were expected to view them

Twining the two different approaches of GIS viewsheds and phenomenology within the context of castle studies required a final aim:

1. To define a methodology that could combine the two types of analysis to give a better understanding of the historical human communities in question

10.1 Practical experience of using GIS in a comparative study

To avoid the lure of anecdotal investigation, a large subject population was taken from all across the island of Ireland (Chapter 2: section 2.5). This made it possible to test systematically the archaeological application of GIS viewshed analysis across a wide continuum of circumstances; geographical, political and social. The castles examined included royal, baronial and sub-baronial (knightly) castles. These were located in diverse environments; rural and urban, coastal, in-land, riverine, on hilly terrain and in flat-lowland areas. The only criterion linking the castles was that they retain enough extant fabric to determine the locations of the specific points from which to base an examination of visibility. The inclusion of such a wide range of circumstances and environments has allowed us to see building preferences that transcended the broad variety of builders and topographic situations. This study has utilized viewshed analysis within a valid historical perspective, examining scenes that were a part of daily life during a time period so well documented that we are able to clearly match up motives with actions. Such historical 'voyeurism' allowed us to discover the quite human (and often manipulative) motivations behind the site choices of the subject castles.

While GIS and viewshed technology has been available for use archaeologically since *c.* 1995, it is still nonetheless at a formative stage in the application of the program within historical archaeology. At this time there is not a body of information which clearly points the way for archaeologists attempting a visibility study who lack a background in geographic information science. Without a formulaic system already in place, it is reliant on each 'would-be viewshed analyst' to become familiar with GIS. It was stated in Chapter 6: section 6.1 that one of the major strengths of using GIS is that distances, which might take hours to measure in the field, take only moments to measure in the lab. However, enthusiasm for this particular strength should be tempered with the reminder that the *skills* needed to measure digitally (a basic understanding of the program, and its constraints and limitations) take time to learn in and of themselves. Likewise, for the results obtained through viewshed analysis to be credible and meaningfully interpreted, a period of experimentation is required. While the fundamental concepts of GIS were not difficult to grasp, and the generation of material (although time-consuming) was not arduous, *analysis* of the data produced involved a degree of heuristics that had not been expected. Indeed, a great deal of the analysis came down to basic common sense and a familiarity with the extended landscape surrounding the castles.

Taking GIS skills into the field flagged up a set of issues (Chapter 6: section 6.2). For instance, the 'scrambling' of coordinate points seen at Adare, Carlingford and other castles, illustrated that no matter how much care is taken when obtaining primary sources of coordinate data, coordinates obtained in the field are not necessarily reliable. Realization of data error while in the field is possible *if* the technician is able to bring all the accoutrements of the program with them into the field (i.e. GPS hand-held device and a lap-top computer with GIS software). Otherwise the coordinate errors can *only* be determined in the lab by someone who has a practical knowledge of the site. Field coordinates should always be corroborated by someone who is familiar with the castle within its landscape setting. Within an Irish context, as the border between the Republic of Ireland and Northern Ireland has softened, digital coordinate information has become increasingly available. This is a boon to GIS technicians as it is possible to obtain coordinate data without cost, via OSi and OSNI orthophotography. However, with the amount of variables that go into the creation of a viewshed, interpretation of visibility from a site by anyone who does not have a personal, working knowledge of that site should be considered suspect.

The choice of mapping data (or DEM) upon which to array the coordinate information is likewise significant. Here there are several issues to be considered. First is the resolution or precision of the data expressed in pixel size (i.e. 10m² or 25m² tiles). In an Irish context, Landmap, OSNI and OSi all offer effective and accurate DEMs. Accuracy should be weighed against precision, and it must be determined at the outset of a study how much emphasis is to be placed on precision. The data with the 'tighter' resolution (10m²), which is available from OSNI and OSi can result in viewsheds that are *very* detailed topographically, and highly precise. If the need for precision is significant, the smaller pixel size should be used. However, cost can be a concern, for while DEM data is now available at no charge from OSNI for sites in the north of Ireland, for sites in the Republic, OSi data can be fairly expensive. If money *is* an issue, and it is determined that accurate results but not necessarily high precision are required, then the 25m² pixels from Landmap which are available at no cost for the use of higher education in the UK are an excellent foundation of data.

Second, the comparability of vertical datums from multiple data sets will need to be dealt with if more than one source of DEM is utilized. It will need to be determined if the vertical datums are compatible. If not, and the precision of an absolute elevation value is required, there is a detailed process by which the two can be brought into alignment (this is done mathematically by taking a generalized value of the difference in the two datums through the raster calculator within the GIS software). In the case of this research, the elevations involved were only standards, pre-determined across the study and relative to the sites themselves, not to the island as a whole. It was determined that the absolute elevation values of the terrain and castle features were not critical, and that indeed, the imposition of an absolute value would constitute spurious precision. While this did preclude the use of the total amount of pixels visible as a fixed assessment of visibility, these amounts, translated into percentages of the whole, offered a method of determining the *general* scope of the views at each castle. It is recognized that the percentages themselves were obtained using non-equivalent pixels, and are far from being 'surgically precise'. They *are* nonetheless, entirely accurate and repeatable. They also make it possible to place each site upon a quantitative scale of visibility, giving a numerical value to an experience which is normally neither quantitative nor numerical: the act of 'seeing'. The third and final issue surrounding the choice of DEM data is the fact that there is variability in quality within digital elevation data. The quality variations create the possibility of intrinsic errors in the representation of topography within a DEM. These errors might not be recognised by someone who has not seen the site. Field verification of the digital mapping data to be used is the best way to confirm the validity of results.

Powerful analytical tools

Once learned, the ability of GIS and the viewshed to gather, analyse, and sort information has the potential to make them powerful tools. This study has pin-pointed two great strengths of GIS, the first of which is the ability to assess and articulate the properties of a site quickly and effectively. This was seen in the assessment of the higher elevation alternative castle sites at seventeen of the twenty subject castles examined in this study. Researchers have noted that castles were not always built on the highest positions available before, but there have been no large scale evaluations to determine how *often* this occurred, the distances between the castle and the surrounding higher points or how the elevations of the actual castle sites may have compared with the elevations of those higher points. Likewise, although we have had the ability to determine distances from a castle site to a landscape feature such as fresh water or a transportation route, to do so physically in the field is time-consuming, and has not been attempted as part of a comparative study of castles. The accuracy and ease of digital measurements using detailed orthophotography or satellite imagery makes it feasible and relatively easy to do large-scale comparative studies of site proximities.

The second (and perhaps most important) strength noted by this study is that, because GIS recognizes only natural topographic features in the landscape, it is able to suggest the *maximum* views possible from a position. The views created by the viewsheds are unhampered by vegetative interference, the presence of other buildings, poor lighting, weather conditions, or the limitations of human visual acuity. In reality, as an expression of the maximum extent of visibility, viewsheds can be unrealistic in scope. This does not mean they are not useful. While the idealistic nature of the two-dimensional viewshed can be seen as both a liability and as an advantage; this study chose to recognize it as a remarkable advantage. Indeed, without encumbrances we are able to view the terrain in a way which is impossible within the modern environment (Gillings and Wheatley 2001: 12); and reveal the topography as it might once have been seen. With the viewshed in its basic form, we might not be able to state unequivocally that something *could* be seen, but we can be quite certain what could *not* be seen. That ability, in and of itself can supply landscape archaeologists with a tremendous amount of knowledge. But it is only one advantage. This study has demonstrated that combining the potential of the maximum view with phenomenology can refine and hone that advantage into an even more powerful tool.

10.2 Twining GIS with phenomenology

The potential of a maximum view becomes realized only when that view is *de-constructed* phenomenologically. We saw a hint of the potential of the maximum view when it illustrated in Chapter 6: sections 6.2 and 6.4 that the catchment or field of vision around a castle most often had an effective radius of only about a kilometre. By relating this to the human viewers, we were able to surmise that while the large scale projective and reflective views may have been possible, apparently they were not considered essential. What appears to have been essential was for the castle to see and to be seen within the *immediate* hinterland, and we are now able to say with confidence that visibility within the near-distance played a role in site choice. By dividing the likely spectators into communities who looked outward or inward (projective or reflective views respectively) we were able to further refine the viewshed. The use of these communities of viewers prompted the need for vantage points representative of the specific communities. This forced a focus from specific locations, again creating catchments of visibility. These particular catchments can in-turn be de-constructed to determine why or how they were significant to a group of individuals. The fact that we have found patterns in the catchments suggests that the community concept is a useful distinction that can allow us to better understand not only siting priorities, but perhaps the communities of individuals themselves.

An in-depth assessment of seigneurial markers was not a facet of this study, and indeed was made difficult by the lack of archaeological work that has been completed on manorial centres in Ireland. However, as shown in Chapter 7: section 7.3, a surprising proportion (90%) of the subject castles exhibited either documentary or circumstantial evidence of elite elements. This suggests that there was certainly a consciousness of the seigneurial environment among these builders. This is significant because it can offer insight into the direction and possible focus of the views from the principal components.

The relationship between the view and the viewers can be seen in two equally possible lights: first, that the castle was built, with seigneurial markers added later within the projective catchments so that they could be viewed from the castle; or second, that the markers themselves offer a *blueprint* for the design of the castle by suggesting the location of a feature in order to maximize the view of the markers from the feature. This research suggests that the second possibility is seen most often in the subject castles. For instance, by analysing the overall viewshed from a site, we are able to determine which directions can see the furthest and have the least interrupted lines-of-sight, especially of approach roads. We can get a sense of where a gate house *should* be sited for the best observation of those roads (as seen at Dundrum Castle). As it would not have been possible to alter the location of fixed geographical positions such as deep water harbours or fords, and difficult (though not impossible) to alter the course of major transportation routes, the sites for gate houses were probably determined by the projective views. Likewise, it is possible to see where a great hall might best be sited in order to offer views of already established seigneurial aspects such as manorial land under cultivation (as at Castleroche), or religious houses (as at Trim Castle). We can also trace patterns in the lines-of-sight. A very good example of this is the projective views from the lord's chambers at Castleroche, whose focal points were the early De Verdon castle of Dún Dealgan, and the town of Dundalk. This finely-tuned focus suggested that even when *percentages* of visibility were not high, visibility may still have been a considerable priority. This makes the question "what lies within the visual catchment area?" an important one to consider. It also suggests that we can use the visual catchment to *predict* the position of a park, etc. or to focus the search for such seigneurial markers. The projective view, as shown by the viewsheds, suggest that the castles were designed *after* the choice of site was made, and were organized specifically to take advantage of particular views. We must appreciate the myriad choices that had to be made in order to assure that the desired vistas were observable from specific chambers or buildings.

10.3 The impact of the research on castle studies

An understanding of siting priorities

Chapters 6 and 7 saw an emerging pattern of commonalities in site choices amongst the castles in this study. The evidence presented in Chapter 7: section 7.1 strongly suggested that the builders of these castles chose sites that: firstly, were on a significant route way; secondly were in close proximity to good pasture or farmland; thirdly, provided a reliable source of fresh water; and fourthly, provided access to a cross road (some combination of road, ford, rivers, ferries or sea lanes). Use of an elevated site or architectural style was seen in at thirteen (65%) of the castles.

However, the evidence seems to suggest that the use of the *highest* possible elevated site was not a priority for siting. And while visibility within the immediate distance did appear to have been a priority, locations which were topographically higher in elevation and which would have made valid alternative castle sites were not utilized, *even* when they would have provided a larger amount of projective and reflective visibility. Instead, convenience to a water source proved more desirable: 84% of the time the castle site was closer to the water source than the alternative site would have been, and 81% of the time the site utilized was closer to a junction of two main roads than the alternative site would have been. This underscores the suggestion that both water and convenience to a cross road were higher priorities than either visibility or elevation.

Along a similar vein, this research has put a new light on the (now discarded) traditional castle lore which stressed the martial aspects of castles. It was demonstrated in Chapter 7: section 7.1 and Chapter 8: section 8.1, that 95% of the subject castles did not meet the two most *basic* requirements for defensibility: a site that was not overlooked geographically, and a site with an internal/secure water supply. While disregard for these basic tenets of defence has been noted on a site-by-site basis for many years, this is the first research that has been able to supply data in Ireland on a broad scale to confirm it. Another point revealed by this study is the fact that the gates actually do monitor the approaches. (An army may not necessarily be reasonable and advance along established routes.) This suggests that the builders of these castles were expecting visitors to do the 'right thing', or 'play fair', and that the gates are aimed at the peaceful visitor. Essentially, these castles were designed for a formal way of *life,* not for war.

Lastly, we have seen in Chapter 7: section 7.2 that Ireland provided a landscape uncluttered with feudal manors, an island full of 'green-field' sites on which lords could establish their *capita.* Still, slightly over half of the subject castles (11 castles or 55%) appropriated sites or place names of previous (Irish) political, religious or mythological power. This was true *even* when such locations were not necessarily defensively secure. This suggests that the psychological, social or economic influence of these sites were considered valuable, and that the possibility of utilising a previous power-base may have been an attractive motive for siting.

An understanding of the projective views

While no definitive source has been found suggesting the required distance that a medieval castle should have been able to observe approaching traffic, in Chapter 8: section 8.1 it was shown that *almost* all of the castle gate houses (fourteen of the fifteen castles with gates, or 93%) appear to have been able to monitor at least *one* of their approaches to a distance of 1km. This implies that, as with the reflective views and the projective views in general, a great deal of emphasis was placed on observation of the approaches within the immediate hinterland. It has been suggested in this study that each castle had a dominant or critical route, which would be defensively important for the castle to be able to watch. It has been shown that slightly over a quarter of the castles with gates (four or 27%) had uninterrupted visibility of their dominant approach stretching from the castle to a distance of 1.75km (two castles could observe their dominant routes much further, to at least 11km). But for the rest of the castles, visibility ranged from being able to observe 1km or so fairly densely, to only patchy visibility of their dominant routes. In fact, most of the castles having gates (eleven or 73%) had to monitor their critical routes with patchy projective views. This does *not* suggest a high degree of emphasis on militarily advantageous visibility.

The possible projective views from the great hall and lord's chamber component were organized into eight categories containing the social and cultural aspects possible within both a rural and an urban context. Analysis of the views from great halls in Chapter 8: section 8.2 suggested that the most common views (37%) from great halls were of a Rural Economic nature (i.e. land under cultivation; pastoral scenes; observable transportation routes both by land or water, which could facilitate the operation of, or be under the control of a manor or estate). This was the wealth of the land which powered the estate. In essence, these were bucolic views which strongly highlighted the power and affluence of the lord. The next most common projective views (26%) included scenes that were of a Rural Sylvan/Dramatic nature (what we would now call aesthetically pleasing or picturesque; mountains; forests; expansive ocean vistas etc.). The third most common projective views (11%) were Urban Settlement (towns planned in the English sense or unplanned; boroughs; villages).

Similarly, it was shown in Chapter 8: section 8.3 that from the lord's chambers the story is much the same. From the council chambers, rural scenes were preferred: the most common views equally split (with 22% each) between Rural Sylvan/Dramatic and Rural Seigneurial (seigneurial markers associated with the land such as parks, rural mills and eel weirs). Rural

Economic was the next most viewed landscape (with 17%). In addition, Urban Settlement (also seen in context of the great hall) and Urban Religious (parish church or religious house founded/funded by the lord) seem to have been of interest (11% each). We have seen however, that the views from these council chambers were often curtailed by a lack of windows, as at Dundrum; the narrow size of those windows, as at Castleroche; or the placement of the windows being above eye-level, as at Trim. When this was not the case, the views were often highly choreographed, framing scenes that were pleasant but perhaps not necessarily stimulating. The result being that usually, the projective views from council chambers seem geared at keeping the attention of those seated within the room on the matters at hand.

Projective views from the lord's private chambers were very similar to those offered from the great halls: Rural Sylvan/Dramatic and Rural Economic being equally common (22% each), with Rural Seigneurial, Urban Settlement, Urban Religious and Urban Seigneurial (urban mills; bridges; ferries) also equally widespread (11%). The undifferentiated lord's chambers offered views that were mostly Rural Sylvan/Dramatic or Rural Economic. In general, the most commonly preferred views from great halls and lord's chambers alike were rural in nature, this, interestingly even though the majority of subject castles with a great hall (ten out of twelve or 83%) or lord's chamber component (eight out of twelve or 67%) were urban.

An understanding of the reflective views

The phenomenological exercises of this research have suggested that producing a psychological impact on the outside viewer was a significant priority of the castle builders in this study. In Chapter 9: Sections 9.1, the personal first impressions upon the first-sighting of each castle were noted (the initial first impressions of a stranger to the area). These were gauged on a scale that was prepared using the presence or absence of specific castle features. This scale included: feelings of a benign nature (no overwhelming emotions), feelings of an intimidated or threatened nature (a feeling of menace), and feelings of being impressed or over-awed socially or politically (by the power and wealth of the builder). The result of this exercise was that, upon first approach as a visitor to the subject castles, the most common emotion was the feeling of being impressed by the power and success of the lord (55%). This was followed by feelings of being intimidated or threatened (30%). Only seldom did a castle appear benign at a first-sighting (15%). This had much to do with the placement of the gate houses (when applicable), which often inserted themselves into the approach.

It was shown in section 9.2 that after spending time examining each castle within its urban context, experiencing the reflective views specifically from the vantage points of parish church and market place (representing the impressions of an individual local to the area), the 'feel' of the castles was somewhat different. While the most frequent reaction was still that the castles were impressive (62% and 55% respectively from parish church and market place), this was followed by the sense that the castles often appeared quite benign within their neighbourhoods (23% and 27% respectively). Feelings of intimidation or threat were the *least* common response from the local perspective (15% and 18% respectively). Again, this had much to do with the location of the reflective components of parish church and market place, which often viewed the castle gates side-on.

What we now know about the priorities of the subject castles

Generally, the builders of the subject castles placed a high priority on environmental features that could facilitate the successful maintenance of an extensive economic base: access to transportation routes, productive land, fresh water, and proximity to a major cross road. They were concerned with visibility, specifically within the near-distance, where their castles could exhibit reflective views of "perfect clarity" (Gillings and Wheatley 2001: 12). An aspect the builders of the castles within this study found attractive was prominence (both topographical and architectural). Use of either an elevated site, prominent architectural style (especially features such as towers that were very tall), or both, was seen in thirteen of the subject castles (65%), so we may conclude that prominence was also a priority. (This said, seven (or 35%) of the castles do not appear to have striven for *any* type of prominence at all.) While an assessment of personal comfort was not part of this research, it was shown in Chapter 8: section 8.3 that of the thirteen castles which retained some form of accommodation, lodging or lord's chamber within the curtain to examine, nine (or 69%) offered rather nice chambers. This certainly suggests that personal comfort was a priority among the builders.

10.4 Strengths and limitations of the method

Strengths

It is noted that the combination of GIS viewshed analysis with a phenomenological approach to visibility can be seen, to an extent, as an exercise in 'objectivity versus humanity'. The methodology required that objective and subjective elements be considered side-by-side, and that there was a constant dialogue between the two approaches throughout the project. It became apparent that if a focus had been placed only on the science of the viewshed, there might have been a tendency to objectify the viewshed beyond its human component. Because of the clutter of the modern environments surrounding castles, had a focus been placed solely on an experiential approach (by assessing visibility using only heuristic methods) much information would have been missed. That the research has been able to avoid these pitfalls speaks to the strength of combining the two approaches in the context of castle studies.

The separation of the viewers into community groups, and then analysing their respective views as the observations of *individuals* proved helpful as a constant reminder of the humanity of the viewers. It also promoted various streams of research by encouraging the assessment of visibility requirements and priorities of multiple groups. The study set out to be scientific and rational, placing emphasis on measurements, percentages and quantifiable evidence. But in many ways, GIS forced the project into an exercise of the imagination, prompting questions such as "could people have actually seen that far?", and "what would they have seen if they *could* see that far?" Repeatedly GIS led the research back to the human side, demanding that phenomenological aspects be considered and reconsidered. When those human aspects were subsequently addressed, they prompted questions answerable by GIS. This created a loop of imagination - question - procedure - facts - imagination. Ultimately GIS *did* act as a tool to give systematic, objective (and measurable) results; it did do what it was expected it to do. And when it came to seeing castles and their landscape settings, GIS in combination with a phenomenological-based method acted as a powerful tool to stimulate and then subsequently inform the imagination. Perhaps this is true about landscape studies as a whole, which often seem to deal more in exercises in the imagination than in measurable data.

A wealth of useful information about the historical human communities which were associated with castles has been gleaned from this first attempt at a comparative study of castles pairing GIS in combination with experiential investigation. This researcher suggests that the future of castle studies can be greatly enhanced by making a visibility assessment part of *all* castle research from this point forward (i.e. GIS and viewshed analysis should be used routinely when examining *any* castle). It is also a conclusion of this study that within the context of the future of archaeology, fundamental GIS skills and viewshed interpretation should be taught as a basic part of archaeology at the University level. An exposure to GIS technology early-on in professional training will perhaps reduce the reticence experienced by archaeologists to employ it. The future holds promise for archaeologists willing to attempt the use of computerization and integrate GIS into their archaeological interpretations.

Limitations

The methodology used for this research is not without its limitations. Specifically, use of GIS and the generation of viable viewsheds are reliant on the availability of DEM data that is of a high quality. It is also dependent on the subject castle having a relatively large amount of standing fabric to analyse. It is necessary to have a basic knowledge of the locations and approximate elevations of the viewing components, as well as the amount of, and the directions which the windows faced. Caution should be taken however with castles which have undergone restoration; the use of this model requires a careful assessment of the viewing components, and creative restoration could conceivably skew the data.

10.5 Further research

This final section suggests five areas where further research might be done. First, the study time frame could be extended to examine the priorities of subsequent castle builders in Ireland, right through the medieval period. Second, the discovery of a rabbit warren within easy line-of-sight at Shanid Castle, Co. Limerick (discussed in Chapter 7: section 7.3) is a strong indication that it might be profitable to seek evidence of the more ephemeral seigneurial aspects (specifically warrens, dove cotes, gardens, fish ponds, deerparks, etc) within the visual catchment suggested by viewshed analysis. Of special interest would be areas visible from the locations of the great hall and lord's private chambers. However, general viewsheds created using the locations of wall walks or battlements of towers as viewing platforms should also be considered, especially when these could have been accessed directly from a lord's chamber.

A third interesting investigation might include an assessment of the visibility of the subject castles using the fourteen water approaches, to determine if these routes had comparable visibility to land approaches. A fourth suggestion might be a broad scale examination of castle architecture to determine if builders may have utilized passive or solar heating. Lastly, further digital experimentation is urged, especially within the software features that are available to make the DEM meaningful to human perception. One of these is '*hillshade'*, which simulates sunlight and shade patterns upon ambient slopes, and can add contour values, suggesting the appearance of relief on the DEM. This in turn can aid in visualizing the relationship between visibility and the terrain (ESRI 2007: http://resources.arcgis.com/glossary/term/552). Along this line, over a decade ago, Harris and Lock suggested the integration of GIS and multi-media (1995:358), and more recently Gillings and Wheatley promoted the incorporation of other digital technologies such as virtual reality into GIS (2001:1). These are all interesting ideas which would stimulate the human perspective of the viewshed; their use within a castle studies context could prove very informative.

Bibliography

Primary documentary sources

Annals of Connacht, 1224-1544. Corpus of Electronic Texts, website: http://www.ucc.ie/celt/published/T100011/index.html, visited on July 28, 2010.

Calendar of Documents relating to Ireland, ed. H.S. Sweetman, 5 volumes. London: Public Record Office.

The Calendar of Justiciary Rolls of Ireland of Edward I 1295-1303, ed. J. Mills. Dublin: H.M.S.O.

Chartae, Priviligia et Immunitates. Dublin: Irish Record Commission.

Expugnatio Hibernicia, The Conquest of Ireland, Giraldus Cambrensis, ed. A.B. Scott and F.X. Martin. Dublin: Royal Irish Academy.

Index and Corrigenda to Irish Pipe Roll of 14 John, 1211-1212, ed. O. Davies and D.B. Quinn. Belfast: Ulster Journal of Archaeology vol. 6, 28-36.

Irish Pipe Roll 46 Henry III, 1261-1262. Dublin: Ordered by the House of Commons to be printed.

Irish Pipe Roll of 16, Edward III. Dublin: Irish Record Commission.

Primary sources for maps and sketches

All castles unless specified:
Map: Taylor, G. and Skinner, A. (1778) 1969. *Maps of the Roads of Ireland*. Shannon: Irish University Press.

Carrickfergus:
Map: *Plan of Carrickfergus* (British Magazine, May 1760)

Map: James O'Kane Map of 1821

Map: 'Kragfargus Towne', *c.*1560 (British Lib., Cotton MS Augustus I ii 42)

Map: Carrickfergus, *c.*1596 (Public Record Office, London, MPF 98 (ex SP 64/1/31)

Map: Carrickfergus, 1685 by Thomas Phillips (National Library of Ireland, MS 3137 (42)

Sketch: Mallory, J.P. and McNeill, T.E. 1991. The Archaeology of Ulster from Colonization to Plantation. Belfast: Queen's University, Belfast, Institute of Irish Studies.

Clonmacnoise:
Map: 1810 by Thomas Townsend titled 'A Map of the Bogs in the western side of District no. 6, Situate in Westmeath and King's Counties', reproduced in *The Heritage of Clonmacnoise*. Tullamore: Environmental Science Unit of Trinity College in Association with County Offaly Vocational Educational Committee.

Greencastle:
Map: 1755 by James Kennedy of County Down.

Kilbolane:
Map: Ordnance Survey Six-Inch County Map of Ireland, survey completed 1841, map engraved 1844, Dublin: HMSO, Cork volume I, sheet I.

Kiltartan:
Map: Ordnance Survey Six-Inch County Map of Ireland, survey completed 1839, map engraved 1840, Dublin: HMSO, Galway volume II, sheet 123.

Sketch: Grose, F. 1793. *The Antiquities of Ireland*. 2 Volumes, London: S. Hooper.

Lea:
Sketch: Grose, F. 1793. *The Antiquities of Ireland*. 2 Volumes, London: S. Hooper.

Nenagh:
Sketch: Gate house from the exterior, an undated drawing from Fleming (James Stark), 9 vols. of Irish Castlellated Structures, 1974 TX (Courtesy of National Library of Ireland), referenced in Hodkinson 1999.

Rinnduin:
Map: Ordnance Survey Six-Inch County Map of Ireland, survey completed 1837, map engraved 1838, Dublin: HMSO, Roscommon volume I, sheet 46.

Trim:
Swift, J. 1999. *Historical Maps of Ireland*. London: PRC Publishing.

Secondary sources

Adams, C. L. 1904. *Castles of Ireland: Some Fortress Histories and Legends*. London: Elliot Stock.

Anderson, J.J. (ed.) 2005. *Sir Gawain and the Green Knight*. London: Orion Publishing Group.

Barnatt, J. and Pierpont, S. 1983. 'Stone Circles: Observatories or Ceremonial centres'. *Scottish Archaeological Review* 2, 101-115.

Barnwell, P. 2007. 'The Power of Peak Castle: Cultural Contexts and Changing Perceptions'. *Journal of the British Archaeological Association* 160, 20-38.

Barry, T. 2008. 'The Study of Medieval Irish Castles: a bibliographic survey'. *Proceedings of the Royal Irish Academy* 108C, 115-136.

Bishop, I.D. 2002. 'Determination of thresholds of visual impact: the case of wind turbines'. *Environment and Planning B: Planning and Design* 29, 707-718.

Bradley, J. 2003. 'The monastic town of Clonmacnoise'. In H.A. King (ed), *Clonmacnoise Studies, Seminar Papers 1998,* 1. Dublin: The Stationary Office, 42-55.

Brown, R. A. 2004. *Allen Brown's English castles.* Woodbridge: Boydell Press.

Buckley, V.M. and Sweetman, P.D. 1991. *Archaeological survey of Co. Louth.* Dublin: Stationary Office.

Chapman H.P. and Gearey, B.R. 2000. 'Palaeoecology and the perception of prehistoric landscapes: some comments on visual approaches to phenomenology'. *Antiquity* 74, 316-319.

Chapman, H. 2009. *Landscape Archaeology and GIS.* Stroud: History Press.

The Complete Road Atlas of Ireland: Scale 1:210,000 1998. Dublin: Ordnance Survey of Ireland.

Conolly, J. and Lake, M. 2006. *Geographical Information Systems in Archaeology.* Cambridge: Cambridge University Press.

Creighton, O. H. 2002. *Castles and landscapes: power, community and fortification in Medieval England.* London: Continuum.

Creighton, O.H. 2010. 'Room with a View: Framing Castle Landscapes', *Château Gaillard* (forthcoming).

Crescioli, M., D Andrea, A. and Niccolucci, F. 2000. 'A GIS-based analysis of the Etruscan cemetery of Pontecagnano using fuzzy logic'. *NATO ASI Series, Series A, Life Sciences* 321, 157-179.

Cruden, J. 1999. 'Kiltartan Castle, Co. Galway: an historical, archaeological, geochemical and geophysical investigation of the extent of its medieval settlement'. Unpublished MA thesis, Belfast: Queen's University Belfast.

Davies, O. and Quinn, D. B. 1941. 'The Irish Pipe Roll of 14 John, 1211-12'. *Ulster Journal of Archaeology* 4, supplement.

Dixon, P. 1996. 'Design in castle-building; the controlling of access to the lord'. *Chateau Gaillard* 18, 47-58.

Dixon, P. and Marshall, P. 1993. 'The Great Tower in the Twelfth Century: The Case of Norham Castle'. *The Archaeological Journal* 150, 410-432.

Dixon, P. and Marshall, P. 2003. 'The Great Tower at Hedingham Castle: a Reassessment'. In R. Liddiard (ed.), *Anglo-Norman Castles,* Woodbridge: The Boydell Press, 297-306.

Duffy, S. 1997. *Ireland in the Middle Ages.* Basingstoke: MacMillan Press Ltd.

Dugdale, W. 1846. J. Caley, H. Ellis and B. Bandinel (eds), *Monasticon Anglicanum* 6 (2). London: James Bohn.

Dunne, L. 2007. 'The castle at Ballingarry, County Tipperary : a fourteenth-century Gaelic castle', In C. Manning (ed.), *From Ringforts to Fortified Houses.* Bray: Wordwell Ltd, 155-170.

Dunraven, Lord, 1865. *Memorials of Adare manor.* Oxford: Parker.

Eadie, G. 2009. 'A new approach to identifying functions in castles; a study of tower houses in Ireland'. Unpublished PhD thesis, Belfast: Queen's University Belfast.

Fahey, J. 1893. *The History and Antiquities of the Diocese of Kilmacduagh.* Dublin: M.H. Gill and Son.

Fanning, T. 1975a. 'A medieval tile pavement from Swords Castle, Co. Dublin'. *Medieval Archaeology* 19, 205-209.

Fanning, T. 1975b. 'An Irish medieval tile pavement: recent excavations at Swords Castle Co. Dublin'. *Journal of the Royal Society of Antiquaries of Ireland* 105, 47-82.

Faulkner, P.A. 1963. 'Castle planning in the fourteenth century'. *The Archaeological Journal* 120, 251-235.

Fisher, P.F. 1992. 'First experiments in viewshed uncertainty: the accuracy of the viewable area'. *Photogrammetric Engineering and Remote Sensing* 58, 345-352.

Fisher, P.F. 1994. 'Probable and fuzzy models of the viewshed operation', In M. Worboys (ed.), *Innovations in GIS* 1. London: Taylor and Francis, 161-175.

Fisher, P.F. 1996. 'Extending the applicability of viewsheds in landscape planning'. *Photogrammetric Engineering and Remote Sensing* 62 (11), 1297-1302.

Fisher, P. and Farrelly, C. 1997. 'Spatial analysis of visible areas from the Bronze Age cairns of Mull'. *Journal of Archaeological Science* 24, 581-592.

Flanagan, M. T. 1999. 'John de Courcy, the first Ulster plantation and Irish church men', In B. Smith (ed.),

Britain and Ireland 900-1300: Insular responses to medieval European change. Cambridge: Cambridge University Press, 154-178.

Furlong, N. 2006. *Diarmait, King of Leinster.* Dublin: Mercier Press.

Gaffney, V., Stancic, Z. and Watson, H. 1995. 'The impact of GIS on archaeology: a personal perspective', In G.R. Lock and Z. Stancic (eds), *Archaeology and Geographic Information Systems: A European Perspective.* London: Taylor and Francis Ltd, 211-279.

Gaffney, V., Stancic, Z. and Watson, H. 1996. 'Moving from catchments to cognition: tentative steps towards a larger archaeological context for GIS', In M. Aldenderfer and H.D.G. Maschner (eds), *Anthropology, Space, and Geographic Information Systems.* New York: Oxford University Press, 132-154.

Gaskell-Brown, C. 1979. 'Excavations at Greencastle, County Down, 1966-1970'. *Ulster Journal of Archaeology* 42, 51-65.

Geissel, H. 2006. *A road on the long ridge: in search of the ancient highway on the Esker Riada.* Newbridge, Co. Kildare: CRS Publications.

Gillings, M. and Wheatley, D. 2001. 'Seeing is not believing: unresolved issues in archaeological visibility analysis'. In B. Slapšak (ed.), *On the good use of geographical information systems in archaeological landscape studies.* Luxembourg: Office for Official Publications of the European Communities, 25-36.

Givens, J. 2008. *Irish Walled Towns.* Dublin: The Liffey Press.

Glasscock, R.E. 1970. 'Moated sites, and deserted boroughs and villages: two neglected aspects of Anglo-Norman settlement in Ireland', In N. Stephens and R.E. Glasscock (eds), *Irish Geographical Studies in honour of E. Estyn Evens.* Belfast: Queen's University of Belfast Geography Department, 162-177.

Gleeson, D. F. and Leask, H. G. 1936. 'The Castle and Manor of Nenagh'. *The Journal of the Royal Society of Antiquaries of Ireland* 6, 247-269.

Gosling, P. 1992. *Carlingford Town: An Antiquarian's Guide.* Newry: Carlingford Lough Heritage Trust.

Hamlin, A.E. 1977. *Dundrum Castle, Co. Down.* Historic Monuments and Buildings Series by the Historic Monuments and Buildings Branch, Department of the Environment for Northern Ireland. Northern Ireland: Her Majesty's Stationery Office.

Hamlin, A.E. 2008. *The Archaeology of Early Christianity in the North of Ireland.* BAR British Series, 460, Oxford: BAR Publishing.

Harbison, S. 1995. 'Rindown Castle: a Royal Fortress in Co. Roscommon'. *Journal of the Galway Archaeological and Historical Society* 47, 138-148.

Harris, T.M. and Lock, G.R. 1995. 'Toward an evaluation of GIS in European archaeology: the past, present and future of theory and applications', In G. Lock and Z. Stancic (eds), *Archaeology and Geographical Information Systems: A European Perspective.* London: Taylor and Francis Ltd, 340-362.

Hennessy, M. 2004. *Trim.* Irish Historic Town Atlas 14, Dublin: Royal Irish Academy.

Higuchi, T. 1983. *The visual and spatial structure of landscapes.* Cambridge, Mass: MIT Press.

Hodkinson, B. 1999. 'Excavations in the Gatehouse of Nenagh Castle, 1996-97'. *Tipperary Historical Journal,* 1999. 162-182.

Hodkinson, B. 2003. 'A Summary of recent work at the Rock of Dunamase, Co. Laois', In J.R. Kenyon and K. O'Conor (eds), *The Medieval Castle in Ireland and Wales.* Dublin: Four Courts Press, 32-49.

Hore, H. 1910. 'Ferns, County Wexford'. *Journal of the Royal Society of Antiquaries of Ireland* 11, 298-315 and 360-1.

Jope, E.M. (ed.) 1966. *An Archaeological Survey of County Down.* Belfast: Northern Ireland Ministry of Finance, Her Majesty's Stationary Office.

Kerrigan, P. M. 1995. *Castles and Fortifications in Ireland.* Cork: The Collins Press.

Killanin, Lord, and Duignan, M. V. 1967. *Shell Guide to Ireland, second edition.* London: Ebury Press.

Knight, J. K. 1991. *Chepstow Castle.* Cadw: Welsh Historic Monuments, Cardiff: Cadw.

Knight, J.K. 1992. 'Excavations at Montgomery Castle'. *Archaeologia Cambrensis* 141, 97-180.

Knox, H T. 1908. *The History of Mayo.* Dublin: Hodges, Figgist Co. Ltd

Kvamme, K.L. 1995. 'A view from across the water: the North American experience in archaeological GIS', In G. Lock and Z. Stancic (eds), *Archaeology and Geographical Information Systems: a European perspective.* London: Taylor and Francis Ltd, 1-10.

Leask, H.G. 1914. 'Proceedings: Swords'. *Royal Society of Antiquaries of Ireland* 44, 257-264.

Leask, H.G. 1936. 'Irish Castles: 1180 – 1310'. *Archaeological Journal* 93, 156-159, 173-175.

Leask, H.G. 1937. *Irish Castles: 1180-1310.* Longman, Brown, Green, and Longman.

Leask, H.G. 1941. *King John's Castle Carlingford, Co. Louth.* A short historical descriptive account reprinted from the 107th Annual Report of the Commissioners of Public Works. Dublin: Stationary Office.

Leask, H. G. 1977. *Irish castles and castellated houses.* Dundalk: Dundalgan Press.

Liddiard, R. 2000, *'Landscapes of Lordship'.* BAR British series, 309. Oxford: BAR Publishing.

Liddiard, R. 2005. *Castles in context power, symbolism and landscape, 1066 to 1500.* Macclesfield: Windgather Press.

Lilley, K.D., Lloyd, C.D. and Trick, S. 2007. 'Designs and designers of medieval 'new towns' in Wales'. *Antiquity* 81, 279-293.

Llobera, M. 2001. 'Building past landscape perception with GIS: understanding topographic prominence'. *Journal of Archaeological Science* 2001 (28), 1005-1014.

Llobera, M. 2003. 'Extending GIS-based visual analysis: the concept of visualscapes'. *International Journal of Geographical Information Science* 17, 1, 25-48.

Llobera, M. 2007. 'Reconstructing visual landscapes'. *World Archaeology* 39, 1, 51-69.

Lock, G.R. and Harris, T.M. 1996. Danebury revisited: an English Iron Age hillfort in a digital landscape'. In M. Aldenderfer and H.D.G. Maschner (eds), *Anthropology, Space, and Geographic Information Systems.* New York: Oxford University Press, 214-240.

Loots, L. 1997. 'The use of Projective and Reflective Viewsheds in the Analysis of the Hellenistic City Defence System at Sagalassos, Turkey'. *Archaeological Comuputing Newsletter* 49, 12-16.

Lowerre, A. 2005. *Placing Castles in the Conquest: Landscape, Lordship and Local Politics in the South-Eastern Midlands, 1066-1100.* BAR British Series, 385. Oxford: BAR Publishing.

Lynn, C.J. 1985. 'Some 13th-century castle sites in the west of Ireland: notes on a preliminary reconnaissance'. *Journal of the Galway Archaeological and Historical Society* 40, 109.

Lynn, C. 1988. *Pieces of the Past: Archaeological excavations by the Department of the Environment for Northern Ireland 1970-1986.* Belfast: Her Majesty's Stationary Office.

MacAlister, R. A. S. 1909. *The memorial slabs of Clonmacnois, King's County: with an appendix on the materials for a history of the monastery.* Dublin: University Press.

MacAlister, R.A.S. 1913. 'The Dominican Church at Athenry'. *Journal of the Royal Society of Antiquaries of Ireland* Part III, 63, 196-223.

MacCotter, P. and Nicholls, K. 1996. *The Pipe Roll of Cloyne: (Rotulus Pipæ Clonensis).* Innygrega: Cloyne Literary and Historical Society.

Malcolm, J. 2007. 'Castles and Landscapes in Uí Fhiachrach Muaidhe, *c.* 1235 - *c.* 1400', In L. Doran and J. Lyttleton (eds), *Lordship in medieval Ireland: Image and reality.* Dublin: Four Courts Press, 193-216.

McCarthy, J. 2007. 'Castles in Space: a study of Irish tower houses'. Unpublished MPhil dissertation, Belfast: Queen's University Belfast.

McCleary, S. 1997. 'The History and Archaeology of Adare Castle'. Unpublished Masters dissertation, Belfast: Queen's University Belfast.

McCormick, F. 1999. 'Early evidence for wild animals in Ireland' in Norbert Benecke (ed.), *The Holocene history of the European vertebrate fauna: modern aspects of research.* Berlin: Rahden, 355-370.

McNeill, C. 1949. 'Calendar of Archbishop Allen's Registrar', *Journal of the Royal Society of Antiquaries of Ireland* extra vol, 175.

McNeill, T.E. 1973. *The History and Archaeology of the Anglo-Norman Earldom of Ulster.* Belfast: Unpublished PhD thesis, Belfast: Queen's University, Belfast.

McNeill, T.E. 1980. *Anglo-Norman Ulster: The History and Archaeology of an Irish Barony, 1177-1400.* Edinburgh: John Donald Publishers LTD.

McNeill, T.E. 1981. *Carrickfergus Castle, County Antrim.* Belfast: Department of the Environment for Northern Ireland, Her Majesty's Stationery Office.

McNeill, T.E. 1990. 'Trim Castle, Co. Meath; the first three generations'. *Archaeological Journal* 147, 308-336.

McNeill, T. E. 1992. *English Heritage book of castles.* London: B.T. Batsford.

McNeill, T.E. 1997. *Castles in Ireland; Feudal Power in a Gaelic World.* London: Rutledge.

McNeill, T. E. 2003. *'Squaring circles: flooring round towers in Wales and Ireland'*, In J.R. Kenyon and K. O'Conor (eds), *The Medieval Castle in Ireland and Wales.* Dublin: Four Courts Press, 96-106.

McNeill, T. E. 2006. 'The view from the top', *Les Cahiers de l'Urbanisme N° Septembre 2006,* 122-127.

Moloney, A. 2003. 'From East and West, Crossing the Bogs at Clonmacnoise', In H.A. King (ed), *Clonmacnoise Studies, Seminar Papers 1998, vol. 1.* Dublin: The Stationary Office, 7-10.

Mullally, E. (ed.) 2002. *The deeds of the Normans in Ireland: La geste des engleis en Yrlande: a new edition of the chronicle formerly known as The Song of Dermot and the Earl.* Dublin: Four Courts Press, 135-138.

Murphy, M. and O'Conor, K. 2006. 'Castles and Deer Parks in Anglo-Norman Ireland'. *Eolas: The Journal of the American Society of Irish Medieval Studies* 1, 53-73.

Myles, F. 1991. 'Archaeological Assessment of the Structure and Fabric of King John's Castle, Limerick'. Unpublished archaeological report commissioned by Shannon Development.

Mytum, H. 2003. 'Surface and geophysical survey at Clonmacnoise: defining the extent of intensive monastic settlement', In H.A. King (ed), *Clonmacnoise Studies, Seminar Papers 1998, vol. 2.* Dublin: The Stationary Office, 35-58.

NIMF (Government of Northern Ireland Ministry of Finance), 1966. *An Archaeological Survey of County Down.* Belfast: HMSO.

Nolan, J.P. 1900-01. 'Galway Castles and Owners in 1574'. *Journal of the Galway Archaeological and Historical Society* I part II, 109-123.

O'Brien, K. and Fenlon, J. 2002. *Trim Castle, Co. Meath.* Dúchas The Heritage Service Handbook.

O'Conor, K. and Manning, C. 2003. 'Introduction', In H.A. King (ed.), *Clonmacnoise Studies, Seminar Papers 1998,* vol. 2. Dublin: The Stationary Office, 137-165.

O'Conor, K. 1996. 'Dunamase Castle'. *Journal of Irish Archaeology* 7, 97-115.

Ó Drisceoil, C. 2002. 'Recycled ringforts: The evidence from archaeological excavations for the conversion of pre-existing monuments to motte castles in medieval Ireland'. *Journal of the Co. Louth Archaeological and Historical Society* 25, 189-201.

Ogburn, D. 2006. 'Assessing the level of visibility of cultural objects in past landscapes'. *Journal of Archaeological Science* 33, 405-413.

O'Keeffe, T. 2001. 'Concepts of 'castle' and the construction of identity in medieval and post-medieval Ireland'. *Irish Geography* 34 (1), 69-88.

O'Keeffe, T and Coughlan, M. 2003. 'The chronology and formal affinities of the Ferns donjon, Co. Wexford', In J.R. Kenyon and K. O'Conor (eds), *The Medieval Castle in Ireland and Wales: essays in honour of Jeremy Knight.* Dublin: Four Courts Press, 133-148.

Oliver, W.K. 1997. 'A study of the archery defences at Castleroche, Carlingford Castle and Dunamase Castle; were they designed to fulfil a specific defensive plan, or were they simply included as a result of fashion?' Unpublished thesis, Belfast: Queen's University, Belfast.

Oliver, W.K. 1999. 'An investigation into the original function of the outer enclosure at Castle Roche, Co. Louth, using only surveying techniques: both physical and geophysical'. Unpublished MSc thesis, Belfast: Queen's University.

Orpen, G. 1907. 'Athlone Castle: Its Early History, with Notes on Some Neighbouring Castles'. *The Journal of the Royal Society of Antiquaries of Ireland* 37, No. 3, 257-276.

Orpen, G.H. 1911. *Ireland Under the Normans 1169-1216,* vol. I. Oxford: University Press.

Orpen, G.H. 1920. *Ireland Under the Normans 1216-1333,* vol. III. Oxford: University Press.

Orpen, G.H. 1913-1921. 'The Earldom of Ulster: introductory to the inquisitions of 1333, and inquisitions concerning Carrickfergus and Antrim'. In *RSAI,* ser 6, 3 (1913); 4 (1914); 5 (1915); 10 (1920); 11 (1921).

OPW Office of Public Works, Republic of Ireland: non-dated pamphlet on Athenry Castle.

O'Sullivan, A. and Boland, D. 2000. *The Clonmacnoise Bridge: an early medieval river crossing in County Offaly.* Archaeology Ireland Heritage Guide 11. Bray: Archaeology Ireland Ltd.

Otway-Ruthven, A.J. 1968. *A History of Medieval Ireland.* London: Benn, New York: Barnes and Noble.

Phillips, J.J. 1883. *The Annals and Archaeology of Dundrum County Down Ireland: its Knights Templars' Donjon, a unique Anglo-Irish Castle.* Belfast: Marcus Ward and Co. Ltd Printers.

Pilcher, J. and Hall, V. 2001. *Flora Hibernica.* Cork: The Collins Press.

Potterton, M. 2005. *Medieval Trim: History and Archaeology.* Dublin: Four Courts Press.

Quinn, D.B. 1943. 'Index and Corrigenda to Irish Pipe Roll of 14 John', *Ulster Journal of Archaeology* 6, 28-36.

Reeves, W. 1992. *Ecclesiastical Antiquities of Down, Connor, and Dromore: consisting of a taxation of those dioceses, compiled in the year 1306.* Dublin: Braid Books and Mayola Books.

Reeves-Smyth, T. 1983. 'Landscapes in Paper: Cartographic Sources for Irish Archaeology', In T. Reeves-Smyth and F. Hamond (eds) *Landscape*

Archaeology in Ireland, 1983. BAR British Series, 116. Oxford: B.A.R., 119-177.

Robinson, P. 1986. *Carrickfergus.* Irish Historic Towns Atlas 2. Dublin: Royal Irish Academy.

Roulston, W. 2004. 'The Provision, Building and Architecture of Anglican Churches in the North of Ireland, 1600-1740, vol. I'. Unpublished PhD thesis, Belfast: Queen's University, Belfast.

Ruffell, A., Geraghty, L., Brown, C. and Barton, K. 2004. 'Ground-penetrating Radar Facies as an aid to Sequence Stratigraphic Analysis: Application to the archaeology of Clonmacnoise Castle, Ireland'. *Archaeological Prospection* 11, 247-262.

Rynne, E. 1961. 'Was Desmond Castle, Adare, erected on a Ringfort?'. *North Munster Antiquarian Journal* 8, 193-202.

Salzman, L.F. 1952. *Building in England: Down to 1540.* Oxford: Clarendon Press.

Smith, B. 1993. 'A County Community in Early Fourteenth-Century Ireland: The Case of Louth'. *The English Historical Review* 108, no. 428, 561-588.

Smith, B. 1999. *Colonisation and Conquest in Medieval Ireland: the English in Louth, 1170-1330.* Cambridge Studies in Medieval Life and Thought: Fourth Series. Cambridge: Cambridge University Press.

Speight, S. 2008. 'Castles as Past Culture: Living with Castles in the Post-Medieval World'. *Château Gaillard* 23, 385-394.

Sweetman, P.D. 1979. 'Archaeological excavations at Ferns Castle, Co. Wexford'. *Proceedings of the Royal Irish Academy* 79, section C, 217-45.

Sweetman P.D. 1980. 'Archaeological Excavations at King John's Castle, Limerick'. *Proceedings of the Royal Irish Academy* 80, C, No. II, 207-233.

Sweetman, D. 1991. *Medieval Castles of Ireland.* Cork: The Collins Press.

Sweetman, P.D. 1992. 'Aspects of early thirteenth century castles in Leinster'. *Château Gaillard* 15, 325-33.

Sweetman, P.D. 1998, 'The development of Trim Castle in light of recent research'. *Château Gaillard* 18, 223-230.

Sweetman, D. 1999. *Medieval Castles of Ireland.* Cork: The Collins Press.

Swift, J. 1999. *Historical Maps of Ireland.* London: PRC Publishing.

Taylor, G. and Skinner, A. (1778) 1969. *Maps of the Roads of Ireland.* Shannon: Irish University Press.

Thomas, A. 1992. *The Walled Towns of Ireland.* Dublin: Irish Academic Press.

Tilley, C. 1993. 'Art, Architecture, Landscape (Neolithic Sweden)', In B. Bender (ed.), *Landscape: Politics and Perspectives.* Oxford: Berg Publishers, 49-84.

Tilley, C. 1994. *A Phenomenology of Landscape: Places, Paths and Monuments.* Oxford: Berg Publishers.

Tubridy, M. 1987. 'The Story of Mongan Bog', In M. Turbridy (ed.), *The Heritage of Clonmacnoise.* Tullamore: Environmental Science Unit of Trinity College in Association with County Offaly Vocational Educational Committee, 79.

Waterman, D.M. 1964. 'The Water Supply of Dundrum Castle, Co. Down', *Ulster Journal of Archaeology* 27, 136-138.

Wheatley, D. 1995. 'Cumulative viewshed analysis: a GIS-based method for investigating intervisibility, and its archaeological application'. In G. Lock and Z. Stancic (eds) *Archaeology and Geographical Information Systems: A European Perspective.* London: Taylor and Francis, 171-186.

Wheatley, D. and Gillings, M. 2000. 'Vision, perception and GIS: developing enriched approaches to the study of archaeological visibility', In G. Lock (ed.), *Beyond the Map: Archaeology and Spatial Technologies.* NATO Science Series. Amsterdam: IOS Press, 1-27.

Wheatley, D. and Gillings, M. 2002. *Spatial technology and archaeology: the archaeological applications of GIS.* New York: Taylor and Francis.

Wiggins, K. 2000. *Anatomy of a Siege: King John's Castle: Limerick, 1642.* Bray: Wordwell Ltd.

Wiggins, K. 2009. *Excavations at King John's Castle, Limerick 1990 – 1998.* Castle Studies Group 23rd Annual Conference, unpublished brochure.

General internet resources used

Website: http://www.ucc.ie/celt/published/G100011/index.html, visited on July 27, 2010, '*Annals of Connacht.* 1224-1544'.

Website: http://www.pvv.ntnu.no/~madsb/home/war/vegetius/, visited on July 15, 2010, Vegetius, F.R. 'The Military Institutions of the Romans', translated from the Latin by Lieutenant J. Clarke, 1767, Copyright Expired, Etext version by Brevik, M. 2001.

Website: http://www.wwwestra.com/horses/history_travel.htm, visited on July 14, 2010, Westra, K, 'The Three Corner Farm, Horses: an informational site '.

Website: www.nmr.mgh.harvard.edu/.../doc/HST583-Lab1.html, visited on June 10, 2010, from Harvard School of Medicine, 'Illustration of the visual cortex'.

Website: http://resources.arcgis.com/glossary, visited in June 2007, ESRI glossary of GIS vocabulary, 'Description of GIS viewshed'.

Website: http://resources.arcgis.com/glossary/term/552, visited on July 28, 2010 'Description of *hillshade'*.

Website: http://en.wikipedia.org/wiki/File:Les_Tr%C3%A8s_Riches_Heures_du_duc_de_Berry_mars.jpg, visited on June 9, 2010, 'Le Tres Riches Heures du duc de Berry: March', within public domain.

Website: http://upload.wikimedia.org/wikipedia/commons/e/e9/Les_Tr%C3%A8s_Riches_Heures_du_duc_de_Berry_juin.jpg, visited on June 9, 2010, 'Le Tres Riches Heures du duc de Berry: June', within public domain.

Website: http://en.wikipedia.org/wiki/File:Les_Tr%C3%A8s_Riches_Heures_du_duc_de_Berry_juillet.jpg, visited on June 9, 2010, 'Le Tres Riches Heures du duc de Berry: July', within public domain.

Website: http://upload.wikimedia.org/wikipedia/commons/c/c5/Les_Tr%C3%A8s_Riches_Heures_du_duc_de_Berry_aout.jpg, visited on June 9, 2010, 'Le Tres Riches Heures du duc de Berry: August', within public domain.

Website: http://en.wikipedia.org/wiki/File:Les_Tr%C3%A8s_Riches_Heures_du_duc_de_Berry_septembre.jpg, visited on June 9, 2010, 'Le Tres Riches Heures du duc de Berry: September', within public domain.

Website: http://en.wikipedia.org/wiki/File:Les_Tr%C3%A8s_Riches_Heures_du_duc_de_Berry_octobre.jpg, visited on June 9, 2010, 'Le Tres Riches Heures du duc de Berry: October', within public domain.

Website: http://cn.wikipedia.org/wiki/File:Towrlndn.JPG, visited on June 9, 2010, 'Digital reproduction of Louis d'Orleans held prisoner in the White Tower', within public domain.

Website: http://resources.arcgis.com/content/search-result?searchKeyWord=artifact&clearCacheTime=1279122123&searchTaxonomy=&searchProduct=&searchRC=All&searchCollectionType=0, visited on July 14, 2010, ArcGis 9.2 Desktop Help.

Website: http://eprints.soton.ac.uk/28803/, visited on September 13, 2007,

Gillings, M. and Wheatley, D. 2001. 'Seeing is not believing: unresolved issues in archaeological visibility analysis'. In B. Slapšak (ed.) *On the good use of geographical information systems in archaeological landscape studies*, Luxembourg: Office for Official Publications of the European Communities, 25-36.

Website: http://www.dnr.state.mn.us/mis/gis/tools/arcview/extensions/DNRGarmin/DNRGarmin.html, visited on July 24, 2010.

Site specific online resources used

Adare Castle, Co. Limerick:
Website: http://ims0.osiemaps.ie/website/publicviewer/main.aspx#V1,546949,646800,7,
visited on November 2, 2009, Ordnance Survey Ireland, OSi Shop Map Viewer.

Athenry Castle, Co. Galway:
Website: http://ims0.osiemaps.ie/website/publicviewer/main.aspx#V1,550392,728001,8, visited on March 24, 2009, Ordnance Survey Ireland, OSi Shop Map Viewer.

Website: http://www.athenryheritagecentre.com/history_castle.htm, visited on March 27, 2009, Rynne, E. 'Athenry Castle'.

Website http://www.excavations.ie/Pages/Details.php?Year=&County=Galway&id=3343, visited on March 27, 2009, Papazian, C. 'Athenry Castle'

Athlone Castle, Co. Westmeath:
Website: http://ims0.osiemaps.ie/website/publicviewer/main.aspx#V1,603947,741399,7,
visited on March 17, 2009, Ordnance Survey Ireland, OSi Shop Map Viewer.

Carrickfergus Castle, Co. Antrim:
Website: http://www.nimap.net/Mapviewer/Default.aspx, visited on Januray 12, 2010, Ordnance Survey Ireland, OSi Shop Map Viewer.

Castleroche Castle, Co. Louth:
Website: http://ims0.osiemaps.ie/website/publicviewer/main.aspx#V1,699085,811839,7, visited on April 7, 2009, Ordnance Survey Ireland, OSi Shop Map Viewer.

Clonmacnoise Castle, Co. Offaly:
Website: http://ims0.osiemaps.ie/website/publicviewer/main.aspx#V1,600821,730623,8,

visited on July 27, 2009, Ordnance Survey Ireland, OSi Shop Map Viewer.

Dunamase Castle, Co. Laois:
Website:
http://ims0.osiemaps.ie/website/publicviewer/main.aspx#V1,653234,697930,5,
visited on Januray 23, 2009, Ordnance Survey Ireland, OSi Shop Map Viewer.

Website: http://www.clanomore.com/dunarch.htm, visited on December 3, 2008,
Hodkinson, B.J., 'The Sources for the History of Dunamase Castle in the Medieval Period'.

Website:
http://www.excavations.ie/Pages/Details.php?Year=&County=Laois&id=2048, visited on January 19, 2009, Hodkinson, B. 1996, 'Dunamase'.

Dundrum Castle, Co. Down:
Website: http://www.nimap.net/Mapviewer/Default.aspx, visited on September 26, 2009, Northern Ireland Map Web NImapWeb.

Ferns Castle, Co. Wexford:
Website:
http://ims0.osiemaps.ie/website/publicviewer/main.aspx#V1,701684,649889,8,
visited on October 8, 2009, Ordnance Survey Ireland, OSi Shop Map Viewer.

Website:
http://www.excavations.ie/Pages/Details.php?Year=&County=Wexford&id=875, visited on October 19, 2007, Walsh, C., 1999: 876, 'Castlelands, Ferns'.

Greencastle, Co. Down:
Website:
http://ims0.osiemaps.ie/website/publicviewer/main.aspx#V1,724500,811787,7, visited on September 7, 2009, Ordnance Survey Ireland, OSi Shop Map Viewer.

Website: http://www.nimap.net/Mapviewer/Default.aspx, visited on March 22, 2010, Northern Ireland Map Web NImapWeb.

Ward, K. 2005. CAF Fieldwork Monitoring Report:
http://www.qub.ac.uk/schools/CentreforArchaeologicalFieldworkCAF/Reports/MonitoringReports/Filetoupload,64475,en.pdf

Kilbolane Castle, Co. Cork:
Website:
http://ims0.osiemaps.ie/website/publicviewer/main.aspx#V1,542953,621301,8, visited on October 15, 2009, Ordnance Survey Ireland, OSi Shop Map Viewer.

Kiltartan Castle, Co. Galway:
Website:
http://ims0.osiemaps.ie/website/publicviewer/main.aspx#V1,545877,704763,7, visited on November 11, 2009, Ordnance Survey Ireland, OSi Shop Map Viewer.

Lea Castle, Co. Laois:
Website:
http://ims0.osiemaps.ie/website/publicviewer/main.aspx#V1,657027,712097,8, visited on December 28, 2009, Ordnance Survey Ireland, OSi Shop Map Viewer.

Limerick Castle, Co. Limerick:
Website:
http://ims0.osiemaps.ie/website/publicviewer/main.aspx#V1,557698,657791,8, visited on April 11, 2009, Ordnance Survey Ireland, OSi Shop Map Viewer.

Nenagh Castle, Co. Tipperary:
Website:
http://ims0.osiemaps.ie/website/publicviewer/main.aspx#V1,586648,679350,8, visited on March 12, 2009, Ordnance Survey Ireland, OSi Shop Map Viewer.

Rinnduin Castle, Co. Roscommon:
Website:
http://ims0.osiemaps.ie/website/publicviewer/main.aspx#V1,600588,754314,8, visited on February 11, 2009, Ordnance Survey Ireland, OSi Shop Map Viewer.

Trim Castle, Co. Meath:
Website:
http://ims0.osiemaps.ie/website/publicviewer/main.aspx#V1,680195,756691,7, visited on January 23, 2009, Ordnance Survey Ireland, OSi Shop Map Viewer.

www.ingramcontent.com/pod-product-compliance
Lightning Source LLC
LaVergne TN
LVHW072255300926
849160LV00018B/1670
* 9 7 8 1 4 0 7 3 1 0 7 4 9 *